AMINO ACIDS IN FARM ANIMAL NUTRITION

MEDWAY CAMPUS LIBRARY

This book is due for return or renewal on the last date stamped below,
but may be recalled earlier if needed by other readers.
Fines will be charged as soon as it becomes overdue.

the
UNIVERSITY
of
GREENWICH

AMINO ACIDS IN FARM ANIMAL NUTRITION

Edited by

J.P.F. D'MELLO

The Scottish Agricultural College
West Mains Road
Edinburgh EH9 3JG
UK

CAB INTERNATIONAL

CAB INTERNATIONAL   Tel: Wallingford (0491) 832111
Wallingford   Telex: 847964 (COMAGG G)
Oxon OX10 8DE   E-mail:cabi@cabi.org
UK   Fax: (0491) 833508

A catalogue entry for this book is available from the British Library.

ISBN 0 85198 881 4

Typeset by Colset Pte Ltd, Singapore
Printed and bound in the UK at Biddles Ltd, Guildford

Contents

Contributors

BAKER, D.H. *Department of Animal Sciences and Division of Nutritional Sciences, University of Illinois, 290 Animal Sciences Laboratory, 1207 West Gregory Drive, Urbana, Illinois 61801, USA*

BALDWIN, R.L. *Department of Animal Science, University of California, Davis, California 95616–8521, USA*

BATTERHAM, E.S. (Deceased) *NSW Agriculture, Pig Research Unit, Wollongbar Agricultural Institute, Wollongbar, NSW 2477, Australia*

BECKETT, J. *Department of Animal Science, University of California, Davis, California 95616–8521, USA*

BUTTERY, P.J. *Department of Applied Biochemistry and Food Science, University of Nottingham, Faculty of Agricultural and Food Sciences, Sutton Bonington, Loughborough, LE12 5RD, UK*

CALVERT, C.C. *Department of Animal Science, University of California, Davis, California 95616–8521, USA*

CHEN, X.B. *The Rowett Research Institute, Greenburn Road, Bucksburn, Aberdeen, AB2 9SB, UK*

COLE, D.J.A. *University of Nottingham, Faculty of Agricultural and Food Sciences, Sutton Bonington, Loughborough, LE12 5RD, UK*

D'MELLO, J.P.F. *The Scottish Agricultural College, West Mains Road, Edinburgh, EH9 3JG, UK*

FISHER, C.   *Leyden Old House, Kirknewton, Midlothian EH27 8DQ, UK*

FULLER, M.F.   *The Rowett Research Institute, Greenburn Road, Bucksburn, Aberdeen, AB2 9SB, UK*

HANIGAN, M.D.   *Department of Animal Science, University of California, Davis, California 95616–8521, USA*

MCNAB, J.M.   *Roslin Institute (Edinburgh), Roslin, Midlothian EH25 9PS, UK*

MOUGHAN, P.J.   *Department of Animal Science, Massey University, Private Bag, Palmerston North, New Zealand*

OLDHAM, J.D.   *Genetics and Behavioural Sciences Department, The Scottish Agricultural College, Bush Estate, Penicuik, Midlothian EH26 0QE, UK*

ØRSKOV, E.R.   *The Rowett Research Institute, Greenburn Road, Bucksburn, Aberdeen, AB2 9SB, UK*

VAN LUNEN, T.A.   *University of Nottingham, Faculty of Agricultural and Food Sciences, Sutton Bonington, Loughborough, LE12 5RD, UK*

WILLIAMS, A.P.   *Cnwc y Deri, Heol Smyrna, Llangain, Carmarthen, Dyfed, SA33 5AD, UK*

WILSON, R.P.   *Department of Biochemistry and Molecular Biology, Mississippi State University, PO Drawer BB, Mississippi State, MS 39762, USA*

Preface

The past two or three decades have been marked by significant advances in our knowledge of the amino acid nutrition of farm animals. These developments have undoubtedly been the result of market-led activities, in particular the industrial manufacture and commercial exploitation of pure methionine and lysine. In Western Europe, the momentum for the extended use of amino acids is gathering pace with the realization that these supplements may have an important role to play in the amelioration of nitrogen pollution from animal wastes. In addition, as global fish stocks decline and with increasing consumer apprehension over the use of animal protein concentrates in livestock nutrition, the prospects for alternative products are being considered. Against this background, a number of companies are now poised to launch into the animal feed industry at least two other amino acids, namely threonine and tryptophan. Further impetus for the exploitation of pure amino acids has recently emerged from the establishment of a link between the incidence of immune hypersensitivity in neonatal animals and dietary protein source and level. Judicious use of amino acid supplements together with overall reductions in antigen loads from specific protein sources are options available for the alleviation of post-weaning stress in calves and piglets.

It is opportune, therefore, to consider recent advances and future developments which may determine the outlook for amino acids in farm animal nutrition. It is hoped that this book will provide the basis for such a reflection. To this end, I have enlisted the expertise of a team of authors internationally recognized for their innovative contributions in the field of amino acid metabolism and requirements of farm animals.

A recurring theme in this book relates to methodological advances, which have been quite striking during the past 25 years. Over this period,

ix

the primacy of ion-exchange chromatography has steadily been eroded by the emergence of high performance liquid chromatography. Ileal digestibility of amino acids has now become the routine technique for measuring bioavailability. An inescapable feature of several investigations has been the consistent demonstration that cereal grains are not only poorly endowed with the critical indispensable amino acids, but this deficit is compounded by reduced ileal digestibility of these amino acids. Other studies have suggested that for a number of heat-processed feedstuffs, amino acids may be absorbed in forms that are inefficiently utilized for growth and optimum feed efficiency. For some raw materials therefore, the fundamental assumption that ileal digestibility reflects availability is being questioned.

The adoption of a new system based on metabolizable protein has renewed focus on amino acid utilization in ruminant animals. However, the proposed framework still lacks the appropriate methodology for calculating requirements and responses in terms of individual amino acids. In addition, any future refinement of the system would need to recognize the role of low-molecular weight peptides, absorbed from the gut, in the supply of tissue nitrogen.

Innumerable studies have been conducted to determine the responses of non-ruminant farm animals to individual amino acids. In the early investigations, progress was thwarted by the preoccupation with determining 'requirements'. Attitudes are now changing, albeit very slowly, as nutritionists realize that it is the response of an animal to graded inputs of an amino acid that is important, rather than a statement of immutable requirements. By and large amino acid responses have been obtained with the graded supplementation technique but an alternative method based on diet dilution has also been proposed. Recent studies with the latter procedure have shown conclusively that in the broiler chick, amino acids are more efficiently utilized when provided as pure supplements rather than as components of intact protein, particularly if poor-quality protein sources are used. Such demonstrations of superior utilization of pure amino acids are, in the main, confined to studies with animals fed ad lib. In the case of the pig, feeding frequency may determine efficiency of utilization.

Continuing dissatisfaction with empiricism has meant that much effort is now being expended in the 'systems approach' involving the development of dynamic computerized models. There is growing recognition that the commercial need for accurate nutritional prediction in animal production cannot be fulfilled solely by empirical methods. The role of models in this respect is no longer a matter for debate. In addition, however, mechanistic models may also assist in identifying important limitations in current knowledge so that rapidly declining resources can be employed more effectively to generate a sound database for the future.

The methodology for dietary protein evaluation is now firmly directed at the development of amino acid patterns in perfect balance as embodied in the 'ideal protein' concept. This approach is now finding wide application, even in the new metabolizable protein system for ruminants. Intragastric infusion of amino acids provides a novel technique for establishing the ideal balance for ruminant animals. General, non-specific amino acid imbalances regularly occur in farm animal nutrition, precipitating adverse effects on food intake and growth. However, recent studies confirm that the efficiency of utilization of limiting amino acids remains unimpaired by such imbalances.

Comparative issues are given some prominence in this book. Traditionally, the amino acid nutrition of farm animals has been considered separately for ruminants and non-ruminants. Such a situation can no longer be justified. If this book helps to improve exchange and integration of information across the species barrier, then it will have served an important function.

Many excellent reviews on various aspects of amino acid metabolism and nutrition have appeared over the past few years but their publication in different journals and texts is perceived as a disadvantage. It is hoped that this book will appeal to final year honours undergraduates and postgraduate students as a coherent synthesis of the existing literature. Authors were asked to provide comprehensive reviews with a critical appraisal of current knowledge. Attempts have, therefore, been made to support chapters with a comprehensive bibliography, wherever possible.

I am grateful to all contributing authors for their cooperation and forbearance with an impatient editor. Any success that this book may enjoy will be entirely due to their efforts and expertise. Finally, I feel that it is incumbent upon me to thank Mr T. Hardwick of CAB INTERNATIONAL for his encouragement during the preparation of this book.

Note in proof: It is with the deepest regret that I record here the death of Dr E.S. Batterham, one of my contributing authors. His distinctive and innovative approach to amino acid research is well recognized. There is no doubt in my mind that his legacy of ideas will continue to inspire further research, and not just in the field of non-ruminant nutrition.

J.P.F. D'Mello
Editor

Amino Acid Metabolism in Farm Animals: An Overview

P.J. Buttery[1] and J.P.F. D'Mello[2]

[1]*Department of Applied Biochemistry and Food Science, University of Nottingham, Faculty of Agricultural and Food Sciences, Sutton Bonington, Loughborough, LE12 5RD, UK:* [2]*The Scottish Agricultural College, West Mains Road, Edinburgh, EH9 3JG, UK.*

Introduction

Dietary protein supply is one of the major factors influencing the productivity of farm animals. Supplementation of the diets of animals with amino acids to enhance the quality of the dietary protein is common practice, particularly in the poultry and pig industries. The importance of amino acids to the industry has been one of the reasons that many aspects of amino acid metabolism in farm animals have been extensively studied.

As is well known there are 20 amino acids commonly found in proteins. In addition there are many more amino acids which are rarely found in proteins. All of the protein-bound amino acids, with the exception of glycine, have an optically active carbon atom to which the amino group and the carboxyl group are attached. In the vast majority of cases it is the L-isomer which is of metabolic and nutritional significance. The main role of amino acids in all living organisms is as the monomeric unit from which proteins are synthesized. In addition amino acids also serve as important sources of energy, especially glucose. Some amino acids are also the precursors of other biologically important compounds, for example adrenaline and some of the bile salts.

Biosynthesis of Amino Acids in Animal Tissues

The metabolism of amino acids has been reviewed many times but several aspects require emphasis in order to understand some of the variations in nutritional requirements for and nutritional responses to amino acids between species.

Table 1.1. Generalization of amino acid requirements of animals.

Not generally synthesized	Synthesized from essential amino acids	Usually non-essential but some animals have a requirement
Lysine	Tyrosine (from phenylalanine)	Arginine if urea cycle absent
Histidine	Cystine (from methionine)	Glycine in uricotelic species
Leucine		Proline if conversion from glutamate deficient
Isoleucine		
Valine		
Methionine		
Threonine		
Tryptophan		
Phenylalanine		

The so-called non-essential amino acids are largely synthesized by transamination of intermediates of glycolysis and the tricarboxylic acid cycle. However, for their synthesis an adequate supply of amino groups must be available. It is possible to feed diets to animals where the supply of amino groups is limiting and as a consequence their growth rate can be increased by supplying non-protein nitrogen in the diet.

Examination of the qualitative requirements of a variety of species of animals shows a remarkable consistency (Table 1.1). Variations between species can usually be traced to minor differences in their metabolism. For example, birds do not have an active urea cycle and therefore require arginine in the diet (see Chapter 10). Rapidly growing mammals will also respond to arginine in the diet not because they do not have an active urea cycle but because the vast majority of the arginine that is synthesized is catabolized in the liver by the active arginase associated with the urea cycle. Consequently not enough leaves the liver to support the rapid growth of the extra-hepatic tissues. Another example of qualitative differences between species is the need for cats to have arginine in the diet. Cats given a mixture of amino acids deficient in arginine rapidly become comatosed as a result of ammonia poisoning. This is readily reversed on supplying arginine or ornithine in the diet; the cat only has a limited capacity to synthesize ornithine from glutamate (Baker and Czarnecki-Maulden, 1991). Poultry often need proline in the diet due to their limited capacity to synthesize it from glutamic acid (Boorman and Lewis, 1971).

Often methionine is the amino acid which limits growth. A significant proportion of dietary methionine is used for the biosynthesis of cysteine via the trans-sulfuration pathway. Because of this pathway dietary recommendations often link cysteine and methionine together. It is however very important to realize that the pathway is not reversible and that it is essential to supply methionine in the diet. Conversely it is often said that because of the presence of this pathway there is no need to supply cysteine in the diet. Baker (1989) illustrates the importance of a detailed consideration of this interconversion in determining the requirements of cysteine and methionine. For example, the methionine requirement for the growing chick can range from 0.27 to 6.0 g kg^{-1} depending on the cysteine content of the diet. He also illustrates the need, when considering the total requirement in g kg^{-1} diet for a pair of interconvertible amino acids, to remember that the molecular weight of the product (e.g. cysteine) and the precursor (e.g. methionine) is rarely the same. Other classic examples in amino acid nutrition where product-precursor relationships need to be taken into account include phenylalanine/tyrosine, cysteine/tyrosine and carnosine/histidine, and these are referred to in several other chapters in this volume.

Utilization of Isomers of Amino Acids

All amino acids used in protein synthesis must be in the L-form but D-amino acids are found naturally, for example in bacterial cell walls. Commercial production of amino acids often involves chemical synthesis although the use of stereospecific fermentative processes is becoming of increasing importance. Often commercial diets are supplemented with crystalline amino acids and it is therefore of importance to be able to predict the utilization of racemic mixtures. Some racemic mixtures are almost as effective as the L-isomer. The differences that do occur (see Table 1.2, which presents data for the chick) can be explained by differences in the metabolism of individual amino acids.

D-amino acids can be catabolized by D-amino acid oxidase to yield keto acids. These keto acids can normally be transaminated to yield the L-isomer of the amino acid. There are no transaminases for lysine and threonine in animal tissues, hence the D-isomers of these amino acids are not nutritionally active. Sometimes the keto acid is readily catabolized hence reducing its potential to be transaminated to form the L-isomer. Presumably this competition between irreversible oxidation and transamination of the keto acid is the reason why some other D-amino acids are not readily interconverted into the L-isomer. This topic is considered further in Chapter 3 of this volume. Inclusion of D-isomers in the diet will, unless they are utilized with the same efficiency as the L-isomer, result in an increase in nitrogen excretion by the animal and potentially an increase in pollution of the environment.

Table 1.2. Utilization of dietary D-amino acids by the growing chick. (From Boorman and Lewis, 1971. Reproduced by permission of Academic Press.)

Amino acid	Nutritional value
Methionine Phenylalanine Leucine Proline	Almost equivalent to the L-isomer
Valine	Approximately half as potent as the L-isomer
Tryptophan Histidine Isoleucine Lysine Threonine Arginine	Of little or no nutritional value

Toxicity of Amino Acids

The balance of a mixture of amino acids in the diet is very unlikely to exactly meet the requirements of each of the animal's tissues. A deficiency of an amino acid is likely to cause a reduction in performance. Excesses of amino acids can also be deleterious. Even small excesses of some amino acids, for example methionine, can cause problems. In addition, an excess of one amino acid, while not toxic in itself, can induce an apparent deficiency of another amino acid. For example a relatively small excess of leucine can cause an apparent deficiency of isoleucine. The susceptibility of an animal to imbalances and excesses of amino acids is influenced by the overall protein supply. Animals fed relatively high levels of protein are more tolerant. Ruminants appear to be particularly sensitive to methionine toxicity, presumably because they do not naturally experience wide variations in the amino acid composition of the digesta entering the intestine. However, a wide range of non-protein amino acids occurs naturally in certain crop plants and particularly in tropical legumes. For example, the aromatic amino acid, mimosine, occurs in *Leucaena leucocephala*, a ubiquitous species yielding timber and palatable forage for ruminants. A structural analogue of arginine, canavanine, is widely distributed in various tropical legumes, including *Canavalia ensiformis*, *Gliricidia sepium* and *Indigofera spicata*. Both amino acids are endowed with potent toxic properties. In temperate regions, the factor causing haemolytic anaemia in cattle and sheep consuming forage brassicas has been identified as a ruminal metabolite of *S*-methylcysteine sulfoxide, an analogue of methionine, distributed throughout the plant. The toxicity of amino acids is discussed in detail in Chapter 4 of this volume.

Arginine Metabolism

As indicated previously, the primary direction of arginine metabolism in mammals is via the urea cycle, enabling the disposal of excess amino acid nitrogen. However, the secondary metabolism of arginine is also of considerable biochemical and physiological significance. Thus, the action of arginine decarboxylase permits many organisms to synthesize putrescine and other polyamines. In addition, putrescine may be synthesized by the action of ornithine decarboxylase (ODC). In animals, however, only the latter pathway is functional. Although the specific functions of polyamines await elucidation, recent studies suggest that these compounds are essential for normal growth and development in all living organisms, and may regulate RNA synthesis and stabilize membrane structures. Polyamine production appears to be an indispensable feature of all tissues actively engaged in protein synthesis. In Chapter 15, Oldham points out that arginine uptake by the mammary gland from the blood supply substantially exceeds the quantities of this amino acid secreted in milk. This is attributed to the need to synthesize non-essential amino acids, particularly proline, within the gland. However, the excess uptake of arginine may also reflect the need for polyamine synthesis by tissues actively synthesizing proteins in the mammary gland. Polyamine synthesis is an important focal point for the action of anti-nutritional factors. Thus in lectin-induced hyperplastic growth of the small intestine, levels of putrescine, spermidine, spermine and cadaverine are markedly enchanced (Pusztai *et al.*, 1993). On the other hand the growth-retarding effect in chicks fed *Canavalia ensiformis* has been attributed to inhibition of polyamine synthesis (Chapter 4). The non-protein amino acid, canavanine, which this legume contains, is metabolized to canaline, a potent inhibitor of ODC (D'Mello, 1993).

A striking feature of arginine metabolism, elucidated recently, relates to the synthesis of nitric oxide (NO). The biosynthesis of NO involves the oxidation of arginine by NADPH and O_2 via the action of NO-synthases with the intermediate production of N^{ω}-hydroxy-arginine. It is now known that NO plays a key role in vasorelaxation, neurotransmission, immunocompetence, male reproductive performance and gut motility (Moncada *et al.*, 1991). It is suggested in Chapter 4 that canavanine may inhibit NO synthesis through its structural antagonism with arginine. Enneking *et al.* (1993) arrived at a similar conclusion from their studies on canavanine-induced feed intake inhibition in pigs.

Catabolism of Amino Acids

The K_m values of the enzymes involved the catabolism of amino acids are higher than those of the amino acid-activating enzymes; for example the K_m for threonine by serine-threonine dehydratase is $8.4–13 \times 10^{-3}$ M and the K_m for the threonyl tRNA synthetase is 4.3×10^{-6} M (Kang-Lee and Harper, 1978). When the supply of an amino acid is low it is used relatively efficiently for protein synthesis. As the supply of amino acid is in excess of that required for protein synthesis amino acid oxidation increases. This phenomenon has been observed for the majority of the so-called essential amino acids. This ensures that when amino acids are in short supply they are preferentially used for body protein synthesis. The activity of most amino acid degradative enzymes increases with the dietary supply of protein. This can occur via an increase in the amount of enzyme present; for example rat liver histidase increases seven-fold when rats are changed from a $180\,\mathrm{g\,kg^{-1}}$ diet to one containing $800\,\mathrm{g\,kg^{-1}}$. Some amino acid catabolizing enzymes exist in active and inactive forms. Phenylalanine hydroxylase, the first and rate-limiting step in the main route of catabolism of phenylalanine, is such an enzyme; it can be activated by substrate and by glucagon. Glucagon increases phosphorylation of the enzyme protein. Phenylalanine increases the activity of the enzyme by allosteric activation.

The increase in oxidation as amino acid supply exceeds the requirements for protein anabolism has often been used to assess the requirement of an animal for an individual amino acid. With some amino acids the inflection point is not always distinct. Methionine is such an amino acid. Kim *et al.* (1983) were able to determine the methionine requirement of pigs using ^{14}C-phenylalanine as an indicator amino acid. At methionine concentrations below those required for maintenance protein is degraded to supplement the deficient amino acid supply and other amino acids such as phenylalanine are in excess. As the methionine supply increases protein anabolism increases and the excess of phenylalanine is reduced and thus phenylalanine oxidation is reduced. Thus by monitoring the oxidation of the indicator amino acid the supply of methionine which exceeds the requirement for protein synthesis can be determined. The use of such techniques has played a major role in determining the amino acid requirements of farm animals, especially the larger species. The results of such studies are referred to many times in this volume. Amino acid catabolism and protein synthesis would appear to be linked processes. The anabolic drive hypothesis of Millward and Rivers (1988) suggests that before their oxidation excesses of amino acids exert a transient anabolic effect on protein deposition. These authors suggest that 'there are advantages in consuming levels of indispensable amino acids in excess of the level required to match identifiable needs'. Although the hypothesis was developed in relation to a more accurate estimation of human amino acid requirements it does suggest that an animal exhibiting a rapid rate of protein synthesis will have an enhanced rate of amino acid oxidation and hence nitrogen excretion. The exact link between amino acid oxidation and protein synthesis is difficult to demonstrate experimentally but if enhanced protein synthesis is by necessity linked with an enhanced catabolism of amino acids then this has major implications in attempts to reduce nitrogen pollution from highly productive farm animals.

Amino Acid Nutrition and Nitrogen Pollution

Increasingly the animal industry is obliged to reduce the effects of intensive production systems on the environment. Until recently consideration of the response of animals to amino acid supply was confined to maximizing the efficiency of outputs like meat and milk but little or no attention was paid to reducing the output of nitrogenous materials in the faeces and the urine. Output in the faeces can be minimized by increasing the digestibility of feeds; for example, anti-nutritional factors in leguminous seeds, such as protease inhibitors, can be removed by heat treatment and potentially by enzyme treatment. Endogenous protein losses also contribute significantly to faecal nitrogen. Enzyme treatment of feedstuffs also appears to be a promising way to reduce the indigestible fibre content of the diet and hence loss of endogenous faecal secretions (Tamminga and Verstegen, 1992).

Reduction in the loss of nitrogen in the urine can be achieved by carefully balancing the dietary supply of amino acids to the requirements of the animal for the desired production response and, as is indicated many times in this volume, these requirements vary with age and physiological status. In addition it is necessary to consider carefully the efficiency with which these amino acids are used. Dietary protein (amino acids) is not used with a constant efficiency; as maximum production is approached the efficiency with which amino acids are used falls and hence the proportion of dietary nitrogen that is excreted increases. This needs to be considered when evaluating the level of production to use in an animal farming enterprise.

Part of the inefficiency with which amino acids are utilized is a consequence of the turnover of proteins. The extent of protein synthesis and degradation per unit of metabolic size in mammals at maintenance is relatively constant, approximately $16 \text{ g kg}^{-0.75}$ (Millward and Garlick, 1977); even in rapidly growing mammals and birds the ratio of protein deposition to protein synthesis rarely exceeds 1:3. This turnover of protein is an essential component of metabolic control but it is energetically expensive, contributing in some cases to 20% of basal heat production (Buttery, 1978). The reutilization of amino acids released on protein catabolism is not 100% efficient but this inefficiency may not be directly related to the extent of protein turnover. The major determinant of the extent of protein degradation will be the concentration of amino acid in the pool associated with the amino acid-catabolizing enzymes not the movement of the amino acid through the pool. The situation is however complicated because of the considerable metabolic control that is exerted over the catabolic enzymes (see above).

The Future

Conventional animal breeding programmes continue to enhance the productivity of animals and it is therefore necessary to continually update knowledge on the response of animals to amino acid supply. Advances in biotechnology potentially provide methods of dramatically enhancing animal performance.

Exogenous growth hormone treatment of mammals receiving an adequate supply of nutrients results in an enhancement of lean and a reduction in fat deposition. This partition towards greater lean deposition is likely to have a marked effect on the requirements for amino acids. Boyd *et al.* (1991) examined how nutrient responses, particularly of pigs, might be influenced by such treatment. The essential components that need to be considered are the effects on the digestibility of the diet, the rate and composition of gain, the efficiency with which absorbed nutrients are used for tissue deposition, the quantity of nutrients required for maintenance and nutrient intake. The main effects of growth hormone treatment would appear to be primarily associated with the use of absorbed nutrients (Boyd and Bauman, 1989) although the reduced intake that can accompany growth hormone treatment could result in a slower rate of passage which may in turn result in a small increase in digestibility (see Verstegen *et al.*, 1990). There is evidence that growth hormone improves the efficiency by which dietary amino acids are partitioned into body protein by up to 20% (see e.g. Campbell *et al.*, 1991). Thus treated animals have to be fed enough protein to enable the increased lean deposition to take place, but the increase in protein requirement will not be linearly related to the enhancement of lean deposition.

Transgenesis is another potential method of enhancing animal productivity. Several species of animals have had extra copies of the growth hormone gene inserted into their genome. While there have been considerable problems in controlling the expression of the inserted growth hormone genes, pigs with an enhanced growth rate and an increased lean to fat ratio have been successfully produced (see Ward and Nancarrow, 1991). Other methods, for instance the use of the immune response and treatment with growth hormone releasing factor, are available to enhance growth hormone activity in farm livestock (Boorman *et al.*, 1992). What is clear is that when and if these animals become a commercial reality it will be necessary to carefully evaluate their response to variation in amino acid supply.

Wool growth is often limited by the supply of cysteine. Methionine or cysteine can be added to the diet but the majority of it will be degraded in the rumen. One method of overcoming this is to protect the amino acids from rumen degradation is such a way that the amino acid is released further down the digestive tract. An alternative approach described in detail by Ward and Nancarrow (1991) is to introduce into the sheep genome the genes that encode the cysteine biosynthetic pathway. The genes for this pathway, which is independent of methionine, are found in prokaryotes and are thus available for inserting into sheep. The genes for the two enzymes involved (serine transacetylase and O-acetylserine sulfhydrylase) together with appropriate promoters have been expressed in mice so there is a realistic chance that they could be incorporated into the genome of the sheep. It has also been suggested that the genes from *Escherichia coli* for lysine and for threonine could be expressed in mammals. The successful exploitation of this technology will without doubt require a fundamental re-evaluation of our knowledge of amino acid responses of farm livestock. It would be very time consuming and expensive to repeat all the studies previously conducted with conventional animals when these genetically modified animals become a commercial reality. This potential need to re-evaluate nutrient responses of animals highlights the importance of developing mechanistic models to describe amino acid metabolism and nutrition (see Chapters 7 and 12).

References

Baker, D.H. (1989) Amino acid nutrition of pigs and poultry. In: Haresign, W. and Cole, D.J.A. (eds) *Recent Advances in Animal Nutrition 1989*. Butterworths, London, pp. 245–260.

Baker, D.H. and Czarnecki-Maulden, G.L.(1991) Comparative nutrition of cats and dogs. *Annual Review of Nutrition* 11, 239–263.

Boorman, K.N. and Lewis, D. (1971) Protein metabolism. In: Bell, D.J. and Freeman, B.M. (eds) *Physiology and Biochemistry of the Domestic Fowl, vol. 1.* Academic Press, London, pp. 339–372.

Boorman, K.N., Buttery, P.J. and Lindsay, D.B. (1992) *The Control of Fat and Lean Deposition*. Butterworths, London.

Boyd, R.D. and Bauman, D.E. (1989) Mechanisms of action for somatotropin in growth. In: Campion, D.R., Hausman, G.J. and Martin, R.J. (eds) *Current Concepts of Animal Growth Regulation*. Plenum, New York, pp. 257–293.

Boyd, R.D., Bauman, D.E., Fox, D.G. and Scanes, C. (1991) Impact of metabolism modifiers on protein accretion and protein and energy requirements of livestock. *Journal of Animal Science* 69 (Suppl. 1), 56–75.

Buttery, P.J. (1978) Amino acids and other nitrogenous compounds. *Comparative Animal Nutrition* 3, 34–79.

Campbell, R.G., Johnson, R.J., Taverner, M.R. and King, R.H. (1991) Interrelationships between exogenous porcine somatotropin (PST) administration and dietary protein and energy intake on protein deposition capacity and energy metabolism of pigs. *Journal of Animal Science* 69, 1522–1531.

D'Mello, J.P.F. (1993) Non-protein amino acids in *Canavalia ensiformis* and hepatic ornithine decarboxylase. *Amino Acids* 5, 212–213.

Enneking, D., Giles, L.C., Tate, M.E. and Davies, R.L. (1993) Canavanine: a natural feed-intake inhibitor for pigs. *Journal of the Science of Food and Agriculture* 61, 315–325.

Kang-Lee, Y.A. and Harper, A.E. (1978) Threonine metabolism in vivo: Effect of threonine intake and prior induction to threonine dehydratase in rats. *Journal of Nutrition* 108, 163–175.

Kim, K.I., McMillan, I. and Bayley, H.S. (1983) Determination of amino acid requirements of young pigs using an indicator amino acid. *British Journal of Nutrition* 50, 369–382.

Millward, D.J. and Garlick, P.J. (1977) The energy cost of growth. *Proceedings of Nutrition Society* 35, 339–349.

Millward, D.J. and Rivers, J.P.W. (1988) The nutritional role of indispensable amino acids and the metabolic basis for their requirements. *European Journal of Clinical Nutrition* 42, 367–393.

Moncada, S., Palmer, R.M.J. and Higgs, E.A. (1991) Nitric oxide: physiology, pathophysiology, and pharmacology. *Pharmacological Reviews* 43, 109–142.

Pusztai, A., Grant, G., Spencer, R.J., Duguid, T.J., Brown, D.S., Ewen, S.W.B., Peumans, W.J., Van Damme, E.J.M. and Bardocz, S. (1993) Kidney bean lectin-induced *Escherichia coli* overgrowth in the small intestine is blocked by GNA, a mannose-specific lectin. *Journal of Applied Bacteriology* 75, 360–368.

Tamminga, S. and Verstegen, M.W.A. (1992) Implications of nutrition of animals on environmental pollution. In: Garnsworthy, P.C., Haresign, W. and Cole, D.J.A. (eds) *Recent Advances in Animal Nutrition*. Butterworths, London, pp. 113–130.

Verstegen, M.W.A., van der Hel, W., Henken, A.M., Huisman, J., Kanis, E., van der Wal, P. and van Weerden, E.J. (1990) Effects of exogenous porcine somatotropin administration on nitrogen and energy metabolism in three genotypes of pigs. *Journal of Animal Science* 68, 1008–1016.

Ward, K.A. and Nancarrow, C.D. (1991) The genetic engineering of production traits in domestic animals. *Experientia* 47, 913–922.

Recent Developments in Amino Acid Analysis   2

A.P. WILLIAMS

Cnwc y Deri, Heol Smyrna, Llangain, Carmarthen, Dyfed, SA33 5AD, UK.

Introduction

Advances in the knowledge of amino acid metabolism in farm animals are very dependent on the accurate and precise determination of amino acids in foods, feeds, digesta, body fluids and tissues. The historical development of amino acid analysis has been adequately reviewed by Tristram and Rattenbury (1981) and it is not intended to deal with this period in any detail except where necessary for comparative purposes. Since 1981 more than 9000 papers have been published on amino acid analysis (Malmer and Schroeder, 1990), many describing methods based on reversed phase chromatography (RPC), most of which achieve high speed, resolution and sensitivity by the use of microparticulate stationary phases with pre-column derivatization. This technique has become increasingly popular for the analysis of pure proteins and foods (Williams, 1988). However, there are many other techniques for the analysis of amino acids of which the traditional one, ion-exchange chromatography (IEC) with post-column derivatization, is still considered by many to be the best for the complex samples described elsewhere in this book. The present review will discuss current developments in amino acid analysis including the equally important techniques used to prepare the samples for analysis since hydrolysis and deproteinization are still considered to be major problems in amino acid analysis (Williams, 1986).

Sample Preparation

Isolation of free amino acids

It is necessary to remove protein from samples such as whole blood, plasma and tissues before the concentrations of free amino acids can be determined. Methods of deproteinization have included the use of picric acid, sulfosalicylic acid (SSA), trichloroacetic acid (TCA), high speed centrifugation, ultrafiltration and equilibrium dialysis (Williams, 1988). The most widely used method is precipitation with SSA. An internal standard can be added to correct for mechanical and other losses. The choice is wide but with modern analysers there may be no more than two or three that are suitable (Williams and Cockburn, 1988). After centrifugation to remove the precipitated protein and adjustment of the pH, which must be close to that of the standard calibration mixture, the sample is ready for analysis. Adjustment of pH must not be done before protein precipitation since it will reduce the amount of protein removed (Hubbard *et al.*, 1988). The procedure is simple and popular but not without its critics (Williams, 1986, 1988), and some clinical studies (Uhe *et al.*, 1991) suggest that SSA results in interferences with RPC methods, e.g. the aspartic acid derivative co-elutes with SSA, which can be avoided by the use of ultrafiltration or acetonitrile. However, Davey and Ersser (1990) found that with human plasma the ultrafiltration membrane pores were frequently blocked and reported that acetonitrile was the best deproteinization reagent for phenylisothiocyanate (PITC) derivatization although it was important to evaporate the acetonitrile prior to derivatization. Similar results were reported by Aristoy and Toldra (1991) in comparative studies on the deproteinization of fresh pork muscle before PITC derivatization although they also found that TCA, perchloric acid and picric acid were satisfactory. If, however, another derivatizing reagent such as *o*-phthalaldehyde (OPA) is used acetonitrile can result in a wide variation in the recovery of individual amino acids from human (Qureshi and Qureshi, 1989) or ovine (Sedgwick *et al.*, 1991) plasma. However, Uhe *et al.* (1991) did not observe these effects with human plasma. These differences may be due to the fact that acetonitrile or methanol and acid precipitants such as TCA may result in proteolysis or incomplete extraction (Graser *et al.*, 1985). The use of internal standards would not be valid in such situations. Ironically both of these studies recommended the use of acids, such as SSA, for deproteinization. This demonstrates the lack of an ideal method and the need for great care in method selection.

Hydrolysis

It is generally accepted (Gehrke *et al.*, 1985; Hunt, 1985; Zumwalt *et al.*, 1987a; Williams, 1988) that hydrolysis is the major limiting factor in the determination of amino acids in foods and feeds. The so-called 'standard method' of heating proteins in 6 M hydrochloric acid (HCl) at 110°C for 24 hours is a compromise, since no one method can ever give satisfactory values for all the amino acids, even for pure proteins. With feeds, digesta and faecal samples the problem is exacerbated (Rowan *et al.*, 1992). In recent years, due to the reduction in analysis times, attempts have been made to solve some of these problems most notably by Bech-Andersen and colleagues in Denmark (see Bech-Andersen *et al.*, 1990). This 'streamlined' procedure developed for the EEC, enabled all the amino acids in feeds, except for tryptophan, to be determined after a preliminary oxidation stage. Unfortunately the collaborative trial (Andersen *et al.*, 1984) carried out on this method, which showed that its repeatability and reproducibility was satisfactory, would probably have to be repeated. Many laboratories in this study would probably have new analysers and some would be based on RPC, on which this hydrolysis procedure has not been tested. Using similar methods Elkin and Griffith (1985) and Gehrke *et al.* (1987) have shown that the complete amino acid content of a feed cannot be accurately determined from one oxidized hydrolysate. More than a decade after its conception the method has still not been adopted by the EEC.

Recent developments have tended to concentrate on automating and reducing the duration of the hydrolysis stage since analysis time has been reduced to less than 30 minutes. Gehrke *et al.* (1985) carried out a detailed study of hydrolysis conditions for pure proteins and animal feeds and showed that a significant reduction in time, from 24 to 4 hours, could be achieved by carrying out the hydrolysis in sealed tubes with 6 M HCl at 145°C. The sealed tubes were heated in an oven or heating block. After analysis by IEC the results agreed well with those obtained after hydrolysis at 110°C and 24 hours. Another manual hydrolysis system is the Reacti-Therm III (Pierce, Rockford, Illinois, USA) in which 24 samples can be hydrolysed in 6 M HCl at temperatures ranging from 110 to 200°C for from 15 minutes to 72 hours. Extended hydrolysis times (48 and 72 hours) are recommended for isoleucine and valine since they are resistant to hydrolysis. Unfortunately prolonged hydrolysis can result in the degradation of threonine and serine. A new approach has been to increase the temperature to 180°C and to reduce the hydrolysis time to as little as 5 minutes. Such a system has been developed by Waters Associates (Milford, Massachusetts, USA) for use with their Pico-Tag RPC analyser. Using this, 48 samples can be hydrolysed in 24 hours at 108°C using standard liquid phase 6 M HCl or in 1 hour at 150°C using vapour phase hydrolysis. In the former HCl is added directly to the sample. With vapour phase hydrolysis tubes containing the sample are sealed into a larger vessel containing HCl. As the temperature rises the HCl vaporizes so that only its vapour comes into contact with the sample. This avoids contamination from amino acids present in all but the highest purity HCl (Williams, 1986). Another vapour phase system has been developed by Applied Biosystems (Warrington, England) for use, on-line, with their model 420 RPC analyser. In this system three samples, on glass frits, are hydrolysed at 155°C for 75 minutes and then derivatized automatically (West and Crabb, 1990). The authors concluded that, for pure proteins, the automated hydrolysis system gave better accuracy and recovery than manual hydrolysis.

Another new development is the use of microwave irradiation in which pure proteins can be hydrolysed in liquid or vapour phase 6 M HCl at 180 ± 5°C in 5–10 minutes using commercial microwave ovens. Chiou and Wang (1989, 1990) used 4 M methanesulfonic acid as well as 6 M HCl in liquid phase and reported good agreement between their results and those obtained by the conventional method. The use of methanesulfonic acid gave reliable results, compared with theoretical values, for methionine, cysteine and tryptophan. It was concluded that 5 minutes at 180°C was the optimum time for hydrolysis since 4 minutes or less gave rise to unknown peaks. The difficulty of accurately calibrating the temperature of microwave ovens was also noted. This was also reported by Gilman and Woodward (1990) in studies in which methionyl human growth hormone was hydrolysed by microwave using 6 M HCl in the vapour phase. Optimum conditions were 180°C for 8–10 minutes although it was concluded that this time might not be sufficient for the complete liberation of valine and isoleucine.

All of these new hydrolysis procedures have been developed for use with pure proteins and care must be taken before applying them to feeds. Even with pure proteins there have been problems with the reliability of some systems and vapour phase hydrolysis has the major disadvantage that volatile additives, which prevent the oxidation or destruction of some amino acids, cannot be used (Pickering and Newton, 1990).

The Chromatographic Separation and the Measurement of Amino Acids

After deproteinization or hydrolysis it is necessary to separate the amino acids from each other and for this column chromatography is the best method. This consists of gas liquid chromatography (GLC) and high performance liquid chromatography (HPLC), the latter consisting of IEC and RPC with either post- or pre-column derivatization. Reviews on these techniques have been published (Labadarios *et al.*, 1984; Hare *et al.*, 1985; Perrett, 1985; Zumwalt *et al.*, 1987b; Williams, 1988; Papadoyannis, 1990) but there is little agreement on which method is the best.

Ion-exchange chromatography (IEC)

The free amino acids are separated on a chromatographic column, mixed with a derivatization agent, passed through a reaction coil and then through a detector such as a spectrophotometer or fluorometer. Sulfonated polystyrene cation exchange resins are used as the stationary phase with aqueous sodium or lithium citrate mobile phases for hydrolysates or physiological fluids respectively. This, when coupled with ninhydrin as the derivatization reagent, is the long-established method considered by many to be the best (Williams, 1988; Pickering, 1989). The major development in IEC has been the reduction in analysis time by improvements in resins to enable it to compete with pre-column derivatization methods based on RPC. Unfortunately this has only been achieved by the use of complex buffer systems and column temperatures (Williams, 1988). Inevitably peak resolution has often suffered and the performance of commercial instruments is often only guaranteed if expensive, ready-to-use buffers and ninhydrin are purchased from the manufacturer. The use of ninhydrin has always been something of a problem but there have been few attempts to improve it other than the development of Trione by Pickering (1981). Although still based on ninhydrin, sulfolane is used instead of 2-methoxyethanol as the solvent. It is claimed that the solution, buffered with lithium instead of sodium acetate, is relatively non-toxic and results in better stability, signal to noise ratio and resolution of peaks and more stable baselines than normal ninhydrin. The claim that it does not form precipitates and blockages in flow lines and reaction coils would be a major improvement since these problems have been exacerbated in some current IEC analysers by the faster analysis times and higher reaction coil temperatures. At least one commercial instrument uses Trione but it is three times as expensive as the normal reagent. Other derivatizing reagents, *o*-phthalaldehyde (OPA), fluorescamine, dabsylchloride (DABS-Cl) and 4-fluoro-7-nitro-2,1,3-benzoxadiazole (NBD-F), have been used to improve the sensitivity obtained with ninhydrin. Problems, particularly with the detection of proline and hydroxyproline, have prevented the widespread use of these reagents although the addition of thiols, such as 3-mercaptopropionic acid (MPA), to OPA may overcome this (Ashworth, 1987; Furst *et al.*, 1989).

Reversed phase chromatography (RPC)

The biggest development in amino acid analysis has been the advent of pre-column derivatization in which the mixture of amino acids is reacted with a reagent to form highly fluorescent or UV-absorbing derivatives and then separated by RPC. RPC is a partition system in which the mobile phase, usually acetate buffer with a gradient of acetonitrile or methanol, is more polar than the stationary phase, usually octadecyl bonded silicas. Many derivatizing reagents have been used, e.g. OPA, DABS-Cl, NBD-F 2,3-naphthalene dialdehyde (NDA), 1-fluoro-2,4-dinitrobenzene (FDNB), dansylchloride (DNS-Cl), phenylisothiocyanate (PITC), 9-fluorenylmethyl-chloroformate (FMOC), 4-N,N-dimethylaminoazobenzene-4'-isothiocya-nate (DABITC), 1-fluoro-2,4-dinitrophenyl-5-L-alanine amide (FDAA), 4-chloro-7-nitro-2,1,3-benzoxadiazole (NBD-Cl), ninhydrin and 1,2,3-peri-naphthindantrione hydrate. The highest sensitivity (subfemtomolar) has been achieved by the use of NDA with laser-induced fluorescence detection (Roach and Harmony, 1987). Thousands of papers have been published on pre-column derivatization and at least two of the methods, e.g. PITC (Godtfredsen and Oliver, 1980; Cohen and Strydom, 1988) and OPA (Alvarez-Coque et al., 1989) have been extensively reviewed ostensibly because many laboratories had difficulty in reproducing published procedures. Commercially, PITC (Waters; Applied Biosystems), DABS-Cl (Beckman) and FMOC (Varian) have been the most popular but none of these methods is perfect. For example Rogerson (1987), in discussing the Varian FMOC system, concluded that the resolution of pre-column systems was inferior to IEC because derivatization with a large organic molecule reduces the unique character of each amino acid. Although this might not be a problem with hydrolysates it would be with physiological samples.

The confusion regarding the choice of reagent has been exacerbated by the continual modification of the original methods. For example, the RPC columns used in the FMOC procedure published by Einarsson et al. (1983) were changed to fused silica columns (Einarsson et al., 1986) to improve the separation of amino acids in physiological fluids. One problem with FMOC is that excess reagent has to be removed by extraction with pentane before RPC. This step was tedious and liable to poor precision since some amino acids were extracted by the pentane. Betner and Foldi (1986, 1988) and Gustavsson and Betner (1990) proposed the use of 1-aminoadamantane (ADAM) instead of pentane, but Malmer and Schroeder (1990) have reported that the precision and recoveries were better with pentane than with ADAM. Godel et al. (1992) have recently suggested that the FMOC method can be improved by the use of OPA, with 3-MPA, to derivatize the primary amino acids, followed by derivatization of the secondary amino acids, which do not react with OPA, with FMOC.

DABS-Cl is a popular derivatizing reagent because its amino acid derivatives are stable at room temperature, in daylight, for up to one month and can be detected in the visible region (Knecht and Chang, 1986). However, in common with other reagents, the original method was not without its problems, particularly with the reproducibility of lysine, tyrosine, histidine, threonine and arginine. Modifications proposed by Vendrell and Aviles (1986) and Stocchi *et al.* (1989) claim to have resolved these problems and to have enabled tryptophan to be estimated in samples with high levels of carbohydrate. Stocchi *et al.* (1989) also described the use of DABITC as a derivatization reagent although this was more suitable for the microsequencing of pure proteins.

PITC is another reagent that forms derivatives that can be measured in the UV. Using RPC it gave comparable results with IEC for pure proteins and peptides (Heinrikson and Meredith, 1984). However, RPC often detected significant amounts of some amino acids in peptides where none existed according to their sequence analysis. Subsequent modifications by Bidlingmeyer *et al.* (1984) seem to have overcome this problem and reduced the analysis time from 50 to 23 minutes. Ebert (1986) carried out an extensive study to optimize the chromatography of the PITC amino acids in a partially successful attempt to solve some of the problems inherent in this method, i.e. stability of the mobile phase, poor resolution of arginine, threonine, alanine and proline and reagent-derived peaks and baseline instability. The procedure, with various other modifications, has been applied to feeds (Beaver *et al.*, 1987), foods and physiological fluids (Bidlingmeyer *et al.*, 1987), foods and faeces (Sarwar *et al.*, 1988) foods and feeds (Hagen *et al.*, 1989) and free amino acids in serum and tissues (Sarwar and Botting, 1990) but with varying success for certain amino acids. Sherwood *et al.* (1990) have described a method for determining free amino acids in plasma and urine using PITC derivatization followed by electrochemical detection which eliminates interferences encountered when UV detection is used for these samples.

FMOC, DABS-Cl and PITC methods have mainly been applied to protein hydrolysates. Most studies on the determination of free amino acids in blood and tissues have used OPA for derivatization Graser *et al.*, (1985). Hydroxyproline and proline can be detected by adding thiols such as 2-mercaptoethanol (MCE), ethanethiol or mercaptopropionic acid (MPA) to the OPA. There have been many modifications to methods using OPA. Hoskins *et al.* (1986) used OPA/MCE followed by electrochemical detection in the analysis of tissues and physiological fluids but warned of the need to use highly purified MCE and water because of high background interference. Furst *et al.* (1989) have suggested that problems with the resolution of certain amino acids could be overcome by replacing methanol with acetonitrile in the mobile phase and using OPA/MPA for derivatization. After comparative studies Furst *et al.* (1989, 1990) concluded that this method was better than FMOC, PITC or DNS-Cl for the determination of free amino acids in biological samples. Unfortunately there are many other comparative studies (Williams, 1988) which come to different conclusions depending on which derivatization procedure the author is trying to promote. McClung and Frankenberger (1988), in a comparative study of the determination of amino acids in protein hydrolysates with OPA/MCE and four other derivatization reagents, considered that there were advantages and disadvantages with all the methods and Kamp (1991) came to similar conclusions. The major problem with RPC is the choice of derivative, which presents an almost insurmountable task (Williams, 1988; McClung and Frankenberger, 1988).

Ideally it would be better if pre- and post-column derivatization could be avoided altogether. It is generally accepted that amino acids in aqueous solution cannot be detected by conventional electrochemical methods. However, Polta and Johnson (1983) reported that amino acids, in 0.25 M NaOH, could be separated by anion-exchange chromatography (AEC) and detected with a platinum (Pt) electrode. Variations of this technique using pulsed coulometric detection (PCD) or potential-sweep PCD with Pt and gold electrodes have been reported by Welch *et al.* (1989). These use complex aqueous mobile phases with AEC and would be difficult to apply to complex mixtures such as physiological fluids. Luo *et al.* (1991) have recently described a similar method using a copper (Cu) electrode, but in common with other electrochemical methods for the determination of underivatized amino acids this method is in need of further development. Another promising, but as yet unfulfilled, method for underivatized amino acids has been described by Levin and Grushka (1985). In this an aqueous solution of Cu acetate, alkyl sulfonate and acetate buffer as the mobile phase is used to separate amino acids by RPC. The Cu complexes of the amino acids are formed *in situ* on the column and detected by UV.

Gas liquid chromatography (GLC)

This is another form of pre-column derivatization since amino acids must be converted to their volatile derivatives before chromatography. However, there are even more derivatives than for RPC which may explain the limited use of GLC in animal nutrition studies. The *N*-trifluoro-acetyl *n*-butyl and *N*-heptafluorobutyryl (HFB) isobutyl esters appear to be the best. With the former it is impossible to separate all the amino acids on one column. All the *N*-HFB isobutyl esters of protein hydrolysates can be separated on a single methylsilicone column in 35 minutes or on a capillary column in 10 minutes (Mackenzie, 1987). Capillary columns are best for physiological fluids, analysis taking 1 hour. Table 2.1 summarizes the main features of GLC compared with IEC and RPC and some of the derivatization procedures used.

Table 2.1. Comparison of methods of amino acid analysis.

Method*	IEC	RPC			GLC
		DABS-Cl	PITC	FMOC	
Derivatization reagent†	Ninhydrin	DABS-Cl	PITC	FMOC	HFB anhydride
Derivatization time (min)	5	30	20	5	30
Solvent extraction	No	Yes	No	Yes	No
Detects imino acids	Yes	Yes	Yes	Yes	Yes
Quantitative yield of derivative	Yes	No	Yes	Yes	Yes
Stable derivatives	Yes	Yes	No	Yes	Yes
Column temperature (°C)	48–70	Ambient	52	Ambient	90–260
Analysis time (min)					
Hydrolysates	30	30	25	30	10–20
Physiological fluids	90	60	60	60	60
Detector	Visible	Visible	UV	Fluorometric	N-specific or electron capture
Sensitivity (pmol)	50	1	<1	0.2	0.01
Disadvantages	Expensive instrumentation and columns; Complex mobile phases and column temperatures; Low sensitivity	Some multiple or unstable derivatives; Complex purification and extraction; Interference from salts and lipids; Short column life; Poor resolution for some amino acids			Complex purification and derivatization; Interference from salts and lipids
Applications	Feed and digesta; blood and urine	Individual amino acids; hydrolysates		Hydrolysates	Hydrolysates, blood

* IEC = ion-exchange chromatography; RPC = reversed phase chromatography; GLC = gas liquid chromatography.
† DABS-Cl = dabsylchloride; PITC = phenylisothiocyanate; FMOC = 9-fluorenylmethylchloroformate; HFB = N-heptafluorobutyryl.

A.P. Williams

The Precision and Accuracy of Amino Acid Analysis

The precision of amino acid analysis between laboratories can be tested by collaborative trials. The results of the earlier trials were reviewed by Williams (1981) and updated by Williams (1988). Since then there have been five collaborative trials published for hydrolysates and one for blood plasma. A summary of these results, in comparison with trials completed in the early 1980s, is given in Table 2.2. Kreienbring (1987) reported the results of a collaborative trial in which five East German laboratories analysed a sample of peas by IEC. Precision was similar to that reported in earlier trials. Miller *et al.* (1989) organized a collaborative trial in which eight fish meals were analysed by eight laboratories, four British, three Norwegian and one South African. Five laboratories used IEC and three used GLC methods. It was concluded that neither of the methods was clearly superior but that a major source of variation in the IEC method was the instability of the ninhydrin reagent.

Table 2.2. Comparison of published results of collaborative trials on the analysis of amino acids in protein hydrolysates and blood plasma. Results are expressed as the means of coefficients of variation (%) of laboratory means.

	Protein hydrolysates		Blood plasma	
	1981–84	1985–91	Williams *et al.* (1980)	Rattenbury and Townsend (1990)
Aspartic acid	5.3	5.0	-	18.1
Threonine	6.8	4.5	21.5	16.4
Serine	5.5	7.5	19.0	15.6
Glutamic acid	7.0	4.6	59.5	14.0
Proline	7.4	6.4	-	30.4
Glycine	5.5	8.1	14.3	10.7
Alanine	5.2	5.2	22.0	12.3
Valine	6.2	6.1	17.5	13.1
Isoleucine	7.5	5.2	12.0	11.5
Leucine	5.0	4.7	15.0	10.7
Tyrosine	13.6	9.1	15.0	18.9
Phenylalanine	7.3	3.8	8.3	11.5
Lysine	6.8	4.9	24.0	17.3
Histidine	9.6	10.9	28.5	33.7
Arginine	7.4	9.2	18.8	22.2
Mean	7.3	6.3	21.2	16.3
Cystine	9.2	9.2	-	-
Methionine	7.5	11.3	57.0	59.1
Tryptophan	19.1	12.3	-	32.9

Most collaborative trials have involved 11 laboratories or less (Williams, 1988). However, Crabb *et al.* (1990) reported the results of a trial in which 36 laboratories from the Association of Biomolecular Resource Facilities took part. This was a particularly interesting study since it was the first time that there had been a direct comparison between IEC and RPC, with roughly half the laboratories using each method with either post- or pre-column derivatization. Most of the laboratories using RPC prepared PITC derivatives. β-Lactoglobulin A was chosen as the protein since its theoretical amino acid composition is well defined. Overall precision was similar by IEC and RPC but showed little improvement on earlier trials and it was concluded that better hydrolysis methodologies were required. There was considerable variation in the length and temperature of the acid hydrolysis procedures used and some laboratories added reducing agents to the HCl, which was sometimes used in the vapour phase and sometimes in the liquid phase. Standardization of hydrolysis and other pre-chromatographic procedures (Miller *et al.*, 1989) is a major problem. Although the precision (capacity of the method to provide the same result on repeated application to the same sample) reported by Crabb *et al.* (1990) was satisfactory the accuracy (extent to which a measurement approaches the true value of the measured quantity) was less good, particularly for arginine, methionine, glycine and isoleucine. The glycine error (+70%) was thought to be due to contamination, methionine (−25%) to oxidation and arginine (+27%) due to incomplete resolution from methionine sulfoxide and dehydroalanine in PITC analyses.

Van der Meer (1990) has reported the results of collaborative trials in which 12 Dutch feed laboratories analysed five feeds by IEC. It was concluded that the precision was adequate for feed science, manufacturing and trade. The results of the Dutch group for mixed feeds were in good agreement with results obtained on the same samples by the German International Analytical Group (IAG). The IAG results were reported by Van der Meer (1990) and have been included in the mean values (1985–91) given in Table 2.2. The Dutch group also analysed a sample of pig faeces but the coefficients of variation (CV) were less good than for the feeds, particularly for lysine (CV 13.9%), histidine (CV 37.8%), arginine (CV 12.5%), cystine (CV 16.9%) and methionine (CV 14.8%). Van der Meer (1990) considered that this was due to the difficulty of hydrolysing faeces and required further investigation.

From Table 2.2 it can be seen that the overall precision (mean CV 6.3%) of the 1985–91 protein hydrolysate trials has improved slightly over the earlier trials. However, for certain individual amino acids – serine, glycine, histidine, arginine and methionine – the precision had deteriorated. In a recently published cooperative, rather than collaborative study in Australia, Davies *et al.* (1992) reported an overall precision (mean CV) of 13.1%. No attempt was made to standardize methodology and the variability of results was attributed to poor quality assurance in the participating laboratories. The poor precision for methionine, cystine and tryptophan continues to give concern. Even in collaborative trials carried out specifically for these amino acids (MacDonald *et al.*, 1985; Sarwar *et al.*, 1985; Allred and MacDonald, 1988) the mean CVs of values between laboratories have been high. For cystine, methionine and tryptophan the CVs were 12.0, 9.5 and 16.6% respectively.

The precision obtained for the determination of free amino acids in blood plasma is much poorer than for hydrolysates although only two collaborative trials have been reported. These are summarized in Table 2.2. In the first, Williams *et al.* (1980), the mean CV was 21.2% when four laboratories analysed the same sample of bovine plasma. Precision was equally poor for the centrally deproteinized plasma and for a sample deproteinized by the participating laboratory. Since satisfactory results were obtained for a standard mixture of amino acids it was concluded that the deproteinizing procedure was a major problem, with the wide range of plasma amino acid concentrations and presence of more peaks in plasma than in protein hydrolysates being contributory factors. A much more extensive trial was organized and reported by Rattenbury and Townsend (1990). Nine different samples of lyophilized bovine plasma (plus one human) and two of liquid urine were analysed over two years by 26 clinical laboratories in Britain in a scheme designed to provide external quality assessment. Of these laboratories 18 used IEC with ninhydrin, four IEC with OPA and four RPC with OPA or dansyl derivatives. From Table 2.2 it can be seen that there were small improvements in the precision compared with that reported by Williams *et al.* (1980) except for tyrosine, phenylalanine, histidine and arginine. The performance of the laboratories did not improve in the second year of the trial although the number of errors through incorrect identification of peaks was reduced. In this respect 23 amino acids and related compounds, not normally found in protein hydrolysates, were detected in plasma. The greater number of peaks in a urine chromatogram was also considered to be a factor in the poorer overall precision obtained (CV 21%). IEC with ninhydrin detection gave the best precision although this was not significantly different from the fluorometric IEC or RPC techniques which had advantages in sensitivity and faster analysis times. In discussing possible reasons for this poor precision obtained for plasma samples Rattenbury and Townsend (1990) concluded that individual expertise may be more important than the method or analyser used.

Methods for Specific Amino Acids

The determination of the concentrations of methionine, cystine and tryptophan in feeds and digesta is important because they are often the most limiting amino acids. Unfortunately they cannot be estimated with any degree of precision (Williams, 1981), partly because they are present in low concentrations and partly because they are extensively degraded during acid hydrolysis.

Methionine and cystine

Methionine and cystine undergo oxidation to multiple derivatives, and controlled oxidation of methionine to methionine sulfone and of cystine to cysteic acid must be carried out with performic acid prior to acid hydrolysis. Currently the method described by Bech-Andersen *et al.* (1990), using IEC for analysis, seems to be the most promising. These authors used phenol as a scavenger, but Barkholt and Jensen (1989) used 3,3'-dithiodipropionic acid which, after IEC, gave satisfactory results for cystine in two pure proteins. RPC methods for oxidized hydrolysates, particularly those with high salt contents (Bech-Andersen *et al.*, 1990), do not appear to be in general use (Allred and MacDonald, 1988). However, an interesting new development is the use of RPC with photolytic electrochemical detection instead of derivatization for the determination of the sulfur and aromatic acids in pure proteins (Dou and Krull, 1990). Near-infrared reflectance spectroscopy (NIRS) has been used by Williams *et al.* (1985) for the determination of methionine in peas. This non-destructive method has the advantage of avoiding hydrolysis but does not appear to have been adopted except for the rapid screening of methionine and certain other amino acids in cereal breeding programmes (Gill *et al.*, 1979; Williams *et al.*, 1984).

Tryptophan

The degradation of tryptophan during acid hydrolysis unfortunately does not result in a measurable end product and it is normally measured after hydrolysis with barium, sodium or lithium hydroxide. The multitude of methods published up to 1981 have been reviewed by Steinhart (1984) and there have been many more since then. Most have involved changes to the hydrolysis conditions; for example, Wong *et al.* (1984) proposed the addition of pyridine borane to 6 M HCl, which reduced tryptophan to dihydro-tryptophan. It is claimed that, for pure proteins, the same hydrolysate can be analysed for tryptophan and all the other amino acids by either IEC or RPC. In an extensive study of the determination of tryptophan in foods, Nielsen and Hurrell (1985) found no difference between lithium and sodium hydroxide hydrolysis and recommended hydrolysis for 20 hours at 110°C followed by RPC using a fluorescence detector. Analysis time was about 30 minutes due to the slow elution of the internal standard 5-methyltryptophan. Results agreed well in comparison with three traditional non-chromatographic methods. Similar RPC procedures have been published by Delhaye and Landry (1986), Landry *et al.* (1988), Slump *et al.* (1991) and Bech-Andersen (1991) with a reduction of the analysis time to less than 10 minutes.

An alternative approach involving the revival of an older method, the acid ninhydrin method, has been described by Pinter-Szakacs and Molnar-Perl (1990). This uses acid hydrolysis of intact foods or feeds in the presence of ninhydrin, with which the tryptophan reacts before it can be degraded. Corrections must be made for tyrosine and it is not clear if tryptophan derivatives and degradation products react with the ninhydrin. Steinhart and Sandmann (1977) developed a rapid fluorometric detection system called the optical multichannel analyser for the determination of tryptophan in plasma. This was an attempt to overcome the lability of tryptophan to photo-oxidation and oxidation, but there were problems and details of an improved system have since been reported (Steinhart and Stutzle, 1984). The new system can apparently be applied to hydrolysates but few details were given.

Other amino acids

From time to time certain unusual amino acids attract considerable interest and other methods have to be developed for their determination. Examples include 3-methyl histidine (proposed as an index of muscle protein turnover), lysinoalanine (formed by alkali treatment of proteins), 2-aminoethyl phosphonic acid and diaminopimelic acid (proposed as markers of protozoal and bacterial protein respectively in ruminant digesta). Methods for these have been discussed by Cockburn and Williams (1984) and Williams (1988) and since interest in most of them has since declined it is not intended to review them here. However, it does illustrate the need to be aware of the existence of such amino acids because, even if they are not of interest, they may interfere with the resolution of other amino acids. A recent example is the detection of a toxic amino acid, gizzerosine, in fish meals. This is formed from histidine and lysine during overheating and causes gizzard erosion in chickens. A method, using RPC after pre-column derivatization with DNS-Cl, has recently been described by Wagener *et al.* (1991). Its formation may also result in a reduction in available lysine.

Measurement of Amino Acid Availability by Chemical Methods

The biological availability of amino acids in feeds is important in monogastric feed formulation to ensure that amino acid requirements for optimum performance are met. The many *in vivo* and *in vitro* methods have been reviewed by Whitacre and Tanner (1989) and Batterham (1992). Of these, growth assays, usually with chicks or rats, are the methods against which all others are usually evaluated although this approach is not particularly reliable (AFRC, 1987). However, these are very expensive and time consuming and there have been considerable efforts to find alternative *in vitro* methods. Microbiological methods using microorganisms such as *Streptococcus zymogenes* were of considerable interest at one time since the total and available content of essential amino acids in feeds could be determined simultaneously (Whitacre and Tanner, 1989). However, they are also expensive and time consuming and it is doubtful if there are any laboratories currently using these methods.

Enzymatic methods, in which the conditions of biological digestion are simulated with proteolytic enzymes, have recently been reviewed by Assoumani and Nguyen (1991). Assoumani *et al.* (1990) developed a technique for the measurement of available lysine based on the use of immobilized enzymes. The use of an α-L-lysine oxidase electrode to monitor the effect of microwave heat treatment on early Maillard reactions in raw soyabeans was proposed. Lysine must first be freed by hydrolysing the soyabeans at constant pH for 7 hours, with trypsin, chymotrypsin and peptidase followed by carboxypeptidase B. After this, analysis is very rapid (1 minute) and highly specific but whether this method is generally applicable to all feeds has yet to be demonstrated.

Marletta *et al.* (1992) measured available cystine and methionine in kidney beans after *in vitro* enzymic digestion. Good agreement was reported for cystine, but not methionine, in comparison with *in vivo* protein digestibilities determined with rats. A satisfactory explanation for this difference could not be suggested.

Chemical methods for the determination of available amino acids have been mainly restricted to lysine and were reviewed by Friedman (1982). The nutritive value of feeds subjected to heat during processing or to adverse storage conditions is often less than predicted from measurements of their total lysine content. Under these conditions, lysine may become nutritionally unavailable if its free ε-amino group reacts with, for example, carbohydrates, forming bonds resistant to digestive tract proteases. The most widely used method used to be that of Carpenter (1960) in which FDNB reacts with the free ε-amino groups of lysine and, after acid hydrolysis, the resultant dinitrophenyl (DNP)-lysine is measured spectrophotometrically. However, since this method is very time consuming, involving a considerable number of manipulations, and is prone to interference from other compounds there have been many attempts to find a better method. These include methods based upon the difference between the total lysine content after acid hydrolysis and the residual lysine content measured after acid hydrolysis of the sample reacted with FDNB, which is assumed to be the available lysine content. This so-called 'difference' method was developed by Roach *et al.* (1967) and later officially adopted by the Association of Official Analytical Chemists (Couch, 1975). However, the method was still time consuming and does not measure available lysine accurately in foods containing early Maillard products. Other reagents that have been used include OPA (Goodno *et al.*, 1981), succinic anhydride and DNS-Cl (Anderson and Erbersdobler, 1982) and 2,4,6-trinitrobenzene sulfonic acid (TNBS) (James and Ryley, 1986). All suffer from problems including long analysis time and interfering compounds and have usually been developed with model systems or pure proteins. Analysis time has been reduced to 25 minutes using TNBS (Tomarelli *et al.*, 1985) and RPC to measure the reaction product ε-TNP-lysine by UV detection at 346 nm. However, the method, used for milk and infant formulas, does not appear to have been used for other feeds. Peterson and Warthesen (1979) have also reported the analysis of available lysine in casein, soya and gluten-glucose mixtures with FDNB followed by analysis of the DNP-lysine by RPC. The analysis took only 15 minutes with a C_{18} reversed phase column and a mobile phase of acetonitrile-0.01 M acetate buffer at pH 4.0. A similar method for milk proteins was described by Wundenberg and Klostermeyer (1983).

Hurrell *et al.* (1979) have suggested the use of dye-binding techniques using Acid Orange 12 to determine the available and total lysine content of a wide range of animal feeds. Using commercial dye-binding equipment the procedure takes 40 minutes. Two of these dye-binding methods were compared with chick growth assays and the FDNB method of Carpenter (1960) in an extensive collaborative trial on eight fish meals reported by Barlow *et al.* (1984). It was concluded that none of the *in vitro* methods could satisfactorily distinguish between normal commercial fish meals in terms of their absolute content of biologically available lysine. However, the dye-binding methods were considered suitable for process control within a single laboratory.

A completely different approach to the determination of available lysine, but again applied only to pure proteins, is the use of ^{19}F nuclear magnetic resonance (NMR) spectroscopy (Cavanaugh, 1988). The proteins are reacted with S-ethyl trifluorothioacetate and rapidly analysed by NMR. Unfortunately the results have been variable and few laboratories are equipped for this technique.

None of the *in vitro* methods are entirely satisfactory and it would appear that, for the moment, it is better to use one of the rapid RPC or dye-binding methods. If differences between feeds of the same type or of known treatment, e.g. heat or formaldehyde treatment of ruminant feeds, are required then these would be acceptable. Table 2.3 shows the advantages and disadvantages of methods used for determining the availability of amino acids in feeds.

Table 2.3. Comparison of methods for the determination of available lysine.

Method	Advantages	Disadvantages	Time required
Biological	Reference method	Slow and costly Low precision Animals required	3 weeks
Microbiological	Other available amino acids determined	Slow with low precision Enzyme pretreatment needed	1 week
Chemical FDNB		Slow Not specific Hazardous reagents Free lysine not detected Early Maillard products not detected	2 days
Dye-binding	Rapid, simple analysis Good precision	Not specific Free lysine not detected Interfering compounds Inability to differentiate heat damage	40 mins
^{19}F-NMR	Rapid analysis	Poor precision Expensive equipment Reducing sugars interfere	2 days
Immobilized enzymes	Rapid analysis Highly specific Detects early Maillard products	Interfering compounds Short membrane life Coenzymes required	6 hours

Conclusions

It is clear that there is no perfect procedure for the determination of total or available amino acids in either pure proteins or in feeds and physiological fluids. Hydrolysis of proteins, the deproteinization of physiological fluids and the determination of methionine, cystine and tryptophan remain major problems. Other, more general, problems are often overlooked, either because they were published some time ago or because they formed a minor part of the paper in which they appeared (Williams,

1986). Current developments in the chromatography of amino acids, particularly by RPC, although improving the speed and sensitivity of the methods have done little to improve their precision and accuracy. Indeed RPC, with its multitude of pre-column derivatization reagents, presents the analyst with the major problem of choosing the best method. Such confusion exists that some authors have published their own modifications twice or three times. It is necessary to read the small print very carefully in any paper claiming to have improved an existing RPC method! There have been many studies, usually comparing RPC with IEC and occasionally GLC, in which good agreement between methods and the theoretical values is claimed but not always adequately demonstrated. There is probably no ideal pre-column derivatization reagent and it is advisable to choose an RPC system that can use any of the reagents (Kochar and Christen, 1989).

The choice between IEC, RPC and GLC is also difficult and much depends on the applications, sensitivity and urgency required and the expertise available. The traditional IEC method with post-column ninhydrin derivatization is ideal for feed and digesta hydrolysates and the complex mixture of free amino acids in plasma and urine, especially as there is usually no shortage of sample. It is often suggested that the faster analysis times of RPC are a considerable advantage although the times taken for sample preparation and data handling cannot be ignored. Few data are available on the workload of laboratories other than that of Rattenbury and Townsend (1990), who showed that the number of plasma samples analysed annually by 26 clinical laboratories ranged from 50 to 1250 (mean 505). This suggests the recent considerable effort to reduce the time of the chromatography stage is misplaced. Recommendations by AFRC (1987) on the characterization of the nitrogenous constituents of feedstuffs suggest that it would be better if existing methods of amino acid analysis were applied with greater care. The need 'to employ suitably qualified staff and to both allow time and develop procedures for internal checking of results' was also recommended, as was increased communication and collaborative tests between laboratories. In this respect the establishment of external quality assessment schemes such as that described by Rattenbury and Townsend (1990) for clinical laboratories involved in plasma amino acid analysis should be encouraged. Similar schemes for protein hydrolysates already exist in Germany, Holland and the USA. Unfortunately the organization recommended by AFRC (1987) to fulfil this role for protein hydrolysates in the UK, the AFRC Amino Acid Analyser Users Group, has since been disbanded.

References

AFRC (Agricultural and Food Research Council) (1987) Technical Committee on Responses to Nutrients No. 2, Characterisation of feedstuffs: nitrogen. *Nutrition Abstracts and Reviews, Series B: Livestock Feeds and Feeding* 57, 713–736.

Allred, M.C. and MacDonald, J.L. (1988) Determination of sulfur amino acids and tryptophan in foods and feed ingredients: collaborative study. *Journal of the Association of Official Analytical Chemists* 71, 603–606.

Alvarez-Coque, M.C.G., Hernandez, M.J.M., Camanas, R.M.V. and Fernandez, C.M. (1989) Formation and instability of *o*-phthalaldehyde derivatives of amino acids. *Analytical Biochemistry* 178, 1–7.

Andersen, S., Mason, V.C. and Bech-Andersen, S, (1984) EEC collaborative studies on a streamlined hydrolysate preparation method for amino acid determinations in feedstuffs. *Zeitschrift für Tierphysiologie, Tierernährung und Futtermittelkunde* 51, 113–129.

Anderson, T.R. and Erbersdobler, H.F. (1982) A comparison of methods for the rapid determination of available lysine in model proteins. *South African Food Review* 9, 33–35.

Aristoy, M.-C. and Toldra, F. (1991) Deproteinization techniques for HPLC amino acid analysis in fresh pork muscle and dry-cured ham. *Journal of Agricultural and Food Chemistry* 39, 1792–1795.

Ashworth, R.B. (1987) Ion-exchange separation of amino acids with post column orthophthalaldehyde detection. *Journal of the Association of Official Analytical Chemists* 70, 248–252.

Assoumani, M.B. and Nguyen, N.P. (1991) Enzyme modelling of protein digestion and L-lysine availability. In: Fuller, M. (ed.) In vitro *Digestion For Pigs and Poultry*. CAB International, Wallingford, Oxon, pp. 86–104.

Assoumani, M.B., Nguyen, N.P., Lardinois, P.F., Van Bree, J., Baudichau, A. and Bruyer, D.C. (1990) Use of a lysine oxidaze electrode for lysine determination in Maillard model reactions and in soyabean meal hydrolysates. *Lebensmittel Wissenschaft und Technologie* 23, 322–327.

Barkholt, V. and Jensen, A.L. (1989) Amino acid analysis: determination of cysteine plus half-cystine in proteins after hydrochloric acid hydrolysis with a disulfide compound as additive. *Analytical Biochemistry* 177, 318–322.

Barlow, S.M., Collier, G.S., Jurtiz, J.M., Burt, J.R., Opstvedt, J. and Miller, E.L. (1984) Chemical and biological assay procedures for lysine in fish meals. *Journal of the Science of Food and Agriculture* 35, 154–164.

Batterham, E.S. (1992) Availability and utilization of amino acids for growing pigs. *Nutrition Research Reviews* 5, 1–18.

Beaver, R.W., Wilson, D.M., Jones, H.M. and Haydon, K.D. (1987) Amino acid analysis in feeds and feedstuffs using precolumn phenylisothiocyanate derivatization and liquid chromatography – Preliminary study. *Journal of the Association of Official Analytical Chemists* 70, 425–428.

Bech-Andersen, S. (1991) Determination of tryptophan with HPLC after alkaline hydrolysis in autoclave using α-methyl-tryptophan as internal standard. *Acta Agricultura Scandinavica* 41, 305–309.

Bech-Andersen, S., Mason, V.C. and Dhanoa, M.S. (1990) Hydrolysate preparation for amino acid determinations in feed constituents. 9. Modifications to oxidation and hydrolysis conditions for streamlined procedures. *Zeitschrift für Tierphysiologie, Tierernährung und Futtermittelkunde* 63, 188–197.

Betner, I. and Foldi, P. (1986) New automated amino acid analysis by HPLC precolumns derivatization with fluorenylmethyloxycarbonylchloride. *Chromatographia* 22, 381–387.

Betner, I. and Foldi, P. (1988) The FMOC-ADAM approach to amino acid analysis. *LC-GC International* 2, 44–53.

Bidlingmeyer, B., Cohen, S.A. and Tarvin, T.L. (1984) Rapid analysis of amino acids using precolumn derivatization. *Journal of Chromatography* 336, 93–104.

Bidlingmeyer, B., Cohen, S.A., Tarvin, T.L. and Frost, B. (1987) A new, rapid, high sensitivity analysis of amino acids in food type samples. *Journal of the Association of Official Analytical Chemists* 70, 241–247.

Carpenter, K.J. (1960) The estimation of available lysine in animal protein foods. *Biochemical Journal* 77, 604–610.

Cavanaugh, J.R. (1988) Elimination of lactose interference in the determination of available lysine using fluorine-19 nuclear magnetic resonance spectroscopy. *Journal of Dairy Science* 71, 1147–1151.

Chiou, S.-H. and Wang, K.-T. (1989) Peptide and protein hydrolysis by microwave irradiation. *Journal of Chromatography* 491, 424–431.

Chiou, S.-H. and Wang, K.-T. (1990) A rapid and novel means of protein hydrolysis by microwave irradiation using Teflon-Pyrex tubes. In: Villafranca, J.J. (ed.) *Current Research in Protein Chemistry: Techniques, Structure and Function*. Academic Press, London, pp, 3–10.

Cockburn, J.E. and Williams, A.P. (1984) The simultaneous estimation of the amounts of protozoal, bacterial and dietary nitrogen entering the duodenum of steers. *British Journal of Nutrition* 51, 111–132.

Cohen, S.A. and Strydom, D.J. (1988) Amino acid analysis utilizing phenylisothiocyanate derivatives. *Analytical Biochemistry* 174, 1–16.

Couch, J.R. (1975) Collaborative study of the determination of available lysine in proteins and feeds. *Journal of the Association of Official Analytical Chemists* 58, 599–601.

Crabb, J.W., Ericsson, L., Atherton, D., Smith, A.J. and Kutny, R. (1990) A collaborative amino acid analysis study from the Association of Biomolecular Resource Facilities. In: Villafranca, J.J. (ed.) *Current Research in Protein Chemistry: Techniques, Structure and Function*. Academic Press, London, pp. 49–61.

Davey, J.F. and Ersser, R.S. (1990) Amino acid analysis of physiological fluids by high-performance liquid chromatography with phenylisothiocyanate derivatization and comparison with ion-exchange chromatography. *Journal of Chromatography* 528, 9–23.

Davies, R.L., Baigent, D.R., Levitt, M.S., Mollah, Y., Rayner, C.J. and Frensham, A.B. (1992) Accuracy and precision in amino acid analysis. *Journal of the Science of Food and Agriculture* 59, 423–436.

Delhaye, S. and Landry, J. (1986) High-performance liquid chromatography and ultraviolet spectrophotometry for quantitation of tryptophan in barytic hydrolysates. *Analytical Biochemistry* 159, 175–178.

Dou, L. and Krull, I.S. (1990) Determination of aromatic and sulfur-containing amino acids, peptides and proteins using high-performance liquid chromatography with photolytic electrochemical detection. *Analytical Chemistry* 62, 2599–2606.

Ebert, R.F. (1986) Amino acid analysis by HPLC: optimized conditions for chromatography of phenylthiocarbamyl derivatives. *Analytical Biochemistry* 154, 431–435.

Einarsson, S., Josefsson, B. and Lagerkvist, S. (1983) Determination of amino acids with 9-fluorenylmethyl chloroformate and reversed-phase high-performance liquid chromatography. *Journal of Chromatography* 282, 609–618.

Einarsson, S., Folestad, S., Josefsson, B. and Lagerkvist, S. (1986) High-resolution reversed-phase liquid chromatography system for the analysis of complex solutions of primary and secondary amino acids. *Analytical Chemistry* 58, 1638–1643.

Elkin, R.G. and Griffith, J.E. (1985) Hydrolysate preparation of amino acids in sorghum grains: effect of oxidative pretreatment. *Journal of the Association of Official Analytical Chemists* 68, 1117–1121.

Friedman, M. (1982) Chemically reactive and unreactive lysine as an index of browning. *Diabetes* 31, 5–14.

Furst, P., Pollack, L., Graser, T.A., Godel. H. and Stehle, P. (1989) HPLC analysis of free amino acids in biological material – an appraisal of four pre-column derivatization methods. *Journal of Liquid Chromatography* 12, 2733–2760.

Furst, P., Pollock, L., Graser, T.A., Godel, H. and Stehle, P. (1990) Appraisal of four pre-column derivatization methods for the high-performance liquid chromatographic determination of free amino acids in biological materials. *Journal of Chromatography* 499, 557–569.

Gehrke, C.W., Wall, L.S Sr, Absheer, J.S., Kaiser, F.E. and Zumwalt, R.W. (1985) Sample preparation for chromatography of amino acids: acid hydrolysis of proteins. *Journal of the Association of Official Analytical Chemists* 68, 811–821.

Gehrke, C.W., Rexroad, P.R., Schisla, R.M., Absheer, J.S. and Zumwalt, R.W. (1987) Quantitative analysis of cystine, methionine, lysine, and nine other amino acids by a single oxidation-4 hour hydrolysis method. *Journal of the Association of Official Analytical Chemists* 70, 171–174.

Gill, A.A., Starr, C. and Smith, C.B. (1979) Lysine and nitrogen measurements by infra-red reflectance analysis as an aid to barley breeding. *Journal of Agricultural Science (Cambridge)* 93, 727–733.

Gilman, L.B. and Woodward, C. (1990) An evaluation of microwave heating for the vapor phase hydrolysis of proteins. 1. Comparison to vapor phase hydrolysis for 24 hours. In: Villafranca, J.J. (ed.) *Current Research in Protein Chemistry: Techniques, Structure and Function.* Academic Press, London, pp. 23–26.

Godel, H., Seitz, P. and Verhoef, M. (1992) Automated amino acid analysis using combined OPA and FMOC-CI precolumn derivatization. *LG-GC International* 5, 44–49.

Godtfredsen, S.E. and Oliver, R.W.A. (1980) On the analysis of phenylthiohydantoin amino acids by high performance liquid chromatography. *Carlsberg Research Communications* 45, 35–46.

Goodno, C.C., Swaisgood, H.E. and Catignani, G.L. (1981) A fluorimetric assay for available lysine in proteins. *Analytical Biochemistry* 115, 203–211.

Graser, T.A., Godel, H.G., Albers, S., Foldi, P. and Furst, P. (1985) An ultra rapid and sensitive high-performance liquid chromatographic method for determination of tissue and plasma free amino acids. *Analytical Biochemistry* 151, 142–152.

Gustavsson, B. and Betner, I. (1990) Fully automated amino acid analysis for protein and peptide hydrolysates by precolumn derivatization with 9-fluorenyl-methyl chloroformate and 1-aminoadamantane. *Journal of Chromatography* 507, 67–77.

Hagen, S.R., Frost, B. and Austin, J. (1989) Precolumn phenylisothiocyanate derivatization and liquid chromatography of amino acids in food. *Journal of the Association of Official Analytical Chemists* 72, 912–916.

Hare, P.E., St John, P.A. and Engel, M.H. (1985) Ion-exchange separation of amino acids In: Barrett, G.C. (ed.) *Chemistry and Biochemistry of the Amino Acids*. Chapman & Hall, London, New York, pp. 415–425.

Heinrikson, R.L. and Meredith, S.C. (1984) Amino acid analysis by reverse-phase high-performance liquid chromatography: precolumn derivatization with phenylisothiocyanate. *Analytical Biochemistry* 136, 65–74.

Hoskins, J.A., Holliday, S.B. and Davies, F.F. (1986) Problems in the analysis of low levels of amino acids in physiological fluids and tissues using *o*-phthalaldehyde derivatization and reversed-phase high-performance liquid chromatography with electrochemical detection. *Journal of Chromatography* 375, 129–133.

Hubbard, R.W., Chambers, J.G., Sanchez, A., Slocum, R. and Lee, P (1988) Amino acid analysis of plasma: studies in sample preparation. *Journal of Chromatography* 431, 163–169.

Hunt, S. (1985) Degradation of amino acids accompanying *in vitro* protein hydrolysis. In: Barrett, G.C. (ed.) *Chemistry and Biochemistry of the Amino Acids*. Chapman & Hall, London, New York, pp. 376–398.

Hurrell, R.F., Lerman, P. and Carpenter, K.J. (1979) Reactive lysine in foodstuffs as measured by a rapid dye-binding procedure. *Journal of Food Science* 44, 1221–1231.

James, N.A. and Ryley, J. (1986) The rapid determination of chemically reactive lysine in the presence of carbohydrates by a modified trinitrobenzene sulfonic acid procedure. *Journal of the Science of Food and Agriculture* 37, 151–156.

Kamp, R.M. (1991) High-sensitivity amino acid analysis using high-performance liquid chromatography and precolumn derivatization. *LC-GC International* 4, 40–46.

Knecht, R. and Chang, J.Y. (1986) Liquid chromatographic determination of amino acids after gas-phase hydrolysis and derivatization with dimethylamino azobenzenesulfonyl chloride. *Analytical Chemistry* 58, 2375–2379.

Kochar, S. and Christen, P. (1989) Amino acid analysis by high-performance liquid chromatography after derivatization with 1-fluoro-2,4-dinitrophenyl-5-L-alanine amide. *Analytical Biochemistry* 178, 17–21.

Kreienbring, F. (1987) Further results of the comparative determination of amino acids. *Die Nahrung* 9, 855–862.

Labadarios, D., Moodie, I.M. and Shephard, G.S. (1984) Gas chromatographic analysis of amino acids in physiological fluids: a critique. *Journal of Chromatography* 310, 223–231.

Landry, J., Delhaye, S. and Viroben, G. (1988) Tryptophan content of feedstuffs as determined from three procedures using chromatography of barytic hydrolysates. *Journal of Agricultural and Food Chemistry* 36, 51–52.

Levin, S. and Grushka, E. (1985) Reversed-phase liquid chromatographic separation of amino acids with aqueous mobile phases containing copper ions and alkylsulfonates. *Analytical Chemistry* 57, 1830–1835.

Luo, P., Zhang, F. and Baldwin, R.P. (1991) Constant potential amperometric detection of underivatized amino acids and peptides at a copper electrode. *Analytical Chemistry* 63, 1702–1707.

MacDonald, J.L., Krueger, M.W. and Keller, J.H. (1985) Oxidation and hydrolysis determination of sulfur amino acids in food and feed ingredients: collaborative study. *Journal of the Association of Official Analytical Chemists* 68, 826–829.

Mackenzie, S.L. (1987) Gas chromatographic analysis of amino acids as the N-heptafluorobutyryl isobutyl esters. *Journal of the Association of Official Analytical Chemists* 70, 151–160.

Malmer, M.F. and Schroeder, L.A. (1990) Amino acid analysis by high-performance liquid chromatography with methanesulfonic acid hydrolysis and 9-fluorenyl-methylchloroformate derivatization. *Journal of Chromatography* 514, 227–239.

Marletta, L., Carbonaro, M. and Carnovale, E. (1992) *In vitro* protein and sulfur amino acid availability as a measure of bean protein quality. *Journal of the Science of Food and Agriculture* 59, 497–504.

McClung, G. and Frankenberger, W.T. Jr (1988) Comparison of reverse-phase high-performance liquid chromatographic methods for pre-column derivatized amino acids. *Journal of Liquid Chromatography* 11, 613–646.

Miller, E.L., Juritz, J.M., Barlow, S.M. and Wessels, J.P.H. (1989) Accuracy of amino acid analysis of fish meals by ion-exchange and gas chromatography. *Journal of the Science of Food and Agriculture* 47, 293–310.

Nielsen, H.K. and Hurrell, R.E. (1985) Tryptophan determination of food proteins by h.p.l.c. after alkaline hydrolysis. *Journal of the Science of Food and Agriculture* 36, 893–907.

Papadoyannis, I.N. (1990) HPLC in the analysis of amino acids. In: Papadoyannis, I.N. (ed.) *HPLC in Clinical Chemistry*. Marcel Dekker, New York, pp. 97–154.

Perrett, D. (1985) Liquid chromatography of amino acids and their derivatives. In: Barrett, G.C. (ed.) *Chemistry and Biochemistry of the Amino Acids*. Chapman & Hall, London, New York, pp. 426–461.

Peterson, W.R. and Warthesen, J.J. (1979) Total and available lysine determinations using high-pressure liquid chromatography. *Journal of Food Science* 44, 994–997.

Pickering, M.V. (1981) Ninhydrin reagent for use in amine and amino acid analysis. *U.S. Patent* 4, 274, 833.

Pickering, M.V. (1989) Ion-exchange chromatography of free amino acids. *LC-GC International* 2, 25–29.

Pickering, M.V. and Newton, P. (1990) Amino acid hydrolysis: old problems, new solutions. *LC-GC International* 3, 22–26.

Pinter-Szakacs, M. and Molnar-Perl, I. (1990) Determination of tryptophan in unhydrolyzed food and feedstuffs by the acid ninhydrin method. *Journal of Agricultural and Food Chemistry* 38, 720–726.

Polta, J.A. and Johnson, D.C. (1983) The direct electrochemical detection of amino acids at a platinum electrode in an alkaline chromatographic effluent. *Journal of Liquid Chromatography* 6, 1727–1743.

Qureshi, G.A. and Qureshi, A.R. (1989) Determination of free amino acids in biological samples: problems of quantitation. *Journal of Chromatography* 491, 281–289.

Rattenbury, J.M. and Townsend, J.C. (1990) Establishment of an external quality-assessment scheme for amino acid analysis: results from assays of samples distributed in two years. *Clinical Chemistry* 36, 217–224.

Roach, A.G., Sanderson, P. and Williams, D.R. (1967) Comparison of methods for the determination of available lysine value in animal and vegetable protein sources. *Journal of the Science of Food and Agriculture* 18, 274–278.

Roach, M.C. and Harmony, M.D. (1987) Determination of amino acids at subfemtomole levels by high-performance liquid chromatography with laser-induced fluorescence detection. *Analytical Chemistry* 59, 411–415.

Rogerson, G. (1987) Amino acid analysis by HPLC with pre-column derivatization. *International Analyst* 4, 19–26.

Rowan, A.M., Moughan, P.J. and Wilson, M.N. (1992) Effect of hydrolysis time on the determination of amino acid composition of diet, ileal digesta and faeces samples and on the determination of dietary amino acids digestibility coefficients. *Journal of Agricultural and Food Chemistry* 40, 981–985.

Sarwar, G. and Botting, H.G. (1990) Rapid analysis of nutritionally important free amino acids in serum and organs (liver, brain and heart) by liquid chromatography of precolumn phenylisothiocyanate derivatives. *Journal of the Association of Official Analytical Chemists* 73, 470–475.

Sarwar, G., Blair, R., Friedman, M., Gumbmann, M.R., Hackler, L.R., Pellett, P.L. and Smith, T.K. (1985) Comparison of interlaboratory variation in amino acid analysis and rat growth assays for evaluating protein quality. *Journal of the Association of Official Analytical Chemists* 68, 52–56.

Sarwar, G., Botting, H.G. and Peace, R.W. (1988) Complete amino acid analysis in hydrolysates of foods and faeces by liquid chromatography of precolumn phenylisothiocyanate derivatives. *Journal of the Association of Official Analytical Chemists* 71, 1172–1175.

Sedgwick, G.W., Fenton, T.W. and Thompson, J.R. (1991) Effect of protein precipitating agents on the recovery of plasma free amino acids. *Canadian Journal of Animal Science* 71, 953–957.

Sherwood, R.A., Titheradge, A.C. and Richards, D.A. (1990) Measurement of plasma and urine amino acids by high-performance liquid chromatography with electrochemical detection using phenylisothiocyanate derivatization. *Journal of Chromatography* 528, 293–303.

Slump, P., Flissebaalje, T.D. and Haaksman, I.K. (1991) Tryptophan in food proteins: a comparison of two hydrolytic procedures. *Journal of the Science of Food and Agriculture* 55, 493–496.

Steinhart, H. (1984) Summary of the workshop on tryptophan analysis. In: Zebrowska, T., Burazewska, L., Buraczewski, S., Kowalczyk, J. and

Pastuszewska, B. (eds) *Proceedings of the VI International Symposium on Amino Acids*. Polish Scientific Publishers, Warsaw, pp. 434–447.

Steinhart, H. and Sandmann, J. (1977) Determination of tryptophan in plasma with a spectrofluorometer system. *Analytical Chemistry* 49, 950–953.

Steinhart, H. and Stutzle, P. (1984) The application of a spectrofluorometer-optical multichannel analyzer system to the determination of tryptophan. In: Schlossberger, H.G., Kochen, W., Linzen, B. and Steinhart, H. (eds) *Progress in Tryptophan and Serotonin Research*. de Gruyter, Berlin, pp. 103–106.

Stocchi, V., Piccoli, G., Magnani, M., Palma, F., Biagiarelli, B. and Cucchiarini, L. (1989) Reversed-phase high-performance liquid chromatography separation of dimethylaminoazobenzenesulfonyl- and dimethylaminoazobenzene thiohydantoin-amino acid derivatives for amino acid analysis and microsequencing studies at the picomole level. *Analytical Biochemistry* 178, 107–117.

Tomarelli, R.M., Yuhas, R.J., Fisher, A. and Weaber, J.R. (1985) An HPLC method for the determination of reactive (available) lysine in milk and infant formulas. *Journal of Agricultural Food Chemistry* 33, 316–318.

Tristram, G.R. and Rattenbury, J.M. (1981) The development of amino acid analysis. In: Rattenbury, J.M. (ed.) *Amino Acid Analysis*. Ellis Horwood, Chichester, pp. 16–36.

Uhe, A.M., Collier, G.R., McLennan, E.A., Tucker, D.J. and O'Dea, K. (1991) Quantitation of tryptophan and other plasma amino acids by automated pre-column *o*-phthaldialdehyde derivatization high-performance liquid chromatography: improved sample preparation. *Journal of Chromatography* 564, 81–91.

Van der Meer, J.M. (1990) Amino acid analysis of feeds in the Netherlands: four-year proficiency study. *Journal of the Association of Official Analytical Chemists* 73, 394–398.

Vendrell, J. and Aviles, F.X. (1986) Complete amino acid analysis of proteins by dabsyl derivatization and reversed-phase liquid chromatography. *Journal of Chromatography* 358, 401–413.

Wagener, W.W.D., Koch, K.R., Wessels, J.P.H. and Post, P.J. (1991) An investigation into the occurrence of the toxic amino acid gizzerosine in fish meal. *Journal of the Science of Food and Agriculture* 54, 147–152.

Welch, L.E., LaCourse, W.R., Mead, D.A. Jr and Johnson, D.C. (1989) Comparison of pulsed coulometric detection and potential-sweep pulsed coulometric detection for underivatized amino acids in liquid chromatography. *Analytical Chemistry* 61, 555–559.

West, K.A. and Crabb, J.W. (1990) Performance evaluation automatic hydrolysis and PTC amino acid analysis. In: Villafranca, J.J. (ed.) *Current Research in Protein Chemistry: Techniques, Structure and Function*. Academic Press, London, New York, pp. 37–48.

Whitacre, M.E. and Tanner, H. (1989) Methods of determining the bioavailability of amino acids for poultry. In: Friedman, M. (ed.) *Absorption and Utilization of Amino Acids III*. CRC Press, Boca Raton, Florida, pp. 129–141.

Williams, A.P. (1981) Collaborative trials and amino acid analysis. In: Rattenbury, J.M. (ed.) *Amino Acid Analysis*. Ellis Horwood, Chichester, pp. 138–152.

Williams, A.P. (1986) General problems associated with the analysis of amino acids by ion-exchange chromatography. *Journal of Chromatography* 373, 175–190.

Williams, A.P. (1988) Determination of amino acids. In: MaCrae, R. (ed.) *HPLC in Food Analysis*. Academic Press, London, New York, pp. 441–470.

Williams, A.P. and Cockburn, J.E. (1988) Internal standards in amino acid analysis. *Chromatography and Analysis* 1, 9–11.

Williams, A.P., Hewitt, D., Cockburn, J.E., Davies, M.G., Harris, D.A. and Moore, R.A. (1980) A collaborative study on the determination of free amino acids in blood plasma. *Journal of the Science of Food and Agriculture* 31, 474–480.

Williams, P.C., Preston, K.R. and Starkey, P.M. (1984) Determination of amino acids in wheat and barley by near-infrared reflectance spectroscopy. *Journal of Food Science* 49, 17–20.

Williams, P.C., Mackenzie, S.L. and Starkey, P.M. (1985) Determination of methionine in peas by near-infrared reflectance spectroscopy (NIRS). *Journal of Agricultural and Food Chemistry* 33, 811–815.

Wong, W.S.D., Osuga, D.T., Burcham, T.S. and Feeney, R.E. (1984) Determination of tryptophan as the reduced derivative by acid hydrolysis and chromatography. *Analytical Biochemistry* 143, 62–70.

Wundenberg, K. and Klostermeyer, H. (1983) Measurement of reactive lysine by a HPLC method. *Kieler Milchwirtschaftliche Forschungsberichte* 35, 315–317.

Zumwalt, R.W., Absheer, J.S., Kaiser, F.E. and Gerkhe, C.W. (1987a) Acid hydrolysis of proteins for chromatographic analysis of amino acids. *Journal of the Association of Official Analytical Chemists* 70, 147–151.

Zumwalt, R.W., Kuo, K.C.T. and Gerkhe, C.W. (eds) (1987b) *Amino Acid Analysis by Gas Chromatography*. CRC Press, Boca Raton, Florida. Vols 1–3.

Utilization of Precursors for L-Amino Acids

3

D.H. BAKER

Department of Animal Sciences and Division of Nutritional Sciences, University of Illinois, 290 Animal Sciences Laboratory, 1207 West Gregory Drive, Urbana, Illinois, 61801, USA.

Introduction

Since 1980, virtually all amino acids (AA) have been made available as the natural L-isomer. Most AA are made via fermentative synthesis, although some are made via chemical synthesis (e.g. DL-methionine) or from chemical extraction processes (e.g. L-cystine). Chemical synthesis generally results in the DL-racemic mixture (1:1 ratio of D- and L-isomers) while fermentative synthesis and chemical extraction generally yield the L-isomer. Dietary AA or AA precursors must be converted to the L-AA in the animal body in order for protein synthesis to proceed.

As L-AA precursor materials, D-forms of AA that have biological efficacy are converted to L-isomers via a two-step reaction sequence involving oxidation (alpha carbon) to the keto analogue and then transamination of the keto analogue to the L-AA (Fig. 3.1). Some AA analogues are used in clinical applications, particularly for renal patients, and these compounds generally contain a hydroxyl or keto group at the alpha carbon instead of an amino group. For biological efficacy, hydroxyl-substituted AA analogues must first be enzymatically oxidized to the keto analogue, and then transaminated to the L-AA.

The principal barriers to complete utilization of unnatural isomers of AA are (i) inefficient absorption from the gut; (ii) inefficient oxidation to the keto derivative; and (iii) inefficient transamination of the keto derivative to the L-AA. Unnatural AA isomers that are poorly utilized generally are not metabolized. Instead, they are spilled into the urine. D-AA in the food and feed supply of animals and humans are provided by fermentation products (i.e. yogurt, cheese, etc.) in that microbes do contain some D-AA. Under normal dietary conditions, free AA are present in

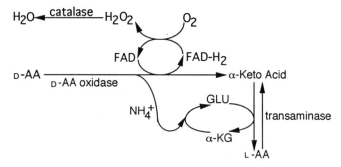

Fig. 3.1. Conversion of D-amino acids to L-amino acids.

very low concentrations in urine, but when poorly utilized D-AA are consumed, aminoaciduria occurs. This is illustrated by D-methionine (poorly utilized by humans) appearing in the urine of human subjects following D-methionine ingestion (Efron *et al.*, 1969; Stegink *et al.*, 1971) and D-tryptophan (poorly utilized by dogs) appearing in the urine of dogs following D-tryptophan ingestion (Czarnecki and Baker, 1982).

The review that follows will summarize the current state of knowledge regarding the biological efficacy of precursor compounds for indispensable AA. Quantitative efficacy estimates represent best estimates based upon the author's review and interpretation of the pertinent literature. In all cases, estimates are based upon growth responses (below the growth requirement of the AA in question) relative to that obtained with the pure L-isomer. Recent evidence suggests that pure L-AA are 100% absorbed from the gut (Liebholz *et al.*, 1986; Nelson *et al.*, 1986; Izquierdo *et al.*, 1988; Han *et al.*, 1990; Chung and Baker, 1992a), although this should not be construed as meaning that the absorbed L-AA are utilized with 100% efficiency. Clearly, when crystalline AA are used as supplements to practical diets, at least twice-daily feeding is required to maximize utilization of the crystalline AA supplements (Batterham, 1984; Baker and Izquierdo, 1985). Table 3.1 summarizes efficacy estimates for many of the AA precursor compounds that have been studied. The information is more complete for laboratory species, particularly chicks and rats, than for pigs, dogs or humans.

Lysine and Threonine

Lysine

The initial steps in lysine degradation in the body involve eliminating the epsilon amino group. Hence, the α-keto analogue of lysine is not formed

Table 3.1. Relative bioavailability (RBV) of amino acid isomers, analogues and precursors. Values are expressed as growth efficacy percentages (molar basis) of the L-isomer, which is assumed in all cases to represent 100% RBV.

Amino acid	Chick	Rat	Mouse	Dog	Human	Pig
Lysine						
L-lysine	100	-	-	-	-	100
D-lysine	0	0	0	-	0	-
Lysinolalanine	0	0	-	-	-	-
γ-glutamyl-L-lysine	-	100	-	-	-	-
Fructosyl-L-lysine	0	0	-	-	-	-
L-homoarginine	0	-	-	-	-	-
Threonine						
L-threonine (2S, 3S)	100	-	-	-	-	100
D-threonine (2R, 3R)	0	0	0	-	0	-
L-allothreonine (2S, 3R)	-	-	-	-	-	-
D-allothreonine (2R, 3S)	-	-	-	-	-	-
Tryptophan						
L-tryptophan	100	-	-	-	-	100
D-tryptophan	20	100	30	35	0	80
N-acetyl-L-tryptophan	-	100	-	-	-	-
N-acetyl-D-tryptophan	-	0	-	-	-	-
Methionine						
L-methionine	100	-	-	-	-	100
D-methionine	90	90	75	100	30	100
DL-methionine	95	95	88	100	65	100
DL-OH-methionine	80	-	70	-	-	100
Keto-methionine	90	-	-	-	-	-
L-methionine sulfone	-	0	0	-	-	-
L-methionine sulfoxide	-	60	85	-	-	-
N-acetyl-L-methionine	100	100	90	100	-	-
N-acetyl-D-methionine	0	0	25	0	-	-
L-homocysteine	65	65	-	-	-	-
D-homocysteine	7	-	-	-	-	-
Cyst(e)ine						
L-cysteine	100	-	-	-	-	100
L-cystine	100	-	-	-	-	100
D-cystine	0	0	0	0	-	-
Keto-cysteine	-	0	-	-	-	-
L-cysteic acid	-	0	-	-	-	-
DL-lanthionine	35	-	35	-	-	-
Glutathione	100	100	-	-	-	-
N-acetyl-L-cysteine	-	100	100	-	-	-
S-methyl-L-cysteine	-	-	0	-	-	-
L-homocysteine	100	-	-	-	-	-

Table 3.1. continued.

Amino acid	Chick	Rat	Mouse	Dog	Human	Pig
D-homocysteine	70	-	-	-	-	-
L-methionine	100	100	-	-	-	100
L-2-oxothiazolidine- 4-carboxylate	80	70	-	-	-	-
Taurine	0	0	-	-	-	-
Arginine						
L-arginine	100	-	-	-	-	100
D-arginine	0	0	-	-	-	-
Keto-arginine	-	0	-	-	-	-
L-ornithine	0	0	-	0	-	0
L-citrulline	90	90	-	90	-	90
Histidine						
L-histidine	100	-	-	-	-	100
D-histidine	10	0	10	-	-	-
Carnosine	100	100	-	-	-	-
Anserine	0	0	-	-	-	-
Balenine (ophidine)	0	0	-	-	-	-
Leucine						
L-leucine	100	-	-	-	-	100
D-leucine	100	50	15	-	-	-
Keto-leucine	100	50	-	-	-	-
L-OH-leucine	100	50	-	-	-	-
D-OH-leucine	100	40	-	-	-	-
Valine						
L-valine	100	-	-	-	-	100
D-valine	70	15	5	-	-	-
Keto-valine	80	50	-	-	-	-
L-OH-valine	80	50	-	-	-	-
D-OH-valine	70	45	-	-	-	-
Isoleucine						
L-isoleucine (2S, 3S)	100	-	-	-	-	100
D-isoleucine (2R, 3R)	0	-	-	-	-	-
L-alloisoleucine (2S, 3R)	0	-	-	-	-	-
D-alloisoleucine (2R, 3S)	60	-	-	-	-	-
L-keto-isoleucine (3S)	85	65	-	-	-	-
D-keto-isoleucine (3R)	0	0	-	-	-	-
L-OH-isoleucine (2S, 3S)	85	65	-	-	-	-
D-OH-isoleucine (2R, 3R)	0	0	-	-	-	-
Phenylalanine						
L-phenylalanine	100	-	-	-	-	100
D-phenylalanine	75	70	-	-	-	-

Table 3.1. *continued.*

Amino acid	Chick	Rat	Mouse	Dog	Human	Pig
Keto-phenylalanine	85	65	-	-	-	-
L-OH-phenylalanine	70	50	-	-	-	-
D-OH-phenylalanine	0	0	-	-	-	-
Fructosyl-L-phenylalanine	0	-	-	-	-	-
Tyrosine						
L-tyrosine	100	-	-	-	-	100
D-tyrosine	100	100	-	-	-	-
L-phenylalanine	100	100	-	-	-	100

in the principal (i.e. saccharopine) pathway of lysine degradation. Small quantities of α-keto-lysine are nonetheless formed in the body, but this occurs via an L-AA oxidase-catalysed pathway, involving oxidative deamination and release of ammonia. There is no transaminase capable of converting α-keto-lysine to L-lysine, and thus D-lysine and the α-keto analogue of lysine have no biological efficacy for animals (Sugahara *et al.*, 1967; Baker, 1986).

L-homoarginine (2-amino-6-quanidinohexanoic acid) is formed when proteins are guanidinated with *o*-methylisourea. Thus, under proper incubation conditions, up to 98% of the lysine residues in a protein, are transformed to homoarginine. Stevens and Bush (1950) suggested that arginase in the rat may act upon homoarginine to form lysine. Schuttert *et al.* (1991), however, observed no conversion of homoarginine to lysine in the small intestine of rats. Recent work in the author's laboratory (Baker and Aoyagi, 1994) using chicks, a species that does not practise coprophagy, confimed that L-homoarginine has no growth promoting activity when added to a lysine-deficient diet.

Lysine is considered the most heat labile of the indispensable AA because of the reactivity of the epsilon amino group. Thus, any exposed free amino groups can react with free carbonyl groups (e.g. reducing sugars such as glucose and lactose) under conditions of high temperature and humidity to form Maillard-bound AA (Carpenter and Booth, 1973; Adrian, 1974; Robbins and Baker, 1980). If Maillard-binding reactions go to completion, fructosyl AA are formed, and AA bound in this fashion have no biological activity (Carpenter and Booth, 1973; Johnson *et al.*, 1977, 1979; Anonymous, 1978; Robbins and Baker, 1980). Heat treatment of proteinaceous food products, particularly if alkaline conditions exist, can also lead to lysinoalanine formation (Sternberg *et al.*, 1975; Finley *et al.*, 1978), a crosslinked AA that has no L-lysine bioactivity (Robbins *et al.*, 1980). Heating of foods can also lead to formation of γ-L-glutamyl-

L-lysine, a compound in which the epsilon amino group of lysine is bound in peptide linkage to the γ-carbon of glutamine (the γ-amide group of glutamine is given off as ammonia). Non-specific peptidases in the gut can hydrolyze the peptide bond in γ-glutamyl-L-lysine, making the lysine in this product fully bioavailable (Waibel and Carpenter, 1972; Raczynski *et al.*, 1975).

Threonine

Threonine is catabolized initially via dehydratase, dehydrogenase or aldolase reactions, and none of these reaction sequences yields the α-keto analogue of threonine. Hence D-threonine has no biological efficacy. Prior to having economically available L-threonine, many early investigations involved use of DL-threonine, which by virtue of having two asymmetric carbon atoms (alpha and beta carbons) gives rise to four potential isomers: L(2S, 3S), D(2R, 3R), L-allo (2S, 3R) and D-allo (2R, 3S). Presumably, those investigators that used DL-threonine actually had a 1:1:1:1 mixture of all four isomers. Utilization of these separate isomers has not been studied, although D-threonine and D-allothreonine (amino group in D position) should have zero bioefficacy. It seems likely that L-allothreonine (hydroxyl group on the beta carbon in the D position) likewise would have zero bioefficacy in that inversion of the hydroxyl group attached to the β-carbon would seem unlikely. Thus, DL-threonine should contain no more than 25% efficacy based upon its 25% content of utilizable L-threonine (Baker, 1986).

Tryptophan

D-tryptophan is utilized well by pigs and rats (du Vigneaud *et al.*, 1932; Baker *et al.*, 1971; Ohara *et al.*, 1980; Arentson and Zimmerman, 1985; Kirchgessner and Roth, 1985; Schutte *et al.*, 1988; Baker and Han, 1993) but poorly by mice, chicks, poults, dogs and humans (Baldwin and Berg, 1949; Morrison *et al.*, 1956; Baker, 1977; Friedman and Gumbmann, 1982; Czarnecki and Baker, 1982). That inefficient gut absorption of D-tryptophan may explain part of the poor utilization in avians is evidenced by the findings of Morrison *et al.* (1956) who observed better utilization of intraperitoneal-injected D-tryptophan than of D-tryptophan given orally. Pigs obtain good activity from D-tryptophan, with values ranging from 60% (Baker *et al.*, 1971) and 70% (Arentson and Zimmerman, 1985) to almost 100% (Kirchgessner and Roth, 1985; Schutte *et al.*, 1988). With humans, the data of Baldwin and Berg (1949) are quite conclusive in suggesting that humans obtain very little, if any, value from D-tryptophan. Acetylated L- but not D-tryptophan appears to be fully active as an

L-tryptophan precursor for rats (du Vigneaud *et al.*, 1932; Baldwin and Berg, 1949).

Sulfur Amino Acids (SAA)

SAA isomers

The D-isomer of methionine is utilized well as an L-methionine precursor, except in apes and humans (Steginik *et al.*, 1971; Kies *et al.*, 1975; Zezulka and Calloway, 1976; Baker, 1977; Cho *et al.*, 1980; Burns and Milner, 1981; Friedman and Gumbmann, 1984; Funk *et al.*, 1990; Chung and Baker, 1992b). Indeed, while the efficacy of D-methionine for pig growth is 100% (Chung and Baker, 1992b), the efficacy of D-methionine for supporting nitrogen balance of adult humans is less than 50% (Kies *et al.*, 1975; Zezulka and Calloway, 1976).

In species obtaining good efficacy from D-methionine, the keto analogue must obviously be formed, and then this must be transaminated to L-methionine. Although keto methionine is not formed in the (major) trans-sulfuration pathway of methionine degradation, an alternative pathway exists wherein keto methionine is formed. Thus, the keto analogue of methionine has been shown to have good L-methionine efficacy in the chick (Harter and Baker, 1977). The keto analogue of cysteine is not produced in metabolism, and thus neither the keto analogue of cysteine (Meister *et al.*, 1954) nor D-cysteine (Baker and Harter, 1978) have L-cysteine bioactivity. An outline of methionine and cysteine metabolism is shown in Fig. 3.2.

Trans-sulfuration involves transfer of the sulfur from methionine to serine, resulting in cysteine biosynthesis (du Vigneaud *et al.*, 1944; du Vigneaud, 1952; Finkelstein and Mudd, 1967). On a molar basis, L-methionine is 100% efficient as a precursor of L-cysteine (Graber and Baker, 1971). The reaction between cysteine and cystine is freely reversible such that both compounds are equal in furnishing cysteine bioactivity for support of protein synthesis. In young, rapidly growing animals, cyst(e)ine can furnish 50% of the requirement for SAA (Graber and Baker, 1971; Sowers *et al.*, 1972; Baker, 1977; Teeter *et al.*, 1978; Halpin and Baker 1984; Hirakawa and Baker, 1985; Chung and Baker, 1992c). For older animals (i.e. adult maintenance), cyst(e)ine can furnish up to 80% of the SAA requirement (Baker *et al.*, 1966; Said and Hegsted, 1970; Fuller *et al.*, 1989).

Hydroxy analogue of methionine

Numerous compounds in the food supply, or produced in metabolism, have SAA sparing activity. The α-hydroxy analogue of methionine

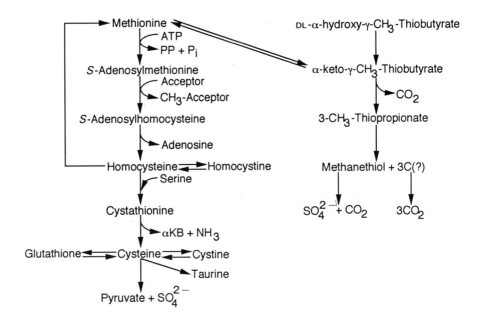

Fig. 3.2. Metabolism of methionine and cysteine.

(OH-Met) is an important commercial product. This compound is made chemically and therefore it is a 1:1 mixture of D- and L-OH-Met (Baker and Boebel, 1980). Considerable controversy (Potter, 1984; Baker, 1986) has surrounded the bioactivity of DL-OH-Met, which is available commercially as the calcium salt (86% of the sulfur in methionine) or the free acid (88% of the sulfur in methionine). It is now well established that two separate enzymes are necessary to convert DL-OH-Met to the α-keto analogue of methionine, a dehydrogenase for D-OH-Met and an oxidase for L-OH-Met (Dibner and Knight, 1984). The keto analogue of methionine is then transaminated to L-methionine, with branched-chain AA being principal amino donors in avians, but glutamine being the principal amino donor in rats (Austic and Rangel-Lugo, 1992). Baker and Boebel (1980) in chicks and Friedman and Gumbmann (1988) in mice showed with purified diets that D-OH-Met is more active than L-OH-Met in stimulating growth of animals fed SAA-deficient diets.

Does DL-OH-Met free acid have 88% L-methionine bioactivity (weight or concentration basis)? This is what one would expect if DL-OH-Met yielded L-Met with 100% efficiency. In avians, Potter (1984) estimated, based upon an extensive literature review, that DL-OH-Met has 75% of the molar activity of L-methionine. This translates to 66% efficacy on a weight or concentration basis (88% $\times$ 0.75 = 66%). Table 3.1 gives

DL-OH-Met as having 80% molar efficacy in chicks (Boebel and Baker, 1982c) but 100% molar efficacy in young pigs (Chung and Baker, 1992b).

Glutathione, taurine and lanthionine

Glutathione (γ-glutamylcysteinylglycine), taurine and lanthionine are found in foods and feeds consumed by animals and humans (Baker *et al.*, 1981; Cho *et al.*, 1984; Wiezbicker *et al.*, 1989; Glass and Czarnecki-Maulden, 1990). Taurine and glutathione (GSH) are constituents of animal-based products, but GSH is also present in some plant-based food products. Lanthionine is a crosslinked SAA that is formed in food products that have been exposed to heat. It is prominent in feáther meal and poultry by-product meals.

Taurine has no SAA-sparing activity in either chicks (Sasse and Baker, 1974) or rats (Martin *et al.*, 1974), although it may be limiting in diets for neonatal infants (Rigo and Senterre, 1977). Glutathione is fully effective on a molar basis in furnishing L-cysteine (Dyer and du Vigneaud, 1936; Harter and Baker, 1977; Boebel and Baker, 1983). Lanthionine in foods can exist in eight different isomeric forms (Jones *et al.*, 1942, 1948; Snow *et al.*, 1976). That absorbed from the gut can be cleaved by cystathionase (Cavallini *et al.*, 1960), releasing either D or L-cysteine. Robbins *et al.* (1980) conducted a quantitative study in chicks that suggested DL-lanthionine has 35% L-cysteine bioactivity.

SAA oxidation products

Methionine and cyst(e)ine in foods are subject to oxidative losses wherein methionine is converted to either methionine sulfoxide or methionine sulfone, and cystine is oxidized to cysteic acid. These compounds are often found in milk-based foods, because hydrogen peroxide is frequently used for sterilization purposes (Fox and Kosikowski, 1967; Yang, 1970; Anderson *et al.*, 1975). Neither L-methionine sulfone nor L-cysteic acid have SAA bioactivity when fed to rats (Anderson *et al.*, 1976), but L-methionine sulfoxide has 60% and 85% bioactivity for rats and mice, respectively.

Homocyst(e)ine

Considerable research has been carried out with homocysteine, an intermediate in the trans-sulfuration pathway (Dyer and du Vigneaud, 1935). This AA is not a constituent of body proteins, but it accumulates in tissues and body fluids of patients afflicted with homocystinuria (Carson *et al.*, 1963; Mudd *et al.*, 1964). Homocysteine has also been implicated as a causative factor in atherosclerosis (McCully and Wilson, 1975; Smolin *et al.*, 1983). In virtually all cases where homocysteine or its oxidation product, homocystine, have been used in nutrition research, the racemic DL mixture has been used, yet the efficacy of the separate isomers was not known. Potentially, homocysteine (or homocystine) could serve as a precursor for either cysteine or methionine (Fig. 3.2). Work with chicks and rats from our laboratory established that D- and L-homocysteine have considerably different bioactivities when serving as an oral source of methionine or cysteine (Harter and Baker, 1978; Baker and Czarnecki, 1985). As might be expected, both isomers of homocysteine are more effective precursors of L-cysteine than of L-methionine (Table 3.1). With diets adequate in cysteine and deficient in methionine, L-homocysteine had 65% the growth-promoting activity of L-methionine, while D-homocysteine was only 7% as effective as L-methionine.

Palatable SAA precursors for food supplementation

While supplementation of animal diets with either DL-methionine or DL-OH-Met has become common, adding bioavailable SAA activity to diets for humans has been held back because: (i) DL-methionine and DL-OH-Met are unpalatable to humans; (ii) L-cysteine is too toxic for routine use; and (iii) L-cystine is relatively insoluble, making it inappropriate for liquified diets (i.e. enteral and parenteral food products). While glutathione (cysteine precursor) is soluble as well as palatable, it is quite expensive. Thus, there is great interest in finding palatable and soluble precursors of methionine or cysteine for human applications. Acetylation of the α-amino group of either L-methionine or L-cysteine results in compounds that are palatable (Balance, 1961; Damico, 1975) and have full SAA bioactivity, although *N*-acetyl-D-methionine has no methionine-sparing activity (Birnbaum *et al.*, 1951; Boggs *et al.*, 1975; Rotruck and Boggs, 1975; Baker, 1979; Burns and Milner, 1981; Baker and Han, 1993). An added virtue of *N*-acetyl-L-methionine is that it is an effective precursor of both methionine *and* cysteine, while *N*-acetyl-L-cysteine can only provide cysteine bioactivity. Moreover, Baker *et al.* (1984) showed that acetylation of the α-amino group of methionine protects methionine from Maillard destruction (i.e. heat and humidity-induced fructosyl-methionine formation).

Meister (1984) has identified a new cysteine precursor, L-2-oxothiazolidine-4-carboxylic acid (OTC), and this compound can serve as an oral precursor of L-cysteine (or GSH). Work in our laboratory has indicated that OTC has 80% and 70% L-cysteine bioactivity in chicks and rats, respectively (Chung *et al.*, 1990).

Arginine

Arginine isomers

Our work with D-arginine showed that neither chicks nor rats could use D-arginine as a precursor for L-arginine (Baker and Boebel, 1981). This confirmed the finding of Sugahara *et al.* (1967) showing no bioefficacy for D-arginine in chicks. The α-keto analogue of arginine is not formed in metabolism, and even if it was, Meister (1954) showed that the α-keto analogue of arginine had no growth-promoting activity in rats.

Urea-cycle precursors

Arginine exemplifies an indispensable AA where *de novo* biosynthesis occurs in mammals but not in avians (Cohen and Hayano, 1946; Tamir and Ratner, 1963; Austic and Nesheim, 1971; Rogers *et al.*, 1972; Easter *et al.*, 1975; Southern and Baker, 1983; Edmonds *et al.*, 1987). Avians lack a mitochondrial source of carbamoyl phosphate synthetase and therefore cannot take up carbamoyl phosphate into the urea cycle to combine with ornithine for citrulline formation.

Arginine is synthesized in the liver from ornithine (derived from glutamate or glutamine conversion in the gut to ornithine, or from arginine degradation in the liver via arginase), but hepatic arginase activity is so high in the liver of mammals that virtually all of the hepatic arginine formed is degraded to ornithine and urea (Szepsi *et al.*, 1970; Czarnecki and Baker, 1984; Edmonds *et al.*, 1987). The kidney, on the other hand, is low in arginase activity such that arginine produced from citrulline in the kidney results in net arginine biosynthesis (Edmonds *et al.*, 1987). Because liver tissue takes up citrulline poorly, orally administered citrulline largely bypasses the liver and instead goes to the kidney where it is converted to arginine. Orally administered ornithine has little or no arginine-sparing activity. Although effectively taken up by kidney as well as liver tissue, ornithine cannot be converted to citrulline in the kidney because the necessary enzyme, ornithine transcarbamoylase, is not present in renal tissue.

In animals fed an arginine-free diet, *de novo* ornithine synthesis must, and does, occur (in intestinal mucosal tissue). Some citrulline is also synthesized in the gut, but in the cat, very little ornithine or citrulline is synthesized because the enzyme pyrroline-5-carboxylate synthetase is very low in mucosal tissue of felids (Rogers and Phang, 1985). This enzyme catalyses two reactions necessary for ornithine synthesis from glutamic acid. Cats thus generally do not survive for even 24 hours when fed an arginine-free diet (Morris and Rogers, 1978). In this case, oral administration of ornithine will ameliorate the hyperammonaemia and keep cats alive, but will not stimulate growth (Morris *et al.*, 1979).

De novo arginine biosynthesis is sufficient to meet the entire arginine requirement of adult gravid and non-gravid swine (Easter *et al.*, 1974, 1975) and of adult humans (Carey *et al.*, 1987). Apparently, in adults of most mammalian species (cats being a clear-cut exception) arginine biosynthesis is sufficient to meet the arginine needs for protein synthesis as well as for synthesis of creatine and polyamines.

Histidine

Histidine isomers

The initial reaction in the primary pathway of histidine degradation is histidase-catalysed deamination, although a small proportion of histidine is also decarboxylated to histamine. The α-keto analogue of histidine is not formed in either of these pathways. Thus, D-histidine would not be expected to have bioactivity as a precursor for L-histidine. Our work with D-histidine indicated zero bioactivity for young rats and only slight utilization in young chicks (Baker and Boebel, 1981), the latter agreeing with the chick work of Sugahara *et al.* (1967) but disagreeing with the work of Fell *et al.* (1959). Earlier rat work by Kamath and Berg (1964) and Nasset and Gatewood (1954) had suggested some utilization of D-histidine by rats.

Carnosine

Animal tissues contain dipeptides of β-alanine and histidine (carnosine) or methylated histidine (anserine, balenine). du Vigneaud *et al.* (1937) showed that L-carnosine was a good precursor of L-histidine in rats, and Robbins *et al.* (1977) later demonstrated that carnosine was 100% effective on a molar basis in furnishing L-histidine. Carnosinase is present in hepatic, renal and intestinal tissue (Hansen and Smith, 1949) such that carnosine consumed either in the food supply (Crush, 1970) or that made available in the body from endogenous turnover is fully effective as a precursor of L-histidine. Methylated carnosines (anserine and balenine) and their methylated histidine components ($1\text{-}CH_3$ and $3\text{-}CH_3$-histidine) have no histidine-sparing activity.

Meat products may contain as much as 0.35% carnosine (Crush, 1970) and there is evidence that the intact dipeptide can be absorbed from the gut (Perry *et al.*, 1967). Unfortunately, a clear-cut biological function for carnosine does not exist, although there is evidence suggesting that it may function in olfaction (Quinn and Fisher, 1977) or as a biological antioxidant (Kohen *et al.*, 1988; Decker and Faraji, 1990). Clearly, histidine deprivation is accompanied by carnosine catabolism to histidine and β-alanine, and this explains why it is difficult to obtain negative growth or nitrogen balance in animals fed a histidine-free diet (Nasset and Gatewood, 1954; Easter and Baker, 1977; Quinn and Fisher, 1977).

Branched-Chain Amino Acids

This group of AA includes the neutral indispensable AA leucine, valine and isoleucine. Utilization of isomers and analogues of these AA takes on added significance in that α-keto analogues of branched-chain AA (BCAA) are used in clinical applications, especially for renal patients.

Leucine

The α-keto analogue of leucine is ketoisocaproic acid (KIC). In normal metabolism, KIC is formed via transamination in the initial step of leucine degradation, and much of the KIC produced (or leucine produced via the reverse reaction) occurs in muscle tissue. Considerable interest exists regarding the biological efficacy of not only KIC, but also the D- and L-isomers of the α-hydroxy analogue of leucine. Both D-leucine and the hydroxy analogues of leucine must be converted first to KIC in order for biological efficacy to occur.

D-leucine is completely efficacious in chicks as an L-leucine precursor (Sugahara *et al.*, 1967; Robbins and Baker, 1977a), but it is only 50% bioactive in rats (Boebel and Baker, 1982a) and only 15% bioactive in mice (Friedman and Gumbmann, 1982). A report by Rose *et al.* (1955a) suggested that humans cannot use D-leucine as a precursor for L-leucine.

The keto analogue of leucine (KIC) is fully efficacious in chicks but only 50% efficacious in rats (Robbins and Baker, 1977a; Boebel and Baker, 1982a; Funk *et al.*, 1987). Both D and L-isomers of the hydroxy analogue of leucine are well utilized by chicks (Robbins and Baker, 1977a), while L-OH-leucine and D-OH-leucine are utilized with 50% and 40% efficiency, respectively, in rats (Boebel and Baker, 1982a; Baker, 1986; Funk *et al.*, 1987).

Valine

Boebel and Baker (1982a) and Funk *et al.* (1987) evaluated the bioefficacy of isomers and analogues of valine for young rats and chicks. D-valine and D-OH-valine were utilized at 70% efficiency in chicks, while the keto analogue of valine (α-ketoisovaleric acid) and L-OH-valine were utilized at 80% efficiency. Rats obtained less efficacy from these compounds: 15% for D-valine, 50% for L-OH-valine and the keto analogue of valine, and 45% for D-OH-valine. Obviously, the enzyme(s) necessary for converting the hydroxy analogues of valine to α-keto-valine must be more active in rats than is the case for the D-AA oxidase required to convert D-valine to its α-keto analogue. Mice appear to use D-valine even less efficiently than rats (Friedman and Gumbmann, 1982), and humans appear to be in the same category as mice (Rose *et al.*, 1955b).

Isoleucine

Considerable controversy surrounds utilization of isoleucine isomers and analogues in that, like threonine, isoleucine has two asymmetric carbon atoms, and therefore this AA has four potential isomers: L $(2S, 3S)$, D $(2R, 3R)$, L-allo $(2S, 3R)$ and D-allo $(2R, 3S)$. While the α-amino group is potentially invertible (D $\rightarrow$ L), the methyl group on the β-carbon is not (Meister, 1951). Thus, D-isoleucine and L-alloisoleucine have no biological activity (Funk and Baker, 1989). D-alloisoleucine, however, has an efficacy value of 60% for chicks. Quantitative information is not available on utilization of pure D-alloisoleucine in other species. It is likely that the DL-isoleucine used by early investigators contained 25% contributions of each of the four isomers. Thus, one can reason that DL-isoleucine should contain no greater than 40% bioactivity, i.e. 25% from L-isoleucine and 15% from D-alloisoleucine. With L-OH-isoleucine, chicks obtain 85% bioactivity and rats 65% from this compound, but D-OH-isoleucine has no bioefficacy for either chicks or rats (Boebel and Baker, 1982a).

The α-keto analogue of isoleucine (α-keto-β-methylvaleric acid, i.e. KMV) has isomeric configuration, and due to its availability commercially, DL-KMV is the compound used in clinical applications. Our work has been conclusive in showing that the D-isomer of KMV is not utilized by either chicks or rats (Boebel and Baker, 1982a; Izquierdo and Baker, 1987; Funk *et al.*, 1987). The L-keto analogue of isoleucine appears to have the same bioefficacy as L-OH-isoleucine. The keto analogue compound currently in use (DL-KMV) has half the efficacy value of L-KMV (Izquierdo and Baker, 1987). Several of the BCAA keto analogues are being used clinically as ornithine salts, and these appear to be as efficacious as sodium or calcium salts of these compounds (Funk *et al.*, 1987).

Phenylalanine and Tyrosine

D-phenylalanine and D-tyrosine are utilized well by both chicks and rats (Rose and Womack, 1946; Fisher *et al.*, 1957; Berg, 1959; Sugahara *et al.*, 1967; Sunde, 1972; Boebel and Baker, 1982b). The keto analogue of both these AA is present as a normal metabolite in the body. Hydroxylation of phenylalanine via phenylalanine hydroxylase yields tyrosine. Studies of the molar efficiency of phenylalanine as a precursor of tyrosine have indicated that the efficiency is 100% in chicks (Sasse and Baker, 1972), rats (Stockland *et al.*, 1971) and pigs (Robbins and Baker, 1977b), and tyrosine can provide between 45 and 50% of the total requirement for aromatic AA (i.e. phenylalanine + tyrosine).

Gaby and Chawla (1976) and Chow and Walser (1975) demonstrated that rats could use the L-α-hydroxy as well as the α-keto analogue (phenylpyruvic acid) of phenylalanine. Quantitative efficacy estimates in chicks and rats were made by Boebel and Baker (1982b), who reported efficacy values of 70% and 50% for L-OH-phenylalanine in chicks and rats, respectively; values for phenylpyruvic acid were 85% and 65%.

Johnson *et al.* (1977, 1979) showed that when phenylalanine reacts with glucose to form fructose-phenylalanine, L-phenylalanine bioactivity is lost completely.

Dispensable Amino Acids

Conclusive quantitative information does not exist concerning utilization of dispensable AA precursors. To grow maximally, chicks require a dietary source of both proline and glycine – to augment that provided via biosynthesis; serine, however, is fully efficacious in replacing the dietary need for glycine, and threonine degradation via the aldolase pathway yields preformed glycine as well (Almquist and Grau, 1944; Baker *et al.*, 1968, 1972; Akrabawi and Kratzer, 1968; Graber and Baker, 1973). Relative to glutamic acid or diammonium citrate, indispensable AA (the classical ten) are inefficient precursors of the amino nitrogen required for dispensable AA biosynthesis (Stucki and Harper, 1961; Allen and Baker, 1974). With chemically defined AA diets, rats, mice and cats require the full array of dispensable AA to grow maximally, but pigs and chicks grow maximally when fed diets containing glutamate, alone, as the source of dispensable amino nitrogen (Rogers and Harper, 1965; Baker *et al.*, 1979; Anderson *et al.*, 1980; Hirakawa *et al.*, 1984; Hirakawa and Baker, 1985; Chung and Baker, 1991).

Sources of nitrogen other than AA can also be used for *de novo* biosynthesis of dispensable AA (Baker, 1992). Through gut (including bacteria) and/or body processes, animals can derive usable nitrogen for dispensable AA synthesis from urea (bacterial urease) and from certain purines (adenine) and pyrimidines (uracil) provided either from the diet or from body turnover (Rose *et al.*, 1949; Featherston *et al.*, 1962; Grimson *et al.*, 1971; Baker and Molitoris, 1974).

References

Adrian, J. (1974) Nutritional and physiological consequences of the Maillard reaction. *World Review Nutrition and Dietetics* 19, 71–122.

Akrabawi, S.S. and Kratzer, F.H. (1968) Effects of arginine or serine on the requirement for glycine by the chick. *Journal of Nutrition* 95, 41–48.

Allen, N.K. and Baker, D.H. (1974) Quantitative evaluation of nonspecific nitrogen sources for the growing chick. *Poultry Science* 53, 258–264.

Almquist, H.J. and Grau, C.R. (1944) The amino acid requirements of the chick. *Journal of Nutrition* 28, 325–332.

Anderson, G.H., Li, G.S.K., Jones, J.O. and Bender, F. (1975) Effect of hydrogen peroxide treatment on the nutritional quality of rapeseed flour fed to weanling rats. *Journal of Nutrition* 105, 317–325.

Anderson, G.H., Ashley, D.V.M. and Jones, J.D. (1976) Utilization of L-methionine sulfoxide, L-methionine sulfone and cysteic acid by the weanling rat. *Journal of Nutrition* 106, 1108–1114.

Anderson, P.A., Baker, D.H., Sherry, P.A. and Corbin, J.E. (1980) Nitrogen requirement of the kitten. *American Journal of Veterinary Research* 11, 1646–1649.

Anonymous (1978) Nutritional implications of the Maillard reaction. *Nutrition Reviews* 36, 28–30.

Arentson, D.E. and Zimmerman, D.R. (1985) Nutritive value of D-tryptophan for the growing pig. *Journal of Animal Science* 60, 474–479.

Austic, R.E. and Nesheim, M.C. (1971) Arginine, ornithine and proline metabolism of chicks: Influence of diet and heredity. *Journal of Nutrition* 101, 1403–1413.

Austic, R.E. and Rangel-Lugo, M. (1992) Metabolism of methionine sources in the chicken. *Proceedings of the Degussa Technical Symposium* (Indianapolis, Indiana), pp. 33–46.

Baker, D.H. (1977) Amino acid nutrition of the chick. In: Draper, H.H. (ed.) *Advances in Nutrition Research, Vol. I.* Plenum Press, New York, pp. 299–335.

Baker, D.H. (1979) Efficacy of the D- and L-isomers of N-acetylmethionine for chicks fed diets containing either crystalline amino acids or intact protein. *Journal of Nutrition* 109, 970–974.

Baker, D.H. (1986) Utilization of isomers and analogs of amino acids and other sulfur-containing compounds. *Progress in Food and Nutrition Science* 10, 133–178.

Baker, D.H. (1992) Applications of chemically defined diets to the solution of nutrition problems. *Amino Acids* 2, 1–12.

Baker, D.H. and Boebel, K.P. (1980) Utilization of D- and L-isomers of methionine and methionine hydroxy analogue as determined by chick bioassay. *Journal of Nutrition* 110, 959–964.

Baker, D.H. and Boebel, K.P. (1981) Utilization of the D-isomers of arginine and histidine by chicks and rats. *Journal of Animal Science* 53, 125–129.

Baker, D.H. and Czarnecki, G.L. (1985) Transmethylation of homocysteine to methionine: efficiency in the rat and chick. *Journal of Nutrition* 115, 1291–1299.

Baker, D.H. and Han, Y. (1993) Bioavailable level (and source) of cysteine determines protein quality of a commercial enteral product: adequacy of tryptophan but deficiency of cysteine for rats fed an enteral product prepared fresh or stored beyond shelflife. *Journal of Nutrition* 123, 541–546.

Baker, D.H. and Harter, J.M. (1978) D-cystine utilization by the chick. *Poultry Science* 57, 562–563.

Baker, D.H. and Izquierdo, O.A. (1985) Effect of meal frequency and spaced crystalline lysine ingestion on the utilization of dietary lysine by chickens. *Nutrition Research* 5, 1103–1112.

Baker, D.H. and Molitoris, B.A. (1974) Utilization of nitrogen from selected purines and pyrimidines and from urea by the young chick. *Journal of Nutrition* 104, 553–557.

Baker, D.H., Becker, D.E., Norton, H.W., Jensen, A.H. and Harmon, B.G. (1966) Quantitative evaluation of the tryptophan, methionine and lysine needs of adult swine for maintenance. *Journal of Nutrition* 89, 441–447.

Baker, D.H., Sugahara, M. and Scott, H.M. (1968) The glycine-serine interrelationship in chick nutrition. *Poultry Science* 47, 1376–1377.

Baker, D.H., Allen, N.K., Boomgaardt, J., Graber, G. and Norton, H.W. (1971) Quantitative aspects of D- and L-tryptophan utilization by the young pig. *Journal of Animal Science* 33, 42–48.

Baker, D.H., Hill, T.M. and Kleiss, A.J. (1972) Nutritional evidence concerning formation of glycine from threonine in the chick. *Journal of Animal Science* 34, 582–586.

Baker, D.H., Robbins, K.R. and Buck, J.S. (1979) Modification of the level of histidine and sodium bicarbonate in the Illinois crystalline amino acid diet. *Poultry Science* 58, 749–750.

Baker, D.H., Blitenthal, R.C., Boebel, K.P., Czarnecki, G.L., Southern, L.L. and Willis, G.M. (1981) Protein-amino acid evaluation of steam-processed feather meal. *Poultry Science* 60, 1865–1872.

Baker, D.H., Bafundo, K.W., Boebel, K.P., Czarnecki, G.L. and Halpin, K.M. (1984) Methionine peptides as potential food supplements: efficacy and susceptibility to Maillard browning. *Journal of Nutrition* 114, 292–297.

Balance, P.E. (1961) Production of volatile compounds related to the flavor of foods from the Strecker degradation of DL-methionine. *Journal of the Science of Food and Agriculture* 12, 532–536.

Baldwin, H.R. and Berg, C.P. (1949) The influence of optical isomerism and acetylation upon the availability of tryptophan for maintenance in man. *Journal of Nutrition* 39, 203–218.

Batterham, E.S. (1984) Utilization of free lysine by pigs. *Pig News and Information* 5, 85–88.

Berg, C.P. (1959) Utilization of the D-amino acids. In: Albanese, A.A. (ed.) *Protein and Amino Acid Nutrition* Academic Press, New York, pp. 57–66.

Birnbaum, S.M., Levinton, L., Kingsley, R.B. and Greenstein, J.P. (1951) Specificity of amino acid acylases. *Journal of Biological Chemistry* 194, 455–470.

Boebel, K.P. and Baker, D.H. (1982a) Comparative utilization of α-keto and D- and L-α-hydroxy analogs of leucine, isoleucine and valine by chicks and rats. *Journal of Nutrition* 112, 1929–1939.

Boebel, K.P. and Baker, D.H. (1982b) Comparative utilization of the isomers of phenylalanine and phenyllactic acid by chicks and rats. *Journal of Nutrition* 112, 367–376.

Boebel, K.P. and Baker, D.H. (1982c) Efficacy of calcium salt and free acid forms of methionine hydroxy analog for chicks. *Poultry Science* 61, 1167–1175.

Boebel, K.P. and Baker, D.H. (1983) Blood and liver concentrations of glutathione, and plasma concentrations of sulfur-containing amino acids in chicks fed deficient, adequate or excess levels of dietary cysteine. *Proceedings of the Society for Experimental Biology and Medicine* 172, 498–501.

Boggs, R.W., Rotruck, J.T. and Damico, R.A. (1975) Acetylmethionine as a source of methionine for the rat. *Journal of Nutrition* 105, 326–330.

Burns, R.A. and Milner, J.A. (1981) Sulfur amino acid requirements of immature beagle dogs. *Journal of Nutrition* 111, 2117–2124.

Carey, G.P., Kime, Z., Rogers, Q.R., Morris, J.G., Hargrove, D., Buffington, C.A. and Brusilow, S.W. (1987) An arginine-deficient diet in humans does not evoke hyperammonemia or orotic aciduria. *Journal of Nutrition* 117, 1734–1739.

Carpenter, K.J. and Booth, V.H. (1973) Damage to lysine in food processing: its measurement and its significance. *Nutrition Abstracts and Reviews* 43, 423–451.

Carson, N.A.J., Cusworth, D.C., Dent, C.E., Field, C.M.B., Neill, D.W. and Westall, R.G. (1963) Homocystinuria: a new inborn error of metabolism associated with mental deficiency. *Archives of Diseases in Childhood* 38, 425–436.

Cavallini, C., DeMarco, D. and Mori, G.G. (1960) The cleavage of cystine by cystathionase and the transsulfuration of hypotaurine. *Enzymologia* 22, 161–173.

Cho, E.S., Johnson, N. and Snider, B.C. (1984) Tissue glutathione as a cyst(e)ine reservoir during cystine depletion in growing rats. *Journal of Nutrition* 114, 1853–1862.

Cho, E.S., Anderson, D.W., Filer, L.J. and Stegink, L.D. (1980) D-methionine utilization in young miniature pigs, adult rabbits, and adult dogs. *Journal of Parenteral and Enteral Nutrition* 4, 544–547.

Chow, K.W. and Walser, M. (1975) Effects of substitution of methionine, leucine, phenylalanine or valine by their α-hydroxy analogs in the diet of rats. *Journal of Nutrition* 105, 372–378.

Chung, T.K. and Baker, D.H. (1991) A chemically defined diet for maximal growth rate in pigs. *Journal of Nutrition* 121, 979–984.

Chung, T.K. and Baker, D.H. (1992a) Apparent and true amino acid digestibility of a crystalline amino acid mixture and of casein: comparison of values obtained with ileal-cannulated pigs and cecectomized cockerels. *Journal of Animal Science* 70, 3781–3790.

Chung, T.K. and Baker, D.H. (1992b) Utilization of methionine isomers and analogs by the pig. *Canadian Journal of Animal Science* 72, 185–188.

Chung, T.K. and Baker, D.H. (1992c) Maximal portion of the young pig's sulfur amino acid requirement that can be furnished by cystine. *Journal of Animal Science* 70, 1182–1187.

Chung, T.K., Funk, M.A. and Baker, D.H. (1990) L-2-oxothiazolidine-4-

carboxylate as a cysteine precursor: efficacy for growth and hepatic glutathione synthesis in chicks and rats. *Journal of Nutrition* 120, 158–165.

Cohen, P.P. and Hayano, M. (1946) The conversion of citrulline to arginine (transamination) by tissue slices and homogenates. *Journal of Biological Chemistry* 166, 239–250.

Crush, K.G. (1970) Carnosine and related substances in animal tissues. *Comparative Biochemistry and Physiology* 34, 3–30.

Czarnecki, G.L. and Baker, D.H. (1982) Utilization of D- and L-tryptophan by the growing dog. *Journal of Animal Science* 55, 1405–1410.

Czarnecki, G.L. and Baker, D.H. (1984) Urea-cycle function in the dog with emphasis on the role of arginine. *Journal of Nutrition* 114, 581–590.

Damico, R. (1975) An investigation of N-substituted methionine derivatives for food supplementation. *Journal of Agricultural and Food Chemistry* 23, 30–33.

Decker, E.A. and Faraji, H. (1990) Inhibition of lipid oxidation by carnosine, a β-alanine-histidine dipeptide. *Journal of the American Oil Chemists Society* 67, 650–652.

Dibner, J.J. and Knight, C.D. (1984) Conversion of 2-hydroxy-4-(methylthio) butanoic acid to L-methionine in the chick: a stereospecific pathway. *Journal of Nutrition* 114, 1716–1723.

du Vigneaud, V. (1952) *Trail of Research in Sulfur Chemistry and Metabolism and Related Fields*. Cornell University Press, Ithaca, New York.

du Vigneaud, V., Sealock, R.R. and Van Etten, C. (1932) The availability of D-tryptophan and its acetyl derivative to the animal body. *Journal of Biological Chemistry* 98, 565–575.

du Vigneaud, V., Sifferd, R.H. and Irving, G.N. (1937) The utilization of L-carnosine by animals on a histidine-deficient diet. *Journal of Biological Chemistry* 117, 589–597.

du Vigneaud, V., Kilmer, G.W., Rachele, J.R. and Cohn, M. (1944) On the mechanism of the conversion *in vivo* of methionine to cystine. *Journal of Biological Chemistry* 155, 645–651.

Dyer, H.M. and du Vigneaud, V. (1935) A study of the availability of D- and L-homocystine for growth purposes. *Journal of Biological Chemistry* 109, 477–480.

Dyer, H.M. and du Vigneaud, V. (1936) The utilization of glutathione in connection with a cystine-deficient diet. *Journal of Biological Chemistry* 115, 543–549.

Easter, R.A. and Baker, D.H. (1977) Nitrogen metabolism, tissue carnosine concentration and blood chemistry of gravid swine fed graded levels of histidine. *Journal of Nutrition* 107, 120–125.

Easter, R.A., Katz, R.S. and Baker, D.H. (1974) Arginine: A dispensable amino acid for postpubertal growth and pregnancy of swine. *Journal of Animal Science* 39, 1123–1128.

Easter, R.A., Katz, R.S. and Baker, D.H. (1975) Nitrogen metabolism and reproductive response of gravid swine fed an arginine-free diet during the last 84 days of gestation. *Journal of Nutrition* 106, 636–641.

Edmonds, M.S., Lowry, K.R. and Baker, D.H. (1987) Urea-cycle metabolism: effects of supplemental ornithine or citrulline on performance, tissue amino acid concentrations and enzymatic activity in young pigs fed arginine-deficient diets. *Journal of Animal Science* 65, 706–716.

Efron, M.L., McPherson, T.C., Shih, V.E., Welsh, C.F. and MacCready, R.A. (1969) D-methioninuria due to DL-methionine ingestion. *American Journal of Diseases of Children* 117, 104–107.

Featherston, W.R., Bird, H.R. and Harper, A.E. (1962) Effectiveness of urea and ammonium nitrogen for the synthesis of dispensable amino acids by the chick. *Journal of Nutrition* 78, 198–206.

Fell, R.V., Wilkinson, W.S. and Watts, A.B. (1959) The utilization by the chick of D- and L-amino acids in liquid and dry diets. *Poultry Science* 38, 1203–1204.

Finkelstein, J.D. and Mudd, S.H. (1967) Transsulfuration in mammals. The methionine-sparing effect of cystine. *Journal of Biological Chemistry* 242, 873–880.

Finley, J.W., Snow, J.T., Johnson, P.H. and Friedman, M. (1978) Inhibition of lysinoalanine formation in food proteins. *Journal of Food Science* 43, 619–621.

Fisher, H., Johnson, D. and Leveille, G.A. (1957) The phenylalanine and tyrosine requirement of the growing chick with special reference to the utilization of the D-isomers of phenylalanine. *Journal of Nutrition* 62, 349–355.

Fox, P.F. and Kosikowski, F.V. (1967) Some effects of hydrogen peroxide on casein and its implications in cheese making. *Journal of Dairy Science* 50, 1183–1188.

Friedman, M. and Gumbmann, M.R. (1982) Bioavailability of D-amino acids in mice. *Federation of American Societies for Experimental Biology Federation Proceedings* 41, 392 (abstract).

Friedman, M. and Gumbmann, M.R. (1984) The utilization and safety of isomeric sulfur-containing amino acids in mice. *Journal of Nutrition* 114, 2301–2310.

Friedman, M. and Gumbmann, M.R. (1988) Nutritional value and safety of methionine derivatives, isomeric dipeptides and hydroxy analogs in mice. *Journal of Nutrition* 118, 388–397.

Fuller, M.F., McWilliam, R., Wang, T.C. and Giles, L.R. (1989) The optimum dietary amino acid pattern for growing pigs. 2. Requirements for maintenance and for protein accretion. *British Journal of Nutrition* 62, 255–267.

Funk, M.A. and Baker, D.H. (1989) Utilization of isoleucine isomers and analogs by chicks. *Nutrition Research* 9, 523–530.

Funk, M.A., Lowry, K.R. and Baker, D.H. (1987) Utilization of the L- and DL-isomers of α-keto-β-methylvaleric acid by rats, and comparative efficacy of the keto analogs of branched-chain amino acids provided as ornithine, lysine and histidine salts. *Journal of Nutrition* 117, 1550–1555.

Funk, M.A. Hortin, A.E. and Baker, D.H. (1990) Utilization of D-methionine by growing rats. *Nutrition Research* 10, 1029–1034.

Gaby, A.R. and Chawla, R.K. (1976) Efficiency of phenylpyruvic and phenyllactic acids as substitutes for phenylalanine in the diet of the growing rat. *Journal of Nutrition* 106, 158–168.

Glass, E.N. and Czarnecki-Maulden, G.L. (1990) Taurine concentration in different food products. *FASEB Journal* 4, A799 (abstract).

Graber, G. and Baker, D.H. (1971) Sulfur amino acid nutrition of the growing chick: quantitative aspects concerning the efficacy of dietary methionine, cysteine and cystine. *Journal of Animal Science* 33, 1005–1011.

Graber, G. and Baker, D.H. (1973) The essential nature of glycine and proline for growing chickens. *Poultry Science* 52, 892–896.

Grimson, R.E., Bowland, J.P. and Milligan, L.P. (1971) Use of nitrogen-15 labelled urea to study urea utilization by pigs. *Canadian Journal of Animal Science* 51, 103–110.

Halpin, K.M. and Baker, D.H. (1984) Selenium deficiency and transsulfuration in the chick. *Journal of Nutrition* 114, 606–612.

Han, Y., Castanon, F., Parsons, C.M. and Baker, D.H. (1990) Absorption and bioavailability of DL-methionine hydroxy analogue compared to DL-methionine. *Poultry Science* 69, 281–288.

Hansen, H.T. and Smith, E.L. (1949) Carnosinase: an enzyme of swine kidney. *Journal of Biological Chemistry* 179, 789–801.

Harter, J.M. and Baker, D.H. (1977) Sulfur amino acid activity of glutathione, DL-α-hydroxy methionine, and α-keto methionine in chicks. *Proceedings of the Society for Experimental Biology and Medicine* 156, 201–204.

Harter, J.M. and Baker, D.H. (1978) Sulfur amino acid activity of D- and L-homocysteine for chicks. *Proceedings of the Society for Experimental Biology and Medicine* 157, 139–143.

Hirakawa, D.A. and Baker, D.H. (1985) Sulfur amino acid nutrition of the growing puppy: determination of dietary requirements for methionine and cystine. *Nutrition Research* 5, 631–642.

Hirakawa, D.A., Olson, L.A. and Baker, D.H. (1984) Comparative utilization of a crystalline amino acid diet and a methionine-fortified casein diet by young rats and mice. *Nutrition Research* 4, 891–895.

Izquierdo, O.A. and Baker, D.H. (1987) Utilization of the L- and DL-isomers of α-keto-methylvaleric acid by chicks. *Nutrition Research* 7, 985–988.

Izquierdo, O.A., Parsons, C.M. and Baker, D.H. (1988) Bioavailability of lysine in L-lysine·HCl. *Journal of Animal Science* 66, 2590–2597.

Johnson, G., Baker, D.H. and Perkins, E.G. (1977) Nutritional implications of the Maillard reaction. I. The availability of fructosephenylalanine to the chick. *Journal of Nutrition* 107, 1659–1664.

Johnson, G., Baker, D.H. and Perkins, E.G. (1979) Nutritional implications of the Maillard reaction. II. The metabolism of fructosephenylalanine in the rat. *Journal of Nutrition* 109, 590–596.

Jones, D.B., Divine, J.P. and Horn, M.J. (1942) A study of the availability of mesolanthionine for the promotion of growth when added to a cystine-deficient diet. *Journal of Biological Chemistry* 146, 571–575.

Jones, D.B., Caldwell, A. and Horn, M.J. (1948) The availability of DL-lanthionine for the promotion in young rats when added to a cystine- and methionine-deficient diet. *Journal of Biological Chemistry* 176, 65–69.

Kamath, S.H. and Berg, C.P. (1964) Antagonism of the D-forms of the essential amino acids toward the promotion of growth by D-histidine. *Journal of Nutrition* 82, 243–248.

Kies, C., Fox, H. and Aprahamian, S. (1975) Comparative value of L-, DL- and D-methionine supplementation of an oat-based diet for humans. *Journal of Nutrition* 105, 809–814.

Kirchgessner, V.M. and Roth, F.X. (1985) Biologische wirksamkeit von DL-tryptophan bei mastschweinen. *Zeitschrift für Tierphysiologie Tierernährung Futtermittelkunde* 54, 135–141.

Kohen, R., Yamamoto, Y., Cundy, K.C. and Ames, B.N. (1988) Antioxidant activity of carnosine, homocarnosine, and anserine present in muscle and brain. *Proceedings of the National Academy of Sciences of the USA* 85, 3175–3179.

Liebholz, J., Love, R.J., Mollah, Y. and Carter, R.R. (1986) The absorption of dietary L-lysine in pigs. *Animal Feed Science and Technology* 15, 141–148.

Martin, W.G., Truex, C.R., Tarka, S.M., Hill, L.J. and Gorby, W.G. (1974) The synthesis of taurine from sulfate. VIII. A constitutive enzyme in mammals. *Proceedings of the Society for Experimental Biology and Medicine* 147, 563–565.

McCully, K.S. and Wilson, R. B. (1975) Homocysteine theory of arteriosclerosis. *Atherosclerosis* 22, 215–227.

Meister, A. (1951) Enzymatic conversion of the stereoisomers of isoleucine to the corresponding α-keto acids. *Nature* 168, 1119.

Meister, A. (1954) Enzymatic transamination reactions involving arginine and ornithine. *Journal of Biological Chemistry* 206, 587–596.

Meister, A. (1984) New aspects of glutathione biochemistry and transport-selective alteration of glutathione metabolism. *Nutrition Reviews* 42, 397–410.

Meister, A., Fraser, P.E. and Tice, S.V. (1954) Enzymatic desulfuration of β-mercaptopyruvate. *Journal of Biological Chemistry* 206, 561–575.

Morris, J.G. and Rogers, Q.R. (1978) Ammonia intoxication in the near-adult cat as a result of a dietary deficiency of arginine. *Science* 1991, 431–432.

Morris, J.G., Rogers, Q.R., Winterrowd, D.L. and Kamikawa, E.M. (1979) The utilization of ornithine and citrulline by the growing kitten. *Journal of Nutrition* 109, 724–729.

Morrison, W.D., Hamilton, T.S. and Scott, H.M. (1956) Utilization of D-tryptophan by the chick. *Journal of Nutrition* 60, 47–63.

Mudd, S.H., Finkelstein, J.D., Irreverre, F. and Laster, L. (1964) Homocystinuria: an enzymatic defect. *Science* 143, 1443–1445.

Nasset, E.S. and Gatewood, V.H. (1954) Nitrogen balance and hemoglobin of adult rats fed amino acid diets low in L- and D-histidine. *Journal of Nutrition* 53, 163–176.

Nelson, T.S., Kirby, L.K. and Halley, J.T. (1986) Digestibility of crystalline amino acids and the amino acids in corn and poultry blend. *Nutrition Reports International* 34, 903–906.

Ohara, I., Otsuka, S., Yugari, Y. and Ariyoshi, S. (1980) Inversion of D-tryptophan to L-tryptophan and excretory patterns in the rat and chick. *Journal of Nutrition* 110, 641–648.

Perry, T.L., Hansen, S., Tischler, B., Bunting, R. and Berry, K. (1967) A new metabolic disorder associated with neurologic disease and mental defect. *New England Journal of Medicine* 277, 1219–1227.

Potter, L.M. (1984) Limiting amino acids in poultry diets. *Proceedings of the Carolina Poultry Nutrition Conference* 9, 33–40.

Quinn, M.R. and Fisher, H. (1977) Effect of dietary histidine deprivation in two rat strains on hemoglobin and tissue concentrations of histidine-containing dipeptides. *Journal of Nutrition* 107, 2044–2054.

Raczynski, G., Snochowski, M. and Buraczewski, S. (1975) Metabolism of E (γ-L-glutamyl)-L-lysine in the rat. *British Journal of Nutrition* 34, 291–296.

Rigo, J. and Senterre, J. (1977) Is taurine essential for neonates? *Biology of the Neonate* 32, 73–76.

Robbins, K.R. and Baker, D.H. (1977a) Comparative utilization of L-leucine and DL-α-hydroxy leucine by the chick. *Nutrition Reports International* 16, 611–615.

Robbins, K.R. and Baker, D.H. (1977b) Phenylalanine requirement of the weanling pig and its relationship to dietary tyrosine. *Journal of Animal Science* 45, 113–118.

Robbins, K.R. and Baker, D.H. (1980) Evaluation of the resistance of lysine sulfite to Maillard destruction. *Journal of Agriculture and Food Chemistry* 28, 25–29.

Robbins, K.R., Baker, D.H. and Norton, H.W. (1977) Histidine status in the chick as measured by growth rate, plasma free histidine and breast muscle carnosine. *Journal of Nutrition* 107, 2055–2061.

Robbins, K.R., Baker, D.H. and Finley, J.W. (1980) Studies on the utilization of lysinoalanine and lanthionine. *Journal of Nutrition* 110, 907–915.

Rogers, Q.R. and Harper, A.E. (1965) Amino acid diets and maximal growth in the rat. *Journal of Nutrition* 87, 267–273.

Rogers, Q.R. and Phang, J.H. (1985) Deficiency of pyrroline-5-carboxylate synthase in the intestinal mucosa of the cat. *Journal of Nutrition* 115, 146–150.

Rogers, W.R., Freedland, R.A. and Symmons, R.A. (1972) The *in vivo* synthesis and utilization of arginine in the rat. *American Journal of Physiology* 223, 236–240.

Rose, W.C. and Womack, M. (1946) The utilization of the optical isomers of phenylalanine, and the phenylalanine requirement for growth. *Journal of Biological Chemistry* 166, 103–110.

Rose, W.C., Smith, L.C., Womack, M. and Shane, M. (1949) The utilization of the nitrogen of ammonium salts, urea, and certain other compounds in the synthesis of nonessential amino acids *in vivo*. *Journal of Biological Chemistry* 181, 307–316.

Rose, W.C., Eades, C.H. Jr. and Coon, M.J. (1955a) The amino acid requirements of man. XII. The leucine and isoleucine requirements. *Journal of Biological Chemistry* 216, 225–234.

Rose, W.C., Wixom, R.L., Lockhart, H.B. and Lambert, G.F. (1955b) The amino acid requirements of man. XV. The valine requirement; summary and final observations. *Journal of Biological Chemistry* 217, 987–995.

Rotruck, J.T. and Boggs, R.W. (1975) Comparative metabolism of L-methionine and N-acetylated derivatives of methionine. *Journal of Nutrition* 105, 331–337.

Said, A.K. and Hegsted, D.M. (1970) Response of adult rats to low dietary levels of essential amino acids. *Journal of Nutrition* 100, 1363–1375.

Sasse, C.E. and Baker, D.H. (1972) The phenylalanine and tyrosine requirements and their interrelationship for the young chick. *Poultry Science* 51, 1531–1536.

Sasse, C.E. and Baker, D.H. (1974) Sulfur utilization by the chick with emphasis on the effect of inorganic sulfate on the cysteine-methionine interrelationship. *Journal of Nutrition* 104, 244–251.

Schutte, J.B., Van Weerden, E.J. and Koch, F. (1988) Utilization of DL- and L-tryptophan in young pigs. *Animal Production* 46, 447–452.

Schuttert, G., Moughan, P.J. and Jackson, F. (1991) *In vitro* determination of the extent of hydrolysis of homoarginine by arginase in the small intestine of the growing rat. *Journal of Agricultural and Food Chemistry* 39, 511–513.

Smolin, L.A., Crenshaw, T.D., Kurtycz, D. and Benevenga, N.J. (1983) Homocyst(e)ine accumulation in pigs fed diets deficient in vitamin B-6: relationship to atherosclerosis. *Journal of Nutrition* 113, 2022–2033.

Snow, J.T., Finley, J.W. and Friedman, M. (1976) Relative reactivities of sulfhydryl groups with *N*-acetyl dehydroalanine and *N*-acetyl dehydroalanine ester. *International Journal of Peptide and Protein Research* 8, 57–64.

Southern, L.L. and Baker, D.H. (1983) Arginine requirement of the young pig. *Journal of Animal Science* 57, 402–412.

Sowers, J.E., Stockland, W.L. and Meade, R.J. (1972) L-methionine and L-cystine requirements of the growing rat. *Journal of Animal Science* 35, 782–788.

Stegink, L.D., Schmitt, J.L., Meyer, P.D. and Kain, P.H. (1971) Effect of diets fortified with DL-methionine on urinary and plasma methionine levels in young infants. *Journal of Pediatrics* 79, 648–655.

Sternberg, M., Kin, C.Y. and Schwende, F.J. (1975) Lysinoalanine: presence in foods and food ingredients. *Science* 190, 992–994.

Stevens, C.M. and Bush, J.A. (1950) New synthesis of α-amino-ε-guanidino-*n* caproic acid (homoarginine) and its possible conversion *in vivo* to lysine. *Journal of Biological Chemistry* 183, 139–147.

Stockland, W.L., Lai, Y.F., Meade, R.J., Sowers, J.E. and Oestemer, G. (1971) L-phenylalanine and L-tyrosine requirements of the growing rat. *Journal of Nutrition* 101, 177–184.

Stucki, W.P. and Harper, A.E. (1961) Importance of dispensable amino acids for normal growth of chicks. *Journal of Nutrition* 74, 377–383.

Sugahara, M., Morimoto, T., Kobayashi, T. and Ariyoshi, S. (1967) The nutritional value of D-amino acid in the chick nutrition. *Agricultural and Biological Chemistry* 31, 77–84.

Sunde, M.L. (1972) Amino acids in avian nutrition. 6. Utilization of D- and DL-amino acids and analogs. *Poultry Science* 51, 44–55.

Szepsi, B., Avery, E.H. and Freedland, R.A. (1970) Role of kidney in gluconeogenesis and amino acid catabolism. *American Journal of Physiology* 219, 1627–1631.

Tamir, H. and Ratner, S. (1963) A study of ornithine, citrulline and arginine synthesis in growing chicks. *Archives of Biochemistry and Biophysics* 102, 259–269.

Teeter, R.G., Baker, D.H. and Corbin, J.E. (1978) Methionine and cystine requirements of the cat. *Journal of Nutrition* 108, 291–295.

Waibel, P.E. and Carpenter, K.J. (1972) Mechanisms of heat damage in proteins. 3. Studies with E (γ-L-glutamyl)-L-lysine. *British Journal of Nutrition* 27, 509–515.

Wiezbicker, G.T., Hegen, T.M. and Jones, D.P. (1989) Glutathione in food. *Journal of Food Composition* 2, 327–337.

Yang, S.F. (1970) Sulfoxide formation from methionine or its sulfide analogs during aerobic oxidation of sulfite. *Biochemistry* 9, 5008–5014.

Zezulka, A.Y. and Calloway, D.H. (1976) Nitrogen retention in men fed isolated soybean protein supplemented with L-methionine, D-methionine, N-acetyl-L-methionine or inorganic sulfate. *Journal of Nutrition* 106, 1286–1291.

Amino Acid Imbalances, Antagonisms and Toxicities

J.P.F. D'MELLO
The Scottish Agricultural College,
West Mains Road, Edinburgh, EH9 3JG, UK.

Introduction

Since amino acids in general enter into a complex variety of metabolic reactions it has been widely assumed that any surplus ingested by farm animals is disposed of without ill-effects. In addition, it has been suggested that the ruminant is endowed with effective detoxification mechanisms by virtue of extensive microbial metabolism of amino acids within the rumen. However, there is now unequivocal evidence to indicate that amino acids may precipitate profound deleterious effects in diverse classes of farm livestock. Adverse effects may arise from the intake of indispensable and dispensable amino acids absorbed in quantities and patterns which are disproportionate to those required for optimum tissue utilization. Alternatively, toxicity may result from the ingestion, and subsequent metabolism, of a wide array of free non-protein amino acids occurring naturally in crop plants and fodder trees. Deleterious effects of amino acids range from depressions in growth, food intake and nutrient utilization to acute neurological aberrations and even death. Precipitation of these effects is determined by factors such as nutritional status and age of the animal, degree of dietary disproportion of amino acids and intrinsic properties and metabolic fate of individual amino acids.

There is compelling evidence to indicate that manifestations of adverse effects of amino acids in farm animals conform with the three categories of imbalances, antagonisms and toxicities embodied in the classification established with the rat (Harper, 1959, 1964; Harper *et al.*, 1970). This system has long existed as a conceptual model, largely confined to the academic domain by virtue of its origin in contrived experiments with laboratory animals. However, there is now enhanced awareness of the wider, practical significance of adverse effects of amino acids in the nutrition of farm livestock. Thus, for example, in an effort to pre-empt the deleterious effects of imbalances, the Agricultural Research Council (1981) proposed a specific pattern for the ideal amino acid composition of dietary protein for pigs. Implicit in this recommendation is the assumption that amino acid imbalances should be avoided if overall efficiency of protein utilization is to be maximized. More recently, Moughan (1991) calculated that, on average, ingested dietary protein was utilized by the growing pig with an efficiency of only 0.30. The low efficiency was attributed, in part, to dietary amino acid imbalance. The extended use of crystalline amino acids as dietary supplements has provided additional focus on the nutritional significance of amino acid imbalances in the practical feeding of pigs and poultry. Bach Knudsen and Jorgensen (1986) concluded that of the two species, the pig is more susceptible to the effects of imbalance when crystalline amino acids are used to supplement the deficiencies of cereal-based diets. This is not to imply that poultry are immune to the effects of amino acid imbalances. Indeed, Waldroup *et al.* (1976) were among early investigators who demonstrated that diets formulated to minimize imbalances and excesses of amino acids were utilized more efficiently by the broiler chick. More recently, investigations at Reading (Morris *et al.*, 1987; Abebe and Morris, 1990a,b; Morris and Abebe, 1990) and elsewhere (Mendonca and Jensen, 1989) indicate that imbalances may increase the requirement of the chick for the first-limiting amino acid, but this effect has not been corroborated by D'Mello (1990).

Antagonisms among structurally related amino acids have also assumed greater importance since the reviews of Harper (1964) and Harper *et al.* (1970). Interactions were originally reported between lysine and arginine and among the branched-chain amino acids, principally in the rat. These antagonisms have now been extended to non-ruminant farm animals and, in addition, other interactions involving several non-protein amino acids naturally present in certain leguminous and brassica plants have also been reported (Austic, 1986; D'Mello, 1989, 1991).

Assessments of toxicity of individual indispensable amino acids in farm animals suggest differences between avian and mammalian species. The unique toxicity of tryptophan in ruminants has also been demonstrated.

This chapter reviews the effects of imbalances, antagonisms and toxicities with respect to their significance in the nutrition of farm livestock. In addressing these issues, importance will be attached to differences between various classes of animals in their sensitivity to disproportionate intakes of indispensable as well as non-protein amino acids. Attention will also be given to biochemical features underlying the adverse effects of amino acids, strategies for mitigation and innate detoxification mechanisms in animals.

Amino Acid Imbalance

The term imbalance has been defined by Harper (1964) as a change in the pattern of amino acids in the diet, precipitating depressions in food intake and growth which are completely alleviated by supplementation with the first-limiting amino acid. The prerequisite for a limiting amino acid in the diet may be satisfied by the use of a deficient protein such as gelatin, but more generally this requirement may be fulfilled by the use of low-protein diets. The definition of imbalance was originally devised as a result of investigations with the rat but is now finding wider application in the nutrition of farm animals. It is, therefore, salutary to recall some of the basic tenets embodied in this class of adverse effects. Two types of imbalance may be recognized (Table 4.1): that caused by the addition of a relatively small quantity of an amino acid to a low-protein diet, and that precipitated by an incomplete mixture of amino acids. In the former case, there is a specific requirement in that the agent precipitating the imbalance should be the second-limiting amino acid (Winje *et al.*, 1954). A more reliable procedure involves the addition of an amino acid mixture devoid of one indispensable amino acid to a low-protein diet limiting in the same amino acid (Pant *et al.*, 1972). Other studies have shown that imbalances may also be created by employing mixtures of the dispensable amino acids (Tews *et al.*, 1979, 1980). These investigations further demonstrated that the most reliable technique of precipitating an imbalance involves the use of amino acids, individually or in mixtures, which compete with the dietary limiting amino acid for transport into the brain.

From experiments employing incomplete amino acid mixtures, Fisher *et al.* (1960) concluded that the chick is as sensitive to an imbalance as the growing rat (Table 4.1). The primary manifestation of adverse effects was a reduction in food intake which, consequently, also reduced intake of the limiting amino acid, leading to reduced growth.

Practical significance of amino acid imbalance

The issue of amino acid imbalance assumed practical significance in poultry nutrition with the studies of Wethli *et al.* (1975) who invoked this phenomenon to explain the inferior utilization, by broiler chicks, of the first-limiting amino acid in low-quality protein sources. In one of their investigations, Wethli *et al.* (1975) designed a series of cereal-based diets containing increasing quantities of groundnut meal to provide crude protein (CP) levels ranging from 120 to 420 g kg^{-1} diet. These diets were formulated with or without supplementary methionine plus lysine. Growth responses were compared with those of chicks fed on a series of control diets containing graded quantities of herring meal to supply CP concentrations in the range 120 to 240 g kg^{-1} diet. Thus the assumed minimal amino

Table 4.1. Effects of amino acid imbalance on growth of rats and chicks fed low-protein diets.

Protein source; dietary level; and amino acid supplements	First-limiting amino acid	Method of precipitating imbalance	Diets	Growth response (proportion of control)	Reference
A. *Rats*					
Egg albumen; 80 g kg^{-1} diet; Thr + Val	His	Addition of second-limiting amino acid (Lys)	Control Imbalanced Corrected	1.00 0.65 0.95	Winje *et al.* (1954)
Casein; 80 g kg^{-1} diet; Met	Trp	Addition of amino acid mixture devoid of Trp	Control Imbalanced Corrected	1.00 0.71 1.10	Pant *et al.* (1972)
B. *Chicks*					
Sesame protein; 110 g kg^{-1} diet; Lys (suboptimal)	Lys	Addition of Lys-free amino acid mixture	Control Imbalanced Corrected	1.00 0.87 1.69	Fisher *et al.* (1960)

Thr = threonine; Val = valine; His = histidine; Lys = lysine; Met = methionine; Trp = tryptophan.

acid needs of the young chick were considered to be satisfied at the high inclusion rates of the two protein sources. As expected, with the unsupplemented groundnut meal diets, growth rates improved as CP concentrations increased up to 360 g kg^{-1} diet, but failed to match those of chicks fed lower levels of CP derived from herring meal. However, supplementation of the groundnut meal diets with methionine plus lysine induced progressive and more efficient liveweight gains at all concentrations of CP up to 270 g kg^{-1} diet. At this level of CP, the supplemented groundnut meal diet supported growth approaching that obtained with the best herring meal control diet containing 210 g CP kg^{-1}. In a further experiment, Wethli *et al.* (1975) observed that diets based on soyabean meal and maize were somewhat inferior to similar diets supplemented with methionine even though the unsupplemented diets, at the higher CP concentrations, satisfied the calculated requirements for the first-limiting amino acid. Of several hypotheses examined, Wethli *et al.* (1975) concluded that the amino acids supplied by low-quality oilseed protein sources were in such disproportion to the needs of the chick as to reduce utilization of the first-limiting amino acid. It was suggested that amino acid imbalances can occur in diets based on conventional ingredients and that pure supplements of the limiting amino acids may be used to correct these imbalances.

Further impetus to the study of imbalances has been provided by the introduction of the diet dilution technique to determine amino acid requirements of poultry. This method, originally devised to determine the methionine requirement of the laying pullet (Fisher and Morris, 1970), has been adapted to investigate the growth responses of broiler chicks to different concentrations of an indispensable amino acid (Gous, 1980; Morris *et al.*, 1987; Abebe and Morris, 1990a,b). The procedure involves the sequential dilution of a high-protein 'summit' diet with an isoenergetic, protein-free mixture. The summit diet is normally formulated to contain a large excess, typically 185% of assumed requirements, of all essential amino acids except the one under test, which is set at a lower level of around 145% of assumed requirements. When the summit diet is blended with the protein-free mixture, the amino acid under test will be first-limiting at all levels of dilution. Although successive dilutions of the summit diet result in different CP concentrations, the dietary amino acid pattern remains constant throughout the diluted series. The method, therefore, relies on the interpretation of responses to different rates of dilution as responses to the first-limiting amino acid and not to changing dietary CP concentrations. At each level of dilution of the summit diet, supplementation with the pure form of the limiting amino acid may be undertaken which, in theory, should elicit responses which are compatible with those obtained merely by diluting the summit diet. This dual approach of dilution and supplementation has been adopted in recent studies on the growth responses of broiler chicks to lysine (Morris *et al.*,

1987) and to tryptophan (Abebe and Morris, 1990b). It will be apparent that the diet dilution technique involves the deliberate creation of an amino acid imbalance in the classical manner established with the rat (Harper *et al.*, 1970). As will be discussed at length later (Chapter 10), substantial disparity has been observed between the growth responses obtained with the dilution of the summit diet and those observed with the supplemented series (D'Mello, 1988, 1992a). This incompatibility of responses has been attributed to the effects of amino acid imbalance in the summit and diluted diets (Abebe and Morris, 1990a) but data by D'Mello (1990) question the validity of this interpretation. Since these issues are associated with matters relating to amino acid utilization by growing chicks, further discussion is reserved until Chapter 10.

Several studies with pigs indicate that amino acid imbalances may occur at the tissue level even though the diet may appear to be in ideal balance. Such imbalances are readily demonstrated on supplementation of cereal-based diets with crystalline amino acids. It has long been recognized that free amino acid supplements are absorbed more rapidly than protein-bound amino acids resulting in an imbalanced supply of amino acids at the sites of protein synthesis (Rolls *et al.*, 1972; Leibholz *et al.*, 1986; Leibholz, 1989). For example, Leibholz *et al.* (1986) observed that the concentration of free lysine in plasma of pigs increased 1–2 hours after feeding a diet containing crystalline lysine, declining thereafter, whereas the circulating concentrations of other amino acids derived from the protein-bound fraction of the diet peaked 2–6 hours after feeding. In pigs fed once-daily, this lack of synchrony in absorption would precipitate an amino acid imbalance at the cellular level. Under these circumstances growth and efficiency of utilization of dietary nitrogen (N) may be impaired, but the deleterious effects may be offset by more frequent feeding. These expectations have been confirmed by Batterham (1974) and Batterham and O'Neill (1978) who observed that the efficiency of utilization of free lysine for growth in pigs was only 0.43 to 0.67 of that recorded with pigs fed the same ration in six equal portions at 3-hourly intervals, whereas no such benefit occurred on feeding the control diet more frequently. Subsequent investigations by Partridge *et al.* (1985) have extended the benefits of increased feeding frequency and lysine supplementation to improvements in N utilization.

Amino acid imbalance and food intake

Accounts of amino acid imbalances conventionally focus on the growth-depressing effects in animals (Harper, 1964; Tews *et al.*, 1979). However, it has been consistently recorded that a predisposing factor is a rapid and marked reduction in food intake. Thus Harper and Rogers (1965) reported that rats fed an imbalanced diet reduced their food intake within 3–6 hours. These results implied that the depression in food intake may be regarded as the primary event responsible for the ensuing retardation of growth. A considerable body of evidence, drawn from studies with both laboratory animals and chicks, supports this premise. If food intake in animals consuming an imbalanced diet is increased by force-feeding (Leung *et al.*, 1964), by insulin injections (Kumta and Harper, 1962), by adjustment of dietary protein to energy ratios (Fisher and Shapiro, 1961) or by exposure of animals to cold environmental temperatures (Klain *et al.*, 1962), commensurate improvements in growth also occur.

The biochemical mechanisms underlying the anorectic effects of imbalanced diets in rats have been described by Harper (1964), but a more explicit account appears in the review of Harper and Rogers (1965). They postulated that surplus amino acids arriving in the portal circulation after consumption of an imbalanced diet stimulate synthesis or suppress breakdown of protein in the liver leading to greater retention of the limiting amino acid relative to that in control groups (Fig. 4.1). The supply of the limiting amino acid for peripheral tissues such as muscle is thereby reduced, although protein synthesis in these tissues continues unimpeded. Eventually, however, the free amino acid patterns of both muscle and plasma become so deranged as to invoke the intervention of the appetite-regulating system to reduce food intake. Growth is reduced as a consequence of the depressed appetite and intake of nutrients. This hypothesis is still accepted as a satisfactory explanation for the effects of amino acid imbalance in the rat (Leung and Rogers, 1987; Tackman *et al.*, 1990) and is thought to have wider application to other mammalian species as well as to poultry (Boorman, 1979).

Alterations in dietary preferences are also a characteristic feature of amino acid imbalance in the rat. Thus when offered a choice, rats consume a balanced diet in preference to an imbalanced one, but more remarkably, in such situations animals will select a protein-free diet incapable of supporting growth instead of an imbalanced diet which allows growth, albeit at a low level (Sanahuja and Harper, 1962; Leung and Rogers, 1987).

A central tenet in the hypothesis outlined by Harper and Rogers (1965) is the association between food intake depression and changes in the tissue patterns of amino acids. In both muscle and plasma, concentrations of the limiting amino acid decline whereas there is an accumulation of those amino acids added to precipitate the imbalance. Since these events occur within a few hours of ingestion of such diets, it has been suggested that changes in plasma amino acid patterns may provide the metabolic signal which ultimately results in anorexia and abnormal feeding behaviour. In subsequent attempts to validate this hypothesis, the role of the first-limiting amino acid has featured prominently. For example initial studies (Leung and Rogers, 1969) indicated that the depression in appetite of rats fed an imbalanced diet may be prevented by the infusion of a small

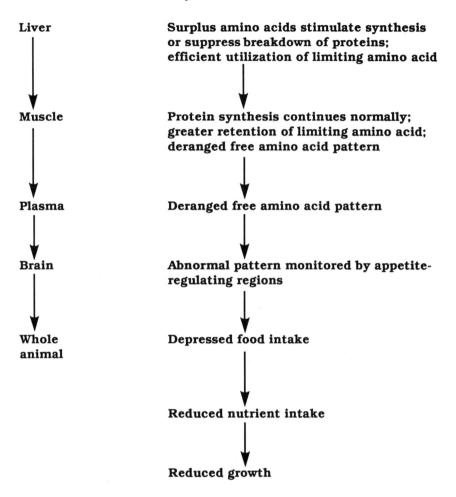

Liver — Surplus amino acids stimulate synthesis or suppress breakdown of proteins; efficient utilization of limiting amino acid

Muscle — Protein synthesis continues normally; greater retention of limiting amino acid; deranged free amino acid pattern

Plasma — Deranged free amino acid pattern

Brain — Abnormal pattern monitored by appetite-regulating regions

Whole animal — Depressed food intake

Reduced nutrient intake

Reduced growth

Fig. 4.1. Effects of amino acid imbalance in the rat; based on the hypothesis of Harper and Rogers (1965).

quantity of the first-limiting amino acid via the carotid artery whereas administration through the jugular vein was ineffective. In later studies, Tobin and Boorman (1979) confirmed that the cockerel fed an imbalanced diet responds in a similar manner to the rat following infusion with the limiting amino acid.

The studies of Leung and Rogers (1969) and others (Fernstrom and Wurtman, 1972) provided the basis of the proposition that food intake and feeding behaviour may be associated with changes in brain uptake and metabolism of critical amino acids. In subsequent studies, therefore, focus shifted towards measurement of brain concentrations of amino acids and their neurotransmitter derivatives. It was soon established that the concentration of the first-limiting amino acid declined more rapidly in cerebral tissues than in plasma (Peng *et al.*, 1972). This observation led to the proposal that the fall in the brain concentration of the limiting amino acid initiates the signal which causes the changes in food intake and dietary choice, although the precise mechanisms remain obscure (Leung and Rogers, 1987). However, the areas of the central nervous system sensitive to amino acid imbalance have been delineated in the rat. From a series of studies involving the insertion of lesions, it appears that specific neural sites are associated with the depression in food intake. These include the anterior prepyriform cortex, the medial amygdala and certain regions of the hippocampus and septum (see Leung and Rogers, 1987). In particular, the sensitivity of the prepyriform cortex to amino acid imbalance has been extensively investigated over several years (Gietzen *et al.*, 1986; Beverly *et al.*, 1990a,b, 1991a). Thus the selection of a protein-free diet in preference to an imbalanced one is reversed if the limiting amino acid is injected directly into the prepyriform cortex. In a more recent study, Beverly *et al.* (1991b) demonstrated that although 2 or 4 nmol injections of threonine were sufficient to enable rats to reverse their selection against a threonine-imbalanced diet, only the lower dose appreciably increased food intake. These results led Beverly *et al.* (1991b) to conclude that the presence of the limiting amino acid in the prepyriform cortex elicits separate effects on dietary selection and on intake of an imbalanced diet, depending upon dose level. Comparable studies on the location and nature of neural areas sensitive to amino acid imbalance have not been conducted with farm animals.

Amino acids may regulate food intake through synthesis and metabolism of neurotransmitters in the brain. There has been considerable work with rats aimed at determining the role of specific neurotransmitters in the hierarchical regulation of food intake in response to intake of protein (Peters and Harper, 1985) and amino acids (Mercer *et al.*, 1989) and to ingestion of imbalanced diets (Leung *et al.*, 1985; Tackman *et al.*, 1990). In one study, the feeding of amino acid imbalanced diets reduced the concentration of noradrenaline (norepinephrine) in the anterior prepyriform cortex of rats (Leung *et al.*, 1985). However, another study led Tackman *et al.* (1990) to conclude that the effects of amino acid imbalances could not be correlated with variations in brain concentrations of serotonin or 5-hydroxindole-3-acetic acid. Their data are supported by those of Harrison and D'Mello (1987) in a study of amino acid imbalance in the chick (Table 4.2). An imbalance caused by the addition of a mixture devoid of tyrosine and phenylalanine to a diet deficient in these amino acids, reduced food intake, but the effects on the concentrations of noradrenaline and dopamine in brain homogenates were minimal.

Table 4.2. The effects of dietary amino acid imbalance on food intake and brain concentrations of neurotransmitters in young chicks. (Adapted from Harrison and D'Mello, 1987.)

Diet	Daily food dry matter intake (g per chick)	Daily weight gain (g per chick)	Neurotransmitters (ng g^{-1} wet weight)	
			Noradrenaline (=norepinephrine)	Dopamine
Basal	16	5	131	212
Basal + imbalancing AA mixture	11	2	109	159
Basal + imbalancing AA mixture + Phe (5 g kg^{-1} diet)	23	14	166	180
Basal + imbalancing AA mixture + Phe (10 g kg^{-1} diet)	25	15	174	198

AA = amino acid mixture lacking phenylalanine (Phe) and tyrosine; basal diet deficient in Phe and tyrosine.

Amino acid imbalance and nutrient utilization

The effect of amino acid imbalance on nutrient utilization has been the subject of some debate. An imbalance would be expected to reduce the overall efficiency of utilization of dietary protein. Experiments with rats (Kumta *et al.*, 1958) confirm this expectation, with N retention efficiency declining from 0.60 to 0.44 on addition of an imbalancing mixture of amino acids to a basal diet containing fibrin. However, in rats pair-fed the basal diet to match intakes of the imbalanced group, efficiency of N retention declined to 0.33, indicating that the effects of imbalance are mediated via a reduction in food intake. Despite these observations, the general consensus is that amino acid imbalances reduce the efficiency of protein utilization in farm animals. Thus Moughan (1991) attributes the low efficiency of protein utilization in pigs partly to dietary amino acid imbalance. In addition, the studies of Partridge *et al.* (1985) demonstrated that imbalances at the tissue level, induced by differential absorption of amino acids from crystalline and protein-bound sources, can reduce the overall efficiency of protein utilization in pigs fed once-daily. Furthermore, Wang and Fuller (1989) demonstrated that manipulation of the composition of a mixture of amino acids simulating the pattern in casein enhanced N retention in pigs through mitigation of deleterious imbalances.

A more contentious issue relates to the effects of amino acid imbalances on efficiency of utilization of the first-limiting amino acid. Salmon (1954) suggested that surplus amino acids causing an imbalance may stimulate catabolic pathways resulting in the degradation of all amino acids, with the inadvertent loss of the limiting amino acid. Initial studies (Florentino and Pearson, 1962) supported this hypothesis in that catabolism of tryptophan in rats was enhanced by feeding excess threonine. However, later investigations failed to provide corroborative evidence. Thus Harper and Rogers (1965) reported that rats fed a threonine-imbalanced diet reduced oxidation of ^{14}C-labelled threonine. In subsequent investigations, Yoshida *et al.* (1966) and Benevenga *et al.* (1968) presented evidence of increased incorporation of the first-limiting amino acid into hepatic proteins of rats fed imbalanced diets. Thus, both whole-animal and biochemical studies with rats have demonstrated enhanced utilization and retention of the limiting amino acid following the administration of imbalanced diets. Despite this evidence, other investigators continue to invoke amino acid imbalance to explain apparent differences in utilization of limiting amino acids in chicks given excess dietary protein (Wethli *et al.*, 1975; Abebe and Morris, 1990a,b). This issue is of sufficient practical significance to merit detailed attention in a later chapter.

Amino Acid Antagonisms

In its simplest form, an amino acid antagonism may be defined as an interaction between structurally similar amino acids resulting in the precipitation of adverse effects. This category of deleterious effects was devised to accommodate the unique actions of excess dietary lysine and leucine in the rat (Harper, 1964). For example, excess lysine specifically impairs the utilization of arginine, while excess leucine depresses the utilization of the other two branched-chain amino acids, isoleucine and valine, even when they are not limiting in the diet. In both antagonisms, mitigation of adverse effects may be effected by appropriate supplementation with specific amino acids, arginine in the former case and isoleucine and valine in the latter interaction. Demonstrations of antagonisms have now been extended to farm animals, including poultry, pigs and preruminant lambs (D'Mello and Lewis, 1970a,b,c; Oestemer _et al._, 1973; Papet _et al._, 1988a). In addition, it is now recognized that antagonisms are not restricted to those caused by lysine and leucine but may be precipitated by a wide range of analogues occurring naturally in crop plants as non-protein amino acids (see D'Mello, 1991). In most cases, the action of the non-protein amino acid analogues is directed at the metabolism and utilization of specific structurally related essential amino acids.

Branched-chain amino acid antagonisms

Interest in the antagonisms involving the branched-chain amino acids (BCAA) has been sustained by the knowledge that maize by-products, sorghum and blood meal contain disproportionate quantities of these amino acids. In addition, maintenance requirements of animals for BCAA may be influenced by these antagonisms. Following the initial demonstration of leucine-induced antagonisms in the rat, much evidence has accumulated to indicate the specificity and complexity of interactions among BCAA in the young chick and turkey poult. In one study D'Mello and Lewis (1970b) showed that dietary supplementation with excess leucine prevented the growth response of chicks to the first-limiting amino acid, methionine, but permitted the response in the presence of supplementary isoleucine. The specificity of the leucine–isoleucine interaction was thus established. However, the results of another experiment led D'Mello and Lewis (1970b) to conclude that the leucine–valine interaction was more potent than that between leucine and isoleucine. This conclusion was based on growth and plasma amino acid data (Table 4.3). The addition of excess leucine to a diet equally but marginally limiting in isoleucine and valine precipitated a severe growth depression in young chicks. Valine supplementation reversed this effect but isoleucine addition failed to elicit a response. Indeed, a combination of excess leucine and supplementary

Table 4.3. Branched-chain amino acid antagonisms in the young chick: effects of excess dietary leucine on growth and plasma amino acid concentrations. (Adapted from D'Mello and Lewis, 1970b.)

Diet	Daily weight gain (g per chick)	Plasma amino acid concentrations (μmol 100 ml^{-1})					
		Val	Ile	Leu	Gly	Lys	Arg
Basal	16	15.7	5.1	12.2	43.2	66.9	16.9
Basal + Val	16	17.8	7.5	12.8	30.9	48.2	18.8
Basal + Ile	15	11.7	11.3	20.2	72.8	73.8	33.0
Basal + Val + Ile	17	-	-	-	-	-	-
Basal + Leu	13	10.7	7.9	37.7	63.0	91.3	23.6
Basal + Leu + Val	15	14.5	7.4	15.2	46.6	61.4	17.8
Basal + Leu + Ile	11	9.3	10.6	40.5	73.5	75.8	27.5
Basal + Leu + Val + Ile	17	-	-	-	-	-	-

Val = valine; Ile = isoleucine; Leu = leucine; Gly = glycine; Lys = lysine; Arg = arginine; basal diet marginally deficient in Val and Ile.

isoleucine impaired growth performance even further. Excess dietary leucine induced a rapid fall in plasma valine concentration which was exacerbated by concomitant supplementation with leucine and isoleucine. The plasma level of isoleucine remained undisturbed on addition of leucine and valine, alone or in combination. The efficacy of valine was further confirmed by its ability to reduce the high plasma levels of leucine, whereas isoleucine addition was ineffective in this respect. The concentrations of most other amino acids in plasma were enhanced marginally on feeding excess leucine. The sensitivity of valine to leucine antagonism and its augmentation by supplementary isoleucine was confirmed in a later study (D'Mello and Lewis, 1970c). Excess leucine depressed plasma valine concentrations from 25.6 to 13.9 μmol 100 ml^{-1} in chicks fed an isoleucine-deficient diet, but combined additions of leucine and isoleucine reduced valine levels further, to 11.7 μmol 100 ml^{-1}. These reductions are of particular significance since dietary valine concentrations were set at adequate levels.

The specificity of the leucine–valine antagonism is further exemplified in studies with turkey poults fed diets supplemented with graded combinations of leucine and valine (D'Mello, 1975). Both leucine and valine accumulated in plasma following supplementation but the extent of accumulation of valine was reduced as dietary concentrations of leucine increased. However, valine supplementation failed to suppress accumulation of leucine in plasma. These studies also illustrated another complex facet of BCAA antagonisms in that leucine supplementation reduced plasma isoleucine concentrations from 14.7 to 6.8 μmol 100 ml^{-1} but valine supplementation with leucine failed to depress plasma isoleucine concentrations any further. The degree to which interactions among BCAA are reversible has not been investigated in any systematic manner, although the dominance of leucine is now apparent.

Despite the primacy of the leucine–valine antagonism, it is possible to devise dietary conditions demonstrating enhanced sensitivity of isoleucine in BCAA interactions. For example, D'Mello (1974) showed that a diet supplemented with a small excess of leucine depressed chick growth, which was only partially alleviated with valine supplements. Examination of the plasma amino acid data indicated that plasma levels of isoleucine had declined from 5.8 to 3.8 μmol 100 ml^{-1} on dietary addition of leucine but plasma isoleucine concentrations declined further, to 2.7 μmol 100 ml^{-1} on providing small excesses of leucine and valine in a combined supplement.

The complexity of interactions among BCAA is further illustrated by the response of laying pullets to excess dietary leucine (Bray, 1970). Egg production and egg yield were reduced by this excess but were restored to satisfactory levels only when isoleucine and valine were added in combination. Failure to recognize the complexity of these interactions may account for the lack of effect of isoleucine alone in alleviating a leucine-induced antagonism in the pig (Oestemer, *et al.*, 1973). Inspection of the plasma amino acid data, however, indicated moderate decreases in circulating levels of both isoleucine and valine, thus suggesting that combined supplements of these amino acids might have been more effective than isoleucine alone.

Antagonisms involving BCAA have been demonstrated in the preruminant lamb by Papet *et al.* (1988a), but a relatively large dose of excess leucine (126 g kg^{-1} diet) is required to depress food intake and plasma concentrations of valine and isoleucine.

From studies, principally with the rat, Harper *et al.* (1984) attributed the leucine-induced changes in plasma levels of isoleucine and valine to increased oxidation of these two amino acids, having discounted any effects emanating from competition during intestinal absorption or kidney reabsorption. Limited studies with the chick support this view. Thus Calvert *et al.* (1982) demonstrated that excess leucine failed to influence excretion of ^{14}C-labelled isoleucine or valine, but markedly increased the oxidation of these amino acids as indicated by enhanced *in vivo* output of $^{14}CO_2$. The catabolism of BCAA is initiated by a reversible aminotransferase reaction. The branched-chain keto acids (BCKA) so formed then undergo irreversible oxidative decarboxylation, catalysed by BCKA dehydrogenase to yield acyl-CoA compounds which are degraded further in a series of reactions analogous to those involved in fatty acid oxidation. Harper *et al.* (1984) suggest that enhanced BCKA oxidation may account for the depletion of plasma isoleucine and valine pools in animals fed excess leucine. Recent studies with preruminant lambs support this view in that marked reductions in plasma concentrations of keto acids derived from isoleucine and valine were recorded in response to excess intake of leucine (Papet *et al.*, 1988a). In subsequent studies Papet *et al.* (1988b) demonstrated that excess leucine increased activities of aminotransferases in the liver and jejunum and also activated BCKA dehydrogenase in the jejunum of lambs.

Excess BCAA may, additionally, induce depletion of brain pools of other amino acids, particularly those which are the precursors of the neurotransmitters. In this regard, Harrison and D'Mello (1986) showed that dietary excesses of three BCAA reduced brain concentrations of noradrenaline, dopamine and 5-hydroxytryptamine in the chick and that the levels of these neurotransmitters may be restored by dietary supplementation with their precursors, phenylalanine and tryptophan (Table 4.4). The significance of these results awaits elucidation. However it is generally conceded that changes in brain metabolism of amino acids and neurotransmitters may be associated with alterations in food intake and feeding behaviour (Leung and Rogers, 1987). Consistent with this hypothesis has been the observation that a substantial element of the adverse effects of

Table 4.4. Brain concentrations of neurotransmitters in chicks fed excess branched-chain amino acids. (Adapted from Harrison and D'Mello, 1986.)

Diet	Neurotransmitters (ng g^{-1} wet weight)		
	Noradrenaline (=norepinephrine)	Dopamine	5-Hydroxytryptamine
Control	296	246	572
Control + BCAA	186	150	357
Control + BCAA + Phe + Trp	269	221	470

BCAA = branched-chain amino acids; Phe = phenylalanine; Trp = tryptophan.

excess leucine arises from the reduction in food intake (Calvert *et al.*, 1982; Papet *et al.*, 1988a) which subjugates effects emanating from oxidative catabolism of isoleucine and valine.

The lysine–arginine antagonism

The considerable variation in the arginine requirements of the chick has provided the impetus for extensive and sustained investigations on the lysine–arginine interaction. In addition, a number of feedstuffs are endowed with adverse ratios of the two amino acids. Evidence of a potent interaction between lysine and arginine in the chick dates back to the studies of Jones (1961) on the toxicity of lysine. Since then much data have accumulated to illustrate specificity, reciprocity and mechanisms of action in this antagonism. The unique specificity of this interaction was tested in several experiments by D'Mello and Lewis (1970a) who designed basal diets which were first limiting in methionine, tryptophan, histidine or threonine, with arginine marginally limiting. Addition of excess lysine to each of these diets precipitated a severe growth depression in chicks which, in every case, was reversed by arginine and not by the amino acid originally limiting in the basal diet. A selection of the results relating to the threonine-deficient diet is shown in Table 4.5, which further illustrates the specific effect of excess lysine in reducing plasma levels of arginine from 16.6 to 7.4 μmol 100 ml^{-1}, whereas plasma threonine concentrations remained unaffected by the surfeit of lysine. Evidence of specificity was also provided by Nesheim (1968) in studies with two strains of chicks differing substantially in their requirements for arginine. Chicks with a high arginine requirement were less able to tolerate dietary excesses of lysine than chicks with a low requirement for arginine. However, a number of dietary factors can reduce or exacerbate the severity of the lysine–arginine antagonism. Excess chloride augments the adverse effects, whereas supplementation

Table 4.5. Specificity of the lysine–arginine antagonism in the young chick: effects of excess dietary lysine on growth and plasma amino acid concentrations. (Adapted from D'Mello and Lewis, 1970a.)

Diet	Daily weight gain (g per chick)	Plasma amino acid concentrations (μmol 100 ml^{-1})						
		Arg	Thr	Lys	Ile	Leu	Tyr	Phe
Basal	13	16.6	26.6	83.8	15.8	22.4	27.8	12.4
Basal + Arg	13	20.0	23.4	91.4	12.0	22.0	27.6	12.0
Basal + Thr	20	13.6	81.6	58.8	12.0	18.6	24.6	10.0
Basal + Arg + Thr	21	28.8	94.8	80.0	15.4	23.6	24.8	12.0
Basal + Lys	8	7.4	33.6	119.8	11.8	19.0	24.2	10.0
Basal + Lys + Arg	13	10.8	40.4	169.6	12.8	23.8	23.4	11.8
Basal + Lys + Thr	10	6.8	133.6	115.0	10.8	18.8	25.0	10.2
Basal + Lys + Arg + Thr	17	7.6	80.0	120.0	9.0	18.2	21.0	8.6

Arg = arginine; Thr = threonine; Lys = lysine; Ile = isoleucine; Leu = leucine; Tyr = tyrosine; Phe = phenylalanine; basal diet first limiting in Thr and second limiting in Arg.

with alkaline salts of monovalent mineral cations reduces or eliminates the potency of this antagonism (see Austic, 1986).

Although the lysine–arginine antagonism has been demonstrated in the rat fed casein diets (Jones et al., 1966), its existence in the pig has been refuted (Edmonds and Baker, 1987). Relatively large excesses $(35 \text{ g kg}^{-1}$ diet) of lysine are required to reduce food intake and efficiency of food utilization. This level of lysine failed to influence arginase activity in any of the tissues examined. Excess lysine depressed plasma arginine but did not affect its concentration in liver, kidney or muscle. On the basis of this evidence, Edmonds and Baker (1987) attributed the adverse effects of excess lysine in the pig to an amino acid imbalance rather than to a specific antagonism.

Reciprocity of the lysine–arginine antagonism is somewhat imperfectly documented. D'Mello and Lewis (1970c) showed that excess arginine depressed growth of chicks fed a lysine-deficient diet, an effect which was reversed by supplementary lysine. However, the specificity and metabolic basis of this effect remain unresolved. In the pig, excess arginine is considered to precipitate its adverse effects through an imbalancing action rather than through a genuine antagonism (Anderson et al., 1984).

The biochemical mechanisms underlying the lysine–arginine antagonism in avian species have now been elucidated. By virtue of their uricotelism, poultry are unable to synthesize arginine and are, therefore, particularly sensitive to this interaction. By far, the most significant contributory factor to the antagonism is the enhanced activity of kidney arginase in chicks fed excess lysine, resulting in increased catabolism of arginine (Austic, 1986). If arginase activity is suppressed by the use of a specific inhibitor, then the susceptibility of chicks to the lysine–arginine antagonism is also attenuated. A second factor implicated in this antagonism is the depression in food intake, presumably arising from lysine-induced disruption of brain uptake and metabolism of other amino acids and their biogenic amines. It should be noted, however, that in the lysine–arginine antagonism, the depression in growth precedes the reduction in food intake (D'Mello and Lewis, 1971). Other mechanisms of secondary importance in chicks fed excess lysine, include enhanced urinary output of arginine and inhibition of hepatic transamidinase activity with consequent reduction in endogenous synthesis of creatine (Austic, 1986).

Antagonisms induced by non-protein amino acids

A wide array of amino acids occurring naturally in unconjugated forms in plants are endowed with the potential to elicit adverse effects in animals. The presence of these non-protein amino acids in economically important species of legumes and brassicas has thwarted attempts to maximize utilization of these plants as sources of food for farm livestock. Non-protein amino acids may occur in all parts of the plant, but the seed is normally the most concentrated source (D'Mello, 1991). In many instances these compounds bear structural analogy with the nutritionally important amino acids or their neurotransmitter derivatives found in the central nervous system of animals. Consequently, manifestations of adverse effects range from reductions in food intake and nutrient utilization to profound neurological disorders and even death (Table 4.6).

The aromatic amino acid, mimosine, contributes significantly to the toxicity of the ubiquitous tropical forage legume, *Leucaena leucocephala*. Mimosine is generally regarded as a structural analogue of tyrosine and its cerebral neurotransmitter derivatives, dopamine and noradrenaline (D'Mello, 1991). However, the effects on brain metabolism of these biogenic amines have yet to be confirmed, and evidence that tyrosine may reverse the deleterious effects of mimosine is equivocal (see D'Mello and Acamovic, 1989). The adverse properties of mimosine are extensive and include disruption of reproductive function, teratogenic effects, loss of hair and wool, and even death. Similar effects may be induced by feeding *Leucaena* to cattle and sheep. Thus the defleecing effects in sheep may be precipitated by the administration of pure mimosine or by the feeding of *Leucaena* forage (Reis *et al.*, 1975).

Manifestations of *Leucaena* toxicity are determined by geographical differences in rumen microbial ecology and are critically dependent upon the rate and extent of bacterial breakdown of mimosine (Fig. 4.2). During this degradation, 3-hydroxy-4(1H)-pyridone (3,4-DHP) is synthesized; this itself is endowed with deleterious properties causing loss of appetite, goitre and reductions of blood thyroxine concentrations (Jones, 1985). Another goitrogen and isomer, 2,3-DHP may also be synthesized in the rumen. Thus the association of mimosine with tyrosine metabolism appears to be mediated indirectly via the two forms of DHP. Some rumen bacteria are capable of detoxifying both forms of DHP to compounds which remain unidentified. Despite these reactions, considerable quantities of mimosine and 3,4-DHP may escape degradation and other derivatives may also be excreted. Ruminants in Australia, the USA and Kenya lack the requisite bacteria involved in the detoxification of the two DHP isomers and, consequently, readily succumb to their goitrogenic effects if high intakes of *Leucaena* are maintained over a protracted period of time (see D'Mello 1992b). On the other hand, in certain other regions where *Leucaena* is indigenous (Central America) or is naturalized (Hawaii and Indonesia), ruminants possess the full complement of bacteria that are required for DHP degradation, which accounts for the absence of *Leucaena* toxicity in these countries (Jones, 1985). However, the transfer of DHP-degrading bacteria to cattle in Australia has been achieved with complete success. Inoculated animals grazing *Leucaena* show markedly higher liveweight gains and serum thyroxine concentrations than untreated *Leucaena*-fed

Table 4.6. Distribution and adverse effects of some non-protein amino acids. (Adapted from D'Mello, 1992b.)

Amino acid	Plant species	Concentration (g kg^{-1} dry weight)	Adverse effects
Aromatic:			
Mimosine	Leucaena leucocephala	145 (seed) 25 (leaf)	Loss of wool; teratogenic effects; organ damage; death
Analogues of sulfur amino acids:			
Se-methylselenocysteine Selenocystathionine Selenomethionine	Astragalus species	?	'Blind staggers'; death
S-methylcysteine sulfoxide	Brassica species	40–60 (leaf)	Haemolytic anaemia; loss of appetite; reduced milk yield; organ damage; death
Arginine analogues:			
Canavanine	Canavalia ensiformis	25–51 (seed)	
	Gliricidia sepium	40 (seed)	
	Robinia pseudoacacia	98 (seed)	Reduced growth and nitrogen utilization
	Indigofera spicata	9 (seed)	
Indospicine	Indigofera spicata	20 (seed)	Teratogenic effects; liver damage
Homoarginine	Lathyrus cicera	12 (seed)	Reduced growth and food intake

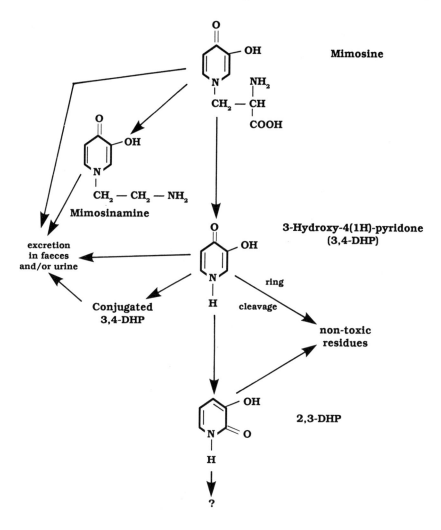

Fig. 4.2. Mimosine metabolism in the ruminant. (From D'Mello, 1991. Reproduced permission of The Royal Society of Chemistry.)

controls (Jones and Megarrity, 1986; Quirk *et al.*, 1988). Dosed cattle rapidly reduce urinary DHP excretion despite a doubling of *Leucaena* intake. The isolation of active DHP-degrading bacteria from faeces of dosed cattle implies that inoculation of just a few animals in a herd may be sufficient to overcome *Leucaena* toxicity (Quirk *et al.*, 1988). This technique thus offers a viable strategy for maximizing utilization of *Leucaena* with the added benefit of improved performance of cattle (Quirk *et al.*, 1990).

Striking analogues of the sulfur-containing amino acids exist naturally in plants (see D'Mello, 1991), particularly in those species where the sulfur atom is replaced by selenium. The debilitating disorders associated with these selenoamino acids are manifestations of acute selenium poisoning. In addition, another analogue, S-methylcysteine sulfoxide (SMCO) occurs in forage and root brassica crops (Table 4.6). The occurrence of this amino acid constitutes a significant deterrent to the exploitation of these crops as fodder for ruminant animals in temperate countries. The adverse effects of SMCO are precipitated following its metabolism by rumen bacteria to dimethyl disulfide (Smith, 1980). A severe haemolytic anaemia appears within 1–3 weeks in animals fed mainly or exclusively on brassica forage. Initial overt indications of the disorder include loss of appetite and reduced milk production, while internal changes include the appearance of refractile, stainable granules (Heinz–Ehrlich bodies) within the erythrocytes and reduced blood haemoglobin concentrations. Extensive organ damage is an accompanying feature of this condition with the liver becoming swollen, pale and necrotic. Critical daily intakes of SMCO range between 15 and 19 g kg^{-1} body weight irrespective of the source of the amino acid. Surviving animals continuing to graze the crop may make spontaneous but incomplete recovery with further fluctuations in blood haemoglobin concentrations. Withdrawal of the forage usually results in the restoration of normal blood composition within 3–4 weeks (Smith, 1980).

Of the three analogues of arginine (Table 4.6), canavanine is more widely distributed and present in higher concentrations in leguminous seeds. Canavanine contributes significantly to the toxicity of *Canavalia ensiformis* (jack bean, JB) for young chicks. Adverse effects may also arise through the synthesis of canaline, a structural analogue of ornithine, by the action of arginase on canavanine. The mammalian metabolism of canavanine corresponds with that of arginine in the urea cycle (see D'Mello, 1991). Since this cycle is non-functional in avian species, they are unable to synthesize arginine and, consequently, readily succumb to the adverse effects of canavanine in jack beans (D'Mello *et al.*, 1989). As shown in Table 4.7, chicks fed JB, autoclaved to denature potent lectins, grow at reduced rates and utilize food and dietary N less efficiently than their control counterparts. Canavanine appears in the serum of JB-fed chicks and, in addition, serum urea concentrations exceed those of control animals. Lysine supplementation serves to exacerbate the effects on growth performance and on food intake and utilization. However, arginine supplementation of the JB diet enhances weight gain and food intake. Since canavanine is not destroyed on autoclaving JB, these responses led D'Mello *et al.* (1989) to propose the existence of a canavanine–arginine interaction analogous to that between lysine and arginine. Similarities exist in several respects. Thus in both antagonisms arginine requirements and urea excretion are enhanced, although the relative proportions of this additional urea

Table 4.7. Whole-animal and metabolic responses of chicks fed a control diet or autoclaved jack bean (AJB) diets containing canavanine. (Adapted from D'Mello *et al.*, 1989.)

Diet	Daily weight gain (g per chick)	Efficiency of food conversion (g gain g^{-1} dry matter intake)	Efficiency of nitrogen retention (g N retained g^{-1} N consumed)	Serum canavanine (mg l^{-1})	Serum urea (mg l^{-1})
Control	35	0.761	0.609	0.0	10.6
AJB basal	22	0.681	0.547	14.2	20.9
AJB basal + Lys	18	0.637	0.508	10.3	27.7
AJB basal + Arg	25	0.698	0.532	12.0	45.8
AJB basal + Lys + Arg	27	0.702	0.540	11.1	42.5

Lys = lysine; Arg = arginine; canavanine content of AJB basal diet: 3.7 g kg^{-1} diet dry matter.

arising from canavanine and arginine remain to be established. The failure
of supplementary arginine to substantially reduce serum concentrations
of the respective antagonists is a feature common to both interactions.
Creatine supplementation markedly improves the efficiencies of dry matter
and N utilization in chicks fed JB (D'Mello *et al.*, 1990). It is noteworthy
that Austic and Nesheim (1972) also reported improvements in the efficiency of food utilization with creatine supplementation in the lysine–
arginine antagonism. However, differences between the two interactions
are also apparent in that arginine enhances the efficiencies of utilization
of food and N in chicks fed excess lysine but not in those fed diets containing canavanine (Table 4.7). D'Mello (unpublished) observed that
homoarginine (Table 4.6) may exacerbate the effects of canavanine-
induced toxicity in chicks fed JB diets, serving to illustrate the diversity of
interactions among the analogues and antagonists of arginine.

   Neurotoxic non-protein amino acids occur in the form of the structurally
related lathyrogens, β-(N)-oxalyl amino alanine and α,γ-diaminobutyric
acid. Their occurrence in legume seeds has been associated with the condition of neurolathyrism in humans, but well-defined neurotoxic effects
have also been observed on administration of the pure forms of these
amino acids to laboratory animals and chicks (see D'Mello, 1991). There
is currently some interest in the use of *Lathyrus sativus* and *Vicia sativa*
grains for poultry (Rotter *et al.*, 1991) but the occurrence of neuro-
lathyrogenic amino acids in these seeds may represent a primary limiting
factor. Ruminants may be able to degrade neurolathyrogens as shown by
the absence of toxicity in wether lambs on feeding *Lathyrus sylvestris* hay
with a relatively high diaminobutyric acid concentration of $12\,g\,kg^{-1}$ dry
weight (Forster *et al.*, 1991).

   There is now overwhelming evidence to indicate that the adverse effects
of the non-protein amino acids are mediated via diverse mechanisms. These
are summarized in Table 4.8 but discussed at some length by D'Mello
(1991). The multimodal action of these amino acids is embodied in the
mechanisms proposed for canavanine. The observation that chicks fed JB
diets containing canavanine increase urea output may reflect enhanced
arginase activity in the kidney. Such an increase might lead to an inadvertent loss of arginine in a manner analogous to that observed in the lysine–
arginine antagonism. Canavanine administration to rats induces substantial
increases in serum and urinary concentrations of ornithine. Hepatic activity
of ornithine decarboxylase is reduced by a factor of five in chicks fed
JB diets containing canavanine (D'Mello, 1993). These effects may be
attributed to the synthesis of canaline, which forms a covalent complex
with pyridoxal phosphate, thereby inhibiting activities of enzymes such as
ornithine decarboxylase which require the vitamin as a cofactor. Ornithine
decarboxylase is a key enzyme in the synthesis of the polyamines involved
in the regulation of cell growth and differentiation. Canavanine may also

Table 4.8. Diverse mechanisms underlying the adverse effects of selected non-protein amino acids.

Amino acid	Biochemical changes	Effects
Canavanine	Enhanced arginase activity	Increased arginine catabolism
	Decreased activity of ornithine decarboxylase following synthesis of canaline	Reduced polyamine synthesis
	Competition with lysine and arginine for transport	Reduced intestinal absorption of lysine and arginine
	Inhibition of transamidinase activity	Reduced creatine synthesis
	Synthesis of aberrant proteins	Enhanced protein turnover*
	Inhibition of nitric oxide synthesis	Impaired immunocompetence*; reduced food intake*
Mimosine	Reduced synthesis of high-tyrosine proteins	Reduced wool strength*
	Reduced DNA synthesis	Inhibition of wool biosynthesis
	Complex formation with pyridoxal phosphate-dependent enzymes	Cystathioninuria
S-methylcysteine sulfoxide	Blockage of sulfhydryl groups	Inactivation of key proteins

* Speculative.

act by competing with arginine and lysine for transport across intestinal cells. A further focal point for toxic action may reside in the ability of canavanine to act, like lysine, by inhibiting transamidinase activity, resulting in reduced creatine synthesis. Canavanine may replace arginine during protein synthesis leading to the synthesis of aberrant proteins with modified functional properties. However, the evidence is equivocal and D'Mello (1991) proposes that canavanyl proteins may be degraded as rapidly as they are formed leading to enhanced overall protein turnover rates, a feature which might contribute to the poor N retention efficiencies of chicks fed JB (Table 4.7). Recent work highlights another potential mechanism underlying canavanine toxicity. Mammalian cells are capable of converting arginine to nitric oxide, an intermediate in the biosynthesis of nitrite and nitrate. Although the metabolic significance of these reactions awaits elucidation, it has been established that canavanine inhibits nitrite and nitrate synthesis in certain cell types (Marletta, 1989). Reduced synthesis of nitric oxide may impair immunocompetence and also cause a depression in food intake through its role in gut motility and neurotransmission (Moncada *et al.*, 1991; Mayer *et al.*, 1993). Canavanine is now recognized to be a potent inhibitor of food intake in pigs (Enneking *et al.*, 1993).

Despite the economic importance of *Leucaena*, information concerning the mechanism of action of mimosine is fragmentary (Table 4.8). Consistent with the action of several structural analogues, synthesis of high-tyrosine protein components of wool is reduced in sheep given intravenous infusions of mimosine (Frenkel *et al.*, 1975). The activities of several pyridoxal phosphate-dependent enzymes may be inhibited through the ability of mimosine to form complexes with the vitamin moiety (see Rosenthal, 1982). In addition both mimosine and 3,4-DHP are known to cause inhibition of wool follicle DNA synthesis *in vitro* (Ward and Harris, 1976).

The mode of action of SMCO requires elucidation although it is acknowledged that its derivative, dimethyl disulfide, inactivates key proteins through blockage of sulfhydryl groups. The reaction with reduced glutathione, a key factor in the production of red blood cells from oxidative injury, represents one mechanism for SMCO toxicity. Brassica-fed ruminants appear to compensate for inactivation of proteins by increasing the synthesis of growth hormone and thyroxine which, in turn, stimulate the production of replacement proteins (Barry *et al.*, 1985).

Amino Acid Toxicity

Unique toxic effects may be precipitated on feeding excess quantities of individual amino acids by virtue of their particular structural or metabolic features. Benevenga and Steele (1984) have reviewed the evidence derived principally from observations with laboratory animals. The acute growth depressions caused by excesses of some individual amino acids may be accompanied by profound and specific lesions in tissues and organs. Toxicities may also be demonstrated in farm livestock, and Baker (1989) has summarized recent data indicating that methionine is the most growth-depressing amino acid when included at $40\,g\,kg^{-1}$ diet. Leucine, isoleucine and valine do not impair growth when added at this level to practical diets for pigs and poultry. Excess threonine depresses growth in chicks but not in pigs, whereas arginine is more toxic to the pig than the fowl. Such instances of toxicity are, in the main, confined to experimental situations. However, Adeola and Ball (1992) reported that dietary excesses of tryptophan or tyrosine administered for 5 days prior to slaughter reduced stress in pigs, a response attributed to increased hypothalamic concentrations of several neurotransmitters.

Of some practical significance is the occasional incidence, under natural conditions, of interstitial pulmonary emphysema and oedema in ruminants subjected to sudden changes in diet. This syndrome is associated with the ruminal production of abnormal quantities of 3-methylindole (skatole) from tryptophan. Oral or intraruminal administration of tryptophan or the indole to cattle can cause acute respiratory distress and pulmonary lesions similar to those seen in the natural disorder (Carlson *et al.*, 1968). Toxicity is mediated after the indole is metabolically activated by a mixed function oxidase to a reactive free radical product capable of initiating tissue damage in the lung. The severity of the syndrome may be reduced by dietary treatments which increase tissue glutathione. Consequently, tryptophan toxicity depends upon the balance between metabolic activation of 3-methylindole and its conjugation with glutathione (Merrill and Bray, 1983).

The unusual amino acid, lysinoalanine, may arise during alkali treatment of protein feedstuffs (Finot, 1983). The feeding of proteins containing lysinoalanine causes a reduction in biological values of diets for rats. In addition renal lesions are induced, characterized by enlargement of the nucleus and cytoplasm and disruption of DNA synthesis. The significance of these observations in the nutrition and metabolism of farm animals requires elucidation since alkali treatment is known to be an effective technique for denaturing antigenically active globular proteins in soyabean meal.

Conclusions

The categories of imbalances, antagonisms and toxicities elucidated in studies with the rat constitute a satisfactory basis for classifying the adverse effects of amino acids in farm livestock. Amino acid imbalances are ubiquitous in conventional diets fed to pigs and poultry, causing reductions in growth performance and efficiency of utilization of dietary N. Imbalances may also occur in pigs at the tissue level due to differential rates of absorption of amino acids from crystalline and intact protein sources. With poultry, it has been observed that high-protein diets based on poor quality ingredients depress utilization of the first-limiting amino acid and this phenomenon has been attributed to the effects of an amino acid imbalance. Antagonisms occur widely in livestock nutrition emanating from adverse ratios of lysine and arginine and of the branched-chain amino acids in some common feedstuffs (see D'Mello, 1978). Antagonisms may also be induced by the non-protein amino acids. These interactions are of particular economic significance since they restrict the exploitation of many crop plants as sources of nutrients for farm animals. Examples of such antagonisms include those caused by the presence of mimosine in the tropical forage legume, *Leucaena leucocephala* and of S-methylcysteine sulfoxide (SMCO) in *Brassica* species grown for winter fodder in temperate countries. In addition to precipitating adverse effects on growth and food utilization, antagonisms generally increase amino acid requirements for maintenance and growth, issues which will be addressed at greater length in Chapter 10.

Although ruminants are normally less susceptible to the effects of imbalances by virtue of extensive microbial metabolism of amino acids, there are notable instances when ruminal degradation serves to precipitate adverse effects through the generation of reactive metabolites from mimosine, SMCO and tryptophan. In the case of the latter two amino acids, reduced glutathione status in the animal may play a critical role in determining toxicity.

The adverse effects of amino acid imbalances are precipitated via reductions in food intake and alterations in the brain uptake and metabolism of amino acids, but the precise mechanisms await elucidation. As regards antagonisms, it is now clear that these are mediated via a diverse array of mechanisms ranging from alterations in the activities of key enzymes to competition with specific indispensable amino acids for transport and protein synthesis.

Animals are able to adapt to disproportionate intakes of certain amino acids as the capacity for amino acid catabolism increases with time. However, the most striking example of degradation is that achieved by some rumen bacteria enabling *Leucaena*-fed cattle and goats in many regions of the tropics to metabolize mimosine to innocuous residues.

References

Abebe, S. and Morris, T.R. (1990a) Note on the effects of protein concentration on responses to dietary lysine by chicks. *British Poultry Science* 31, 255–260.

Abebe, S. and Morris, T.R. (1990b) Effects of protein concentration on responses to dietary tryptophan by chicks. *British Poultry Science* 31, 267–272.

Adeola, O. and Ball, R.O. (1992) Hypothalamic neurotransmitter concentrations and meat quality in stressed pigs offered excess dietary tryptophan and tyrosine. *Journal of Animal Science* 70, 1888–1894.

Agricultural Research Council (1981) *The Nutrient Requirements of Pigs.* Commonwealth Agricultural Bureaux, Farnham Royal, Slough.

Anderson, L.C., Lewis, A.J., Peo, E.R. and Crenshaw, J.D. (1984) Effects of excess arginine with and without supplemental lysine on performance, plasma amino acid concentrations and nitrogen balance of young swine. *Journal of Animal Science* 58, 369–377.

Austic, R.E. (1986) Biochemical description of nutritional effects. In: Fisher, C. and Boorman, K.N. (eds) *Nutrient Requirements of Poultry and Nutritional Research.* Butterworths, London, pp. 59–77.

Austic, R.E. and Nesheim, M.C. (1972) Arginine and creatine interrelationships in the chick. *Poultry Science* 51, 1098–1105.

Bach Knudsen, K.E. and Jorgensen, H. (1986) Use of synthetic amino acids in pig and poultry diets. In: Haresign, W. and Cole, D.J.A. (eds) *Recent Advances in Animal Nutrition.* Butterworths, London, pp. 215–225.

Baker, D.H. (1989) Amino acid nutrition of pigs and poultry. In: Haresign, W. and Cole, D.J.A. (eds) *Recent Advances in Animal Nutrition.* Butterworths, London, pp. 245–260.

Barry, T.N., Manley, T.R., Redekopp, C. and Allsop, T.F. (1985) Endocrine regulation of metabolism in sheep given kale (*Brassica oleracea*) and ryegrass (*Lolium perenne*)-clover (*Trifolium repens*) fresh forage diets. *British Journal of Nutrition* 54, 165–173.

Batterham, E.S. (1974) The effect of frequency of feeding on the utilization of free lysine by growing pigs. *British Journal of Nutrition* 31, 237–242.

Batterham, E.S. and O'Neill, G.H. (1978) The effect of frequency of feeding on the response by growing pigs to supplements of free lysine. *British Journal of Nutrition* 39, 265–270.

Benevenga, N.J. and Steele, R.D. (1984) Adverse effects of excessive consumption of amino acids. *Annual Review of Nutrition* 4, 157–181.

Benevenga, N.J., Harper, A.E. and Rogers, Q.R. (1968) Effects of an amino acid imbalance on the metabolism of the most limiting amino acid in the rat. *Journal of Nutrition* 95, 434–444.

Beverly, J.L. Gietzen, D.W. and Rogers, Q.R. (1990a) Effect of the dietary limiting amino acid in the prepyriform cortex on food intake. *American Journal of Physiology* 259, R709–R715.

Beverly, J.L., Gietzen, D.W. and Rogers, Q.R. (1990b) Effect of the dietary limiting amino acid in the prepyriform cortex on meal patterns. *American Journal of Physiology* 259, R716–R723.

Beverly, J.L., Hrupka, B.J., Gietzen, D.W. and Rogers, Q.R. (1991a) Distribution of the dietary limiting amino acid injected into the prepyriform cortex. *American Journal of Physiology* 260, R525–R532.

Beverly, J.L., Gietzen, D.W. and Rogers, Q.R. (1991b) Threonine concentration in the prepyriform cortex has separate effects on dietary selection and intake of a threonine-imbalanced diet by rats. *Journal of Nutrition* 121, 1287–1292.

Boorman, K.N. (1979) Regulation of protein and amino acid intake. In: Boorman, K.N. and Freeman, B.M. (eds) *Food Intake Regulation in Poultry*. British Poultry Science Ltd, Edinburgh, pp. 87–126.

Bray, D.J. (1970) The isoleucine and valine nutrition of young laying pullets as influenced by excessive dietary leucine. *Poultry Science* 49, 1334–1341.

Calvert, C.C., Klasing, K.C. and Austic, R.E. (1982) Involvement of food intake and amino acid catabolism in the branched chain amino acid antagonism in chicks. *Journal of Nutrition* 112, 627–635.

Carlson, J.R., Dyer, I.A., Johnson, R.J. (1968) Tryptophan-induced interstitial pulmonary emphysema in cattle. *American Journal of Veterinay Research* 29, 1983–1989.

D'Mello, J.P.F. (1974) Plasma concentrations and dietary requirements of leucine, isoleucine and valine: studies with the young chick. *Journal of the Science of Food and Agriculture* 25, 187–196.

D'Mello, J.P.F. (1975) Amino acid requirements of the young turkey: leucine, isoleucine and valine. *British Poultry Science* 16, 607–615.

D'Mello, J.P.F. (1978) Factors affecting amino acid requirements of meat birds. In: Haresign, W. and Lewis, D. (eds) Recent Advances in Animal Nutrition. Butterworths, London, pp. 1–15.

D'Mello, J.P.F. (1988) Dietary interactions influencing amino acid utilization by poultry. *World's Poultry Science Journal* 44, 92–102.

D'Mello, J.P.F. (1989) Toxic amino acids. In: D'Mello, J.P.F., Duffus, C.M. and Duffus, J.H. (eds) *Anti-Nutritional Factors, Potentially Toxic Substances in Plants*. Assocation of Applied Biologists, Warwick, pp. 29–50.

D'Mello, J.P.F. (1990) Lysine utilisation by broiler chicks. *Proceedings of VIII European Poultry Conference, Barcelona, Spain*, pp. 302–305.

D'Mello, J.P.F. (1991) Toxic amino acids. In: D'Mello, J.P.F., Duffus, C.M. and Duffus, J.H. (eds) *Toxic Substances in Crop Plants*. The Royal Society of Chemistry, Cambridge, pp. 21–48.

D'Mello, J.P.F. (1992a) Amino acid supplementation of cereal-based diets for non-ruminants In: D'Mello, J.P.F. and Duffus, C.M. (eds) *Feed Additive and Supplements*, The Scottish Agricultural College, Edinburgh, pp. 1–24.

D'Mello, J.P.F. (1992b) Chemical constraints to the use of tropical legumes in animal nutrition. *Animal Feed Science and Technology* 38, 237–261.

D'Mello, J.P.F. (1993) Non-protein amino acids in *Canavalia ensiformis* and hepatic ornithine decarboxylase. *Amino Acids* 5, 212–213.

D'Mello, J.P.F. and Acamovic, T. (1989) *Leucaena leucocephala* in poultry nutrition – a review. *Animal Feed Science and Technology* 26, 1–28.

D'Mello, J.P.F. and Lewis, D. (1970a) Amino acid interactions in chick nutrition. 1. The interrelationship between lysine and arginine. *British Poultry Science* 11, 299–311.

D'Mello, J.P.F. and Lewis. D. (1970b) Amino acid interactions in chick nutrition.

2. The interrelationship between leucine, isoleucine and valine. *British Poultry Science* 11, 313–323.

D'Mello, J.P.F. and Lewis, D. (1970c) Amino acid interactions in chick nutrition. 3. Interdependence in amino acid requirements. *British Poultry Science* 11, 367–385.

D'Mello, J.P.F. and Lewis, D. (1971) Amino acid interactions in chick nutrition. 4. Growth, food intake and plasma amino acid patterns. *British Poultry Science* 12, 345–358.

D'Mello, J.P.F., Acamovic, T. and Walker, A.G. (1989) Nutritive value of jack beans (*Canavalia ensiformis*) (L.) (DC.) for young chicks: effects of amino acid supplementation. *Tropical Agriculture (Trinidad)* 66, 201–205.

D'Mello, J.P.F., Walker, A.G. and Noble, E. (1990) Effects of dietary supplements on the nutritive value of jack beans (*Canavalia ensiformis*) for the young chick. *British Poultry Science* 31, 759–768.

Edmonds, M.S. and Baker, D.H. (1987) Failure of excess dietary lysine to antagonise arginine in young pigs. *Journal of Nutrition* 117, 1396–1401.

Enneking, D., Giles, L.C., Tate, M.E. and Davies, R.L. (1993) Canavanine: a natural feed-intake inhibitor for pigs. *Journal of the Science of Food and Agriculture* 61, 315–325.

Fernstrom, J.D. and Wurtman, R.J. (1972) Brain serotonin content: physiological regulation by plasma neutral amino acids. *Science* 178, 414–416.

Finot, P.A. (1983) Lysinoalanine in food proteins. *Nutrition Abstracts and Reviews* 53A, 67–80.

Fisher, C. and Morris, T.R. (1970) The determination of the methionine requirement of laying pullets by a diet dilution technique. *British Poultry Science* 11, 67–82.

Fisher, H. and Shapiro, R. (1961) Amino acid imbalance: rations low in tryptophan, methionine or lysine and the efficiency of utilization of nitrogen in imbalanced rations. *Journal of Nutrition* 75, 395–401.

Fisher, H., Griminger, P., Leveille, G.A. and Shapiro, R. (1960) Quantitative aspects of lysine deficiency and amino acid imbalance. *Journal of Nutrition* 71, 213–220.

Florentino, R.F. and Pearson, W.N. (1962) Effect of threonine-induced amino acid imbalance on the excretion of tryptophan metabolites by the rat. *Journal of Nutrition* 78, 101–108.

Forster, L.A., Fontenot, J.P., Perry, H.D., Foster, J.G. and Allen, V.G. (1991) Apparent digestibility and nutrient balance in lambs fed different levels of flatpea hay. *Journal of Animal Science* 69, 1719–1725.

Frenkel, M.J., Gillespie, J.M. and Ries, P.J. (1975) Studies on the inhibition of synthesis of the tyrosine-rich proteins of wool. *Australian Journal of Biological Sciences* 28, 331–338.

Gietzen, D.W., Leung, P.M.B. and Rogers. Q.R. (1986) Norepinephrine and amino acids in prepyriform cortex of rats fed imbalanced amino acid diets. *Physiology of Behaviour* 36, 1071–1080.

Gous, R.M. (1980) An improved method for measuring the response of broiler chickens to increasing dietary concentrations of an amino acid. In: *Proceedings* of *6th European Poultry Conference, Hamburg*, Vol. III, World's Poultry Science Association, Hamburg, pp. 32–39.

Harper, A.E. (1959) Amino acid balance and imbalance. *Journal of Nutrition* 68, 405–418.

Harper, A.E. (1964) Amino acid toxicities and imbalances. In: Munro, H.N. and Allison, J.B (eds) *Mammalian Protein Metabolism, Vol. II*. Academic Press, New York, pp. 87–134.

Harper, A.E. and Rogers, Q.R. (1965) Amino acid imbalance. *Proceedings of the Nutrition Society* 24, 173–190.

Harper, A.E., Benevenga, N.J. and Wohlheuter, R.M. (1970) Effects of ingestion of disproportionate amounts of amino acids. *Physiological Reviews* 50, 428–558.

Harper, A.E., Miller, R.H. and Block, K.P. (1984) Branched-chain amino acid metabolism. *Annual Review of Nutrition* 4, 409–454.

Harrison, L.M. and D'Mello, J.P.F. (1986) Large neutral amino acids in the diet and neurotransmitter concentrations in the chick brain. *Proceedings of the Nutrition Society* 45, 72A.

Harrison, L.M. and D'Mello, J.P.F. (1987) Zinc deficiency, amino acid imbalance and brain catecholamine concentrations in the chick. *Proceedings of the Nutrition Society* 46, 58A.

Jones, J.D. (1961) Lysine toxicity in the chick. *Journal of Nutrition* 73, 107–112.

Jones, J.D., Wolters, R. and Burnett, P.C. (1966) Lysine-arginine-electrolyte relationships in the rat. *Journal of Nutrition* 89, 171–188.

Jones, R.J. (1985) *Leucaena* toxicity and the ruminal degradation of mimosine. In: Seawright, A.A., Hegarty, M.P., James, L.F. and Keeler, R.F. (eds) *Plant Toxicology*. Queensland Poisonous Plants Committee, Yeerongpilly, pp. 111–119.

Jones, R.J. and Megarrity, R.G. (1986) Successful transfer of DHP-degrading bacteria from Hawaiian goats to Australian ruminants to overcome the toxicity of *Leucaena*. *Australian Veterinary Journal* 63, 259–262.

Klain, G.J., Vaughan, D.A. and Vaughan, L.N. (1962) Interrelationships of cold exposure and amino acid imbalances. *Journal of Nutrition* 78, 359–364.

Kumta, U.S. and Harper, A.E. (1962) Amino acid balance and imbalance. IX. Effect of amino acid imbalance on blood amino acid pattern. *Proceedings of the Society for Experimental Biology and Medicine* 110, 512–517.

Kumta, U.S., Harper, A.E. and Elvehjem, C.A. (1958) Amino acid imbalance and nitrogen retention in adult rats. *Journal of Biological Chemistry* 233, 1505–1508.

Leibholz, J. (1989) The utilisation of the free and protein-bound lysine. In: Friedman, M. (ed) *Absorption and Utilisation of Amino Acids, Vol. III*. CRC Press, Boca Raton, pp. 175–187.

Leibholz, J., Love, R.J., Mollah, Y. and Carter, R.R. (1986) The absorption of dietary L-lysine and extruded L-lysine in pigs. *Animal Feed Science and Technology* 15, 141–148.

Leung, P.M.B. and Rogers, Q.R. (1969) Food intake: regulation by plasma amino acid pattern. *Life Sciences* 8, 1–9.

Leung, P.M.B. and Rogers, Q.R. (1987) The effect of amino acids and protein on dietary choice. In: Kawamura, Y. and Kare, M.R. (eds) *Umami: a Basic Taste*. Marcel Dekker, New York, pp. 565–610.

Leung, P.M.B., Rogers Q.R. and Harper, A.E. (1964) Effect of amino acid imbalance on food intake and preference. *Federation Proceedings* 23, 185.

Leung, P.M.B., Gietzen, D.W. and Rogers, Q.R. (1985) Alterations in the amino acid profile of prepyriform cortex from rats fed amino acid-imbalanced diets. *Federation Proceedings* 44, 1523.

Marletta, M.A. (1989) Nitric oxide: biosynthesis and biological significance. *Trends in Biochemical Sciences* 14, 488–492.

Mayer, B., Klatt, P. and Schmidt, K. (1993) Biochemistry and pharmacology of the NO/cGMP signal transducing pathway in cardiovascular and nervous systems. *Amino Acids* 5, 145 (abstract).

Mendonca, C.X. and Jensen, L.S. (1989) Influence of protein concentration on the sulfur-containing amino acid requirement of broiler chickens. *British Poultry Science* 30, 889–898.

Mercer, L.P., Dodds, S.J., Schweisthal, M.R. and Dunn, J.D. (1989) Brain histidine and food intake in rats fed diets deficient in single amino acids. *Journal of Nutrition* 119, 66–74.

Merrill, J.C. and Bray, T.M. (1983) The effect of dietary and sulfur compounds in alleviating 3-methylindole-induced pulmonary toxicity in goats. *Journal of Nutrition* 113, 1725–1731.

Moncada, S., Palmer, R.M.J. and Higgs, E.A. (1991) Nitric oxide: physiology, pathophysiology, and pharmacology. *Pharmacological Reviews* 43, 109–142.

Morris, T.R. and Abebe, S. (1990) Effects of arginine and protein on chicks' responses to dietary lysine. *British Poultry Science* 31, 261–266.

Morris, T.R., Al-Azzawi, K., Gous, R.M. and Simpson, G.L. (1987) Effects of protein concentration on responses to dietary lysine by chicks. *British Poultry Science* 28, 185–195.

Moughan, P.J. (1991) Towards an improved utilization of dietary amino acids by the growing pig. In: Haresign, W. and Cole, D.J.A. (eds) *Recent Advances in Animal Nutrition*. Butterworths, London, pp. 45–64.

Nesheim, M.C. (1968) Kidney arginase activity and lysine tolerance in strains of chickens selected for a high or low requirement of arginine. *Journal of Nutrition* 95, 79–87.

Oestemer, G.A., Hanson, L.E. and Meade, R.J. (1973) Leucine-isoleucine inter-relationship in the young pig. *Journal of Animal Science* 36, 674–678.

Pant, K.C., Rogers, Q.R. and Harper, A.E. (1972) Growth and food intake of rats fed tryptophan-imbalanced diets with or without niacin. *Journal of Nutrition* 102, 117–130.

Papet I., Breuille, D., Glomot, F. and Arnal, M. (1988a) Nutritional and meta-bolic effects of dietary leucine excess in preruminant lambs. *Journal of Nutrition* 118, 450–455.

Papet, I., Lezebot, N., Barre, F., Arnal, M. and Harper, A.E. (1988b) Influence of dietary leucine content on the activities of branched-chain amino acid aminotransferase (EC 2.6.1.42) and branched-chain α-keto acid dehydro-genase (EC 1.2.4.4) complex in tissues of preruminant lambs. *British Journal of Nutrition* 59, 475–483.

96   *J.P.F. D'Mello*

Partridge, I.G., Low, A.G. and Keal, H.D. (1985) A note on the effect of feeding frequency on nitrogen use in growing boars given diets with varying levels of free lysine. *Animal Production* 40, 375–377.

Peng, Y., Tews, J.K. and Harper, A.E. (1972) Amino acid imbalance, protein intake and changes in rat brain and plasma amino acids. *American Journal of Physiology* 222, 314–321.

Peters, J.C. and Harper, A.E. (1985) Adaptation of rats to diets containing different levels of protein: effects on food intake, plasma and brain amino acid concentrations and brain neurotransmitter metabolism. *Journal of Nutrition* 115, 382–398.

Quirk, M.F., Bushell, J.J., Jones, R.J., Megarrity, R.G. and Butler, K.L. (1988) Liveweight gains on *Leucaena* and native grass pastures after dosing cattle with rumen bacteria capable of degrading DHP, a ruminal metabolite from *Leucaena*. *Journal of Agricultural Science, Cambridge* 11, 165–170.

Quirk, M.F., Paton, C.J. and Bushell, J.J. (1990) Increasing the amount of *Leucaena* on offer gives faster growth rates of grazing cattle in South East Queensland. *Australian Journal of Experimental Agriculture* 30, 51–54.

Reis, P.J., Tunks, D.A. and Chapman, R.E. (1975) Effects of mimosine, a potential chemical defleecing agent, on wool growth and the skin of sheep. *Australian Journal of Biological Sciences* 28, 69–84.

Rolls, B.A., Porter, J.W.G. and Westgarth, D.R. (1972) The course of digestion of different food proteins in the rat. 3. The absorption of proteins given alone and with supplements of their limiting amino acids. *British Journal of Nutrition* 28, 283–293.

Rosenthal, G.A. (1982) *Plant Nonprotein Amino and Imino Acids.* Academic Press, New York.

Rotter, R.G., Marquardt, R.R. and Campbell, C.G. (1991) The nutritional value of low lathyrogenic lathyrus (*Lathyrus sativus*) for growing chicks. *British Poultry Science* 32, 1055–1067.

Salmon, W.D. (1954) The tryptophan requirement of the rat as affected by niacin and level of dietary nitrogen. *Archives of Biochemistry and Biophysics* 51, 30–41.

Sanahuja, J.C. and Harper, A.E. (1962) Effect of amino acid imbalance on food intake and preference. *American Journal of Physiology* 202, 165–170.

Smith, R.H. (1980) Kale poisoning: the brassica anaemia factor. *Veterinary Record* 107, 12–15.

Tackman, J.M., Tews, J.K. and Harper, A.E. (1990) Dietary disproportions of amino acids in the rat: effects on food intake, plasma and brain amino acids and brain serotonin. *Journal of Nutrition* 120, 521–533.

Tews, J.K., Kim, Y.W.L. and Harper, A.E. (1979) Induction of threonine imbalance by dispensable amino acids: relation to competition for amino acid transport into brain. *Journal of Nutrition* 109, 304–315.

Tews, J.K., Kim, Y.W.L. and Harper, A.E. (1980) Induction of threonine imbalance by dispensable amino acids: relationships between tissue amino acids and diet in rats. *Journal of Nutrition* 110, 394–408.

Tobin, G. and Boorman, K.N. (1979) Carotid or jugular amino acid infusions and food intake in the cockerel. *British Journal of Nutrition* 41, 157–162.

Waldroup, P.W., Mitchell, R.J., Dayne, J.R. and Hazen, K.R. (1976) Performance of chicks fed diets formulated to minimize excess levels of essential amino acids. *Poultry Science* 55, 243–253.

Wang, T.C. and Fuller, M.F. (1989) The optimum dietary amino acid pattern for growth in pigs. 1. Experiments by amino acid deletion. *British Journal of Nutrition* 62, 77–89.

Ward, K.A. and Harris, R.L.N. (1976) Inhibition of wool follicle DNA synthesis by mimosine and related 4(1H)-pyridones. *Australian Journal of Biological Sciences* 29, 189–196.

Wethli, E., Morris, T.R. and Shresta, T.P. (1975) The effect of feeding high levels of low-quality proteins to growing chickens. *British Journal of Nutrition* 34, 363–373.

Winje, M.E., Harper, A.E., Benton, D.A., Boldt, R.E. and Elvehjem, C.A. (1954) Effect of dietary amino acid balance on fat deposition in the livers of rats fed low protein diets. *Journal of Nutrition* 54, 155–166.

Yoshida, A., Leung, P.M.B., Rogers, Q.R. and Harper, A.E. (1966) Effect of amino acid imbalance on the fate of the limiting amino acid. *Journal of Nutrition* 89, 80–90.

Ideal Amino Acid Patterns | 5

D.J.A. COLE AND T.A. VAN LUNEN
University of Nottingham, Faculty of Agricultural and Food Sciences, Sutton Bonington, Loughborough, LE12 5RD, UK.

It has been suggested that the most important single factor affecting the efficiency of protein utilization for production of meat or eggs is the dietary balance of amino acids. According to Liebig's 'Law of Minimums' the undersupply of a single essential amino acid will inhibit the responses to those in adequate supply. In monogastric animals the supply will be greatly influenced by the diet, but in ruminants the rumen microflora will have a considerable effect on the pattern of amino acids absorbed.

The ideal protein

In order to examine the pattern of dietary amino acids in monogastric animals, the ideal protein provides a simple and effective approach. The development of an ideal protein in pigs has received much attention in recent years. The work of Cole (1978) and Fuller *et al*. (1979) was combined and used by ARC (1981) as the basis for protein requirements in pigs. It had been suggested that the major differences between pigs of different classes (i.e. breed, sex and liveweight) would be the amount of protein that they require according to their potentials for lean meat deposition. The relative amounts of the different essential amino acids needed for 1 g lean or protein deposition would be the same in each case. Thus it was suggested that it should be possible to establish an optimum balance of essential amino acids which, when supplied with sufficient nitrogen for the synthesis of non-essential amino acids, would constitute the ideal protein. Pigs of different classes would require different amounts of the ideal protein but the quality would be the same in each case (Cole, 1978). While such a concept has probably been best developed in pigs, it is equally applicable to other species.

Pigs

Work on the establishment of precise values for the ideal protein was particularly active in the 1970s. On the basis of this, suggestions were made for the balance of essential amino acids in the ideal protein (Table 5.1).

As our knowledge of amino acid requirements increases, so the values used in the original work need to be questioned. However, the original balances have proved to be particularly robust. Nevertheless, some amino acids are worthy of attention, namely threonine, and methionine + cystine.

In the original balance, a value of dietary threonine of 60% of the lysine level received substantial support. More recently evidence has been accumulating that pigs can respond to threonine levels of up to 65–67% of the dietary lysine (Cole, 1993).

While there was considerable variation in the estimates of requirements for methionine + cystine, a value of 50–55% of the dietary lysine supply has generally been adopted. Work at the University of Nottingham (Cole, 1993) has consistently supported a value of 50%. It further suggested that higher requirement values for methionine + cystine may well occur when cystine is providing a substantial part of their joint requirements. As a result it was recommended that dietary methionine should not be allowed to fall below 0.25% of lysine.

An appropriate balance of essential amino acids in the ideal protein would be: lysine, 100; methionine + cystine, 50; threonine, 65–67; tryptophan, 18; isoleucine, 50; leucine, 100; histidine, 33; phenylalanine + tyrosine, 100; valine, 70. Such a balance needs to be accompanied by sufficient N for the synthesis of non-essential amino acids.

Availability

The establishment of an ideal protein in the previous section uses total, rather than available, amino acids. This was a deliberate decision since, as the use of an ideal protein was designed to increase precision, it was felt that the variation in measures of availability would result in loss of accuracy.

Ileal digestibility is now used widely as a measure of availability. In many cases it is a good measure (e.g. Tanksley and Knabe, 1980). However, care needs to be taken in using such an application. For example, it has been shown that lysine retention as a proportion of ileal digestible lysine intake was influenced by dietary lysine concentration (Batterham, *et al.*, 1990). A further example is the case of heat-damaged fish meals. Work at the University of Nottingham (Table 5.2) has shown a reduction in both faecal and ileal digestibility of lysine with such material (Wiseman *et al.*, 1991). However, formulation of diets on the basis of either type of

Table 5.1. The optimum balance of essential amino acids in the ideal protein for pigs (relative to lysine = 100).

Amino acid	Author(s)		
	Cole (1978)	Fuller *et al.* (1979)	ARC (1981)
Lysine	100	100	100
Methionine + cystine	50	53	50
Tryptophan	18	12	15
Threonine	60	56	50
Leucine	100	83	100
Valine	70	63	70
Isoleucine	50	50	55
Phenylalanine + tyrosine	100	96	96
Histidine	40	31.5	–

Table 5.2. Ileal and faecal digestibilities in pigs of lysine in fish meals heated to different temperatures. (From Wiseman *et al.*, 1991.)

Treatment	Ileal digestibility of lysine	Faecal digestibility of lysine
Fish meal untreated	0.92	0.99
3 h at 130°C	0.90	0.95
1.25 h at 160°C	0.84	0.91

digestibility, while giving better performance, did not completely account for the performance problems associated with heat damage. Pigs given diets containing heat-damaged fish meal, based on total, faecal or ileal digestible amino acids, grew at 64%, 83% and 87% respectively of the rate of pigs fed diets containing undamaged fish meal.

Values for the ratios of ileal-digestible essential amino acids are given in Table 5.3. These are based on published work and may not represent the perfect balance.

Availability can be considered in another sense. It has been suggested that synthetic amino acids need to be in phase, at the metabolic sites, with the protein-bound amino acids. To ensure this, it has been suggested that it is necessary to feed more frequently than once per day when using high levels of crystalline lysine (Batterham, 1974; Batterham and O'Neill, 1978)

Table 5.3. Ratios of the ileal-digestible essential amino acids in the ideal protein of pigs used by Yan *et al.* (1986a,b) and Wang and Fuller (1990).

Amino acid	Author(s)		
	Yen *et al.* (1986a)	Yen *et al.* (1986b)	Wang and Fuller (1990)
Lysine	100	100	100
Methionine	37	39	-
Methionine + cystine	52	58	60
Threonine	64	67	66
Tryptophan	19	21	18.5
Isoleucine	73	76	60
Leucine	130	140	111
Phenylalanine + tyrosine	86	95	120
Histidine	43	46	-
Valine	90	97	75

and threonine (Cole, 1993). However, in modern production systems, using ad lib feeding, such problems are unlikely to be encountered.

Poultry

Chickens

The concept of ideal amino acid patterns is supported by work reported by Morris *et al.* (1987) where it was found that for chickens over the range of 140 to 280 g protein kg^{-1} diet, lysine requirement could be expressed as a constant proportion of the protein (5.4%). Boomgardt and Baker (1971, 1973) clearly demonstrated that in diets of varied crude protein content, the chick's requirements for lysine and tryptophan remain constant when expressed as a percentage of dietary protein level. Robbins (1987) reported similar results for threonine. It is important, however, to recognize that such statements may not be true in situations of amino acid oversupply and/or poor amino acid balance. The ratios of amino acids discussed in this chapter assume that appropriate levels of high-quality protein are being used. In practice it is best to use daily intakes of amino acids (i.e. lysine) per animal per day rather than intakes as percentage of the diet or even of the protein.

As with pigs, it appears that the ideal amino acid pattern for chickens is unaffected by genotype. An example of this is found in a report by Han and Baker (1991) where the lysine requirement of fast- and slow-growing broiler genotypes was compared. The amount of lysine required in the diet was the same for both genotypes. The fast-growing birds consumed a greater amount of feed and thus had a higher daily lysine intake.

Ideal amino acid patterns, calculated as a percentage of lysine, for laying hens and broiler chickens are listed in Table 5.4 (calculated from NRC, 1984). It appears that the ideal amino acid pattern for laying hens is slightly different to that of broilers, even during growth. The laying hen has been selected for high egg production and low muscle mass while the broiler chicken has been selected for high muscle mass. This may explain the small differences in pattern between these two types of birds during growth; a greater proportion of dietary amino acids is required for muscle production in the broiler chicken.

Once the hen reaches laying age, dietary amino acids are required for maintenance and egg production. Since maintenance has a relatively small amino acid requirement compared with egg production, the ideal amino acid pattern will be governed to a large extent by the requirement for egg production. In broiler chickens the primary use of dietary amino acids is for muscle growth, which may require a different amino acid pattern for optimum production. Table 5.5 lists the amino acid pattern of whole hens, chicken meat, egg white and egg yolk. The amino acid patterns for each of these products is similar, especially if an average of the values for egg white and yolk is taken. It is interesting to note that the pattern of these products is similar to the ideal amino acid patterns listed in Table 5.4. Any differences which exist are small.

Other Avian Species

The optimum amino acid patterns for turkeys, ducks, pheasants and quail are shown in Table 5.6. The values were calculated from ARC (1975) and NRC (1984) requirement tables.

Turkeys

There does not appear to be much information in the literature relating growth performance of turkeys and dietary amino acids. Since turkeys take much longer than broilers to reach market weight, sex and genotype may have a greater influence on daily amino acid requirements. However, as with pigs and chickens the amino acid pattern remains constant. This is illustrated in Table 5.7 where the NRC (1984) requirement as a percentage of the diet is listed next to the pattern for various ages. Although there are minor differences between ages for specific amino acids, the overall

Table 5.4. Amino acid requirement of layer and broiler chickens (expressed as percent of lysine). (Data based on NRC, 1984.)

Amino acid	Layers				Broilers		
	0-6 weeks	6-14 weeks	14-20 weeks	Laying	0-3 weeks	3-6 weeks	6-8 weeks
Lysine	100.0	100.0	100.0	100.0	100.0	100.0	100.0
Methionine + cystine	70.6	83.3	88.9	85.9	77.5	72.0	70.6
Tryptophan	20.0	23.3	24.4	21.9	19.2	18.0	20.0
Threonine	80.0	95.0	82.2	70.3	66.7	74.0	80.0
Leucine	117.6	138.3	148.9	114.1	112.5	118.0	117.6
Valine	72.9	86.7	91.1	85.9	68.3	72.0	72.9
Isoleucine	70.6	83.3	88.9	78.1	66.7	70.0	70.6
Phenylalanine + tyrosine	117.6	138.3	148.9	125.3	111.7	117.0	117.6
Histidine	30.6	36.7	37.8	25.0	29.2	30.0	30.6
Arginine	117.6	138.3	148.9	106.3	120.0	120.0	117.6

Table 5.5. Amino acid composition of chicken and egg (expressed as percent of lysine). (Data based on FAO, 1970.)

Amino acid	Whole hen	Muscle	Egg white	Yolk
Lysine	100.0	100.0	100.0	100.0
Methionine + cystine	83.1	48.1	95.9	55.2
Tryptophan	20.8	12.9	23.1	20.0
Threonine	73.4	50.0	72.1	72.1
Leucine	126.4	92.6	124.8	111.0
Valine	98.2	64.0	72.5	54.2
Isoleucine	90.1	67.2	77.3	66.9
Phenylalanine + tyrosine	141.7	92.4	142.4	107.5
Histidine	34.9	33.0	35.4	32.9
Arginine	87.4	70.0	86.0	97.7

Table 5.6. Amino acid requirement of growing birds other than chickens (expressed as percent of lysine).

Amino acid	Turkey[1]	Duck[2]	Pheasant[1]	Japanese quail[1]
Lysine	100.0	100.0	100.0	100.0
Methionine + cystine	57.7	83.2	75.0	57.7
Tryptophan	15.4	19.1		16.9
Threonine	60.8	66.3		78.5
Leucine	115.4	132.6		130.0
Valine	72.3	88.8		73.1
Isoleucine	65.4	77.5		75.4
Phenylalanine + tyrosine	107.7	143.8		138.5
Histidine	35.4	43.8		27.7
Arginine	96.2	94.4		96.2

[1] Calculated from NRC (1984).
[2] Calculated from ARC (1975).

Table 5.7. Amino acid requirement of turkeys at various ages (expressed as percent of diet and percent of lysine).

Amino acid	0-4 weeks		5-8 weeks		9-12 weeks		13-16 weeks	
	% Diet[1]	Pattern[2]	% Diet[1]	Pattern[2]	% Diet[1]	Pattern[2]	% Diet[1]	Pattern[2]
Lysine	1.6	100.0	1.5	100.0	1.3	100.0	1.0	100.0
Methionine + cystine	1.05	65.6	0.9	60.0	0.75	57.7	0.65	65.0
Tryptophan	0.26	16.3	0.24	16.0	0.2	15.4	0.18	18.0
Threonine	1.0	62.5	0.93	62.0	0.79	60.8	0.68	68.0
Leucine	1.9	118.8	1.75	116.7	1.5	115.4	1.3	130.0
Valine	1.2	75.0	1.1	73.3	0.94	72.3	0.8	80.0
Isoleucine	1.1	68.8	1.0	67.7	0.85	65.4	0.75	75.0
Phenylalanine + tyrosine	1.8	112.5	1.65	110.0	1.4	107.7	1.2	120.0
Histidine	0.58	36.3	0.54	36.0	0.46	35.4	0.39	39.0
Arginine	1.6	100.0	1.5	100.0	1.25	96.2	1.1	110.0

[1] NRC (1984).
[2] Calculated from NRC (1984).

pattern is constant across ages. However, when requirements are expressed as a percentage of the diet, and thus daily intake values, they tend to decrease with increasing age.

Ducks

The duck, like the turkey, takes longer to reach market weight than the broiler chicken, and it has a tendency to deposit carcass fat. Despite this, the amino acid pattern required for optimum growth will be the same across all ages, only the daily dietary amounts will differ (Table 5.6).

Pheasants and Japanese quail

Values for pheasants are listed in Table 5.6. Because little information is available, many pheasant producers use the requirements for turkeys when formulating diets.

There is more information for Bobwhite and Japanese quail since they are produced commercially for meat and eggs in some parts of the world. The amino acid requirements for quail are listed in Table 5.6, and show a similar pattern to that of the turkey, except for phenylalanine and tyrosine for which the requirement in relation to lysine is higher in the quail than in the turkey.

Ruminants

As in the monogastric animals, ruminants require a nutrient source of essential amino acids. However, because rumen microbes are the primary source of protein they modify the amino acid pattern of the diet. Consequently the supply of amino acids from the rumen to the small intestine for absorption assumes great importance. When microbial protein production is limited, or when amino acid requirements are high, microbial protein may not meet the amino acid requirements for maintenance and production (growth, lactation) (Merchen and Titgemeyer, 1992).

The term 'ruminants' includes young animals whose rumen is not functional for the first few weeks of life. Despite this, the requirement for absorbed amino acids for maintenance and growth will be the same as for older animals. It is the source of the amino acids which will be different. Animals with a functional rumen obtain the majority of their amino acid requirement from rumen microbes, while preruminants require a digestible source of protein which will provide adequate levels of essential amino acids directly to the small intestine.

Table 5.8 lists the amino acid content of rumen bacteria, the primary products of ruminant animals (milk, meat, wool) and the estimated

Table 5.8. Amino acid composition of rumen bacteria, milk, meat and wool compared with the amino acid requirement (expressed as percent of lysine).

Amino acid	Rumen bacteria[1]	Milk[2]	Lamb[2]	Beef[2]	Wool[2]	Requirement[1]
Lysine	100.0	100.0	100.0	100.0	100.0	100.0
Methionine + cystine	50.9	47.6	39.8	44.0	386.7	48.7
Tryptophan	19.2	17.1	13.3	14.3	56.7	13.7
Threonine	66.3	61.0	46.9	50.5	216.7	55.3
Leucine	93.5	124.4	73.5	87.9	313.3	96.9
Valine	65.7	90.2	49.0	58.2	170.0	66.3
Isoleucine	61.9	68.3	46.9	56.0	113.3	62.9
Phenylalanine + tyrosine	114.3	120.7	74.5	91.2	336.7	91.3
Histidine	26.8	36.6	32.7	40.7	26.7	36.4
Arginine	55.4	48.8	62.2	73.6	336.7	33.8

[1] Calculated from Merchen and Titgemeyer (1992).
[2] Calculated from Ørskov (1992).

requirements. All the values are expressed as a percentage of lysine to provide a direct comparison of amino acid composition regardless of protein quantity. It is interesting to note that the amino acid composition of rumen bacteria is similar to that of the products of ruminant animals (other than wool) and thus the requirement. The differences in composition of wool and rumen bacteria may appear to be large; however, it should be noted that synthesis of wool protein per day is small compared with the requirement for maintenance. This is true for all sheep including high wool producing breeds. The wool growth protein requirement will be of the order of 1–2 g N d^{-1}, while tissue maintenance of a 60-kg sheep is 6–8 g N d^{-1} (Ørskov, 1992).

The estimated amino acid requirements listed in Table 5.8 are a compilation of several studies reviewed by Merchen and Titgemeyer (1992). They represent requirements for maintenance and 1 kg d^{-1} gain by a 250 kg steer. The requirements for milk production would be similar except perhaps for leucine and valine, which are present in greater amounts relative to lysine in milk than in meat.

In dairy cows the absorbed amino acid requirement depends on four functions: maintenance, protein deposition in muscle, fetal development and milk production. According to Choi *et al.* (1992) estimates of the ratio of essential amino acids required for maintenance in the cow have not been developed. Even in high-producing dairy cows it is unlikely that amino acids will play a significant role in gluconeogenesis. Amino acid requirements for pregnancy and muscle growth in mature cows appear to be relatively low and can be supplied from normal diets. Therefore, the amino acid composition of milk as listed in Table 5.8 provides an acceptable estimation of the nutrient requirement for lactation, and is similar to the requirement for beef cattle.

Despite the appearance that rumen bacteria provide an absorbable amino acid pattern similar to the requirement, there have been numerous studies which have shown that dietary supplementation or infusion with specific amino acids has increased production of milk and meat. Of the amino acids examined, it appears that in most cases methionine and lysine are the two most limiting amino acids in most ruminant diets. The inclusion of rumen-protected methionine and lysine in corn-based diets improved milk yield by 7.5% (Donkin *et al.*, 1989). Methionine has been identified as the first-limiting amino acid in growing steer calves, followed by lysine and threonine (Richardson and Hatfield, 1978); lysine and methionine are limiting in the rumen bacteria of lambs (Storm and Ørskov, 1984).

The amino acid pattern leaving the rumen will be affected by dietary components, or manipulation of rumen microflora. Most diets fed to ruminants contain some protein that escapes ruminal degradation and provides some amino acids directly to the digestive tract. The quality of this undegraded protein will have an influence on the amino acid supply available to the animal. It is this fact which makes diet formulation for amino acid content for ruminants more difficult than for monogastric animals. It seems likely that a diet comprising ingredients which are readily degraded in the rumen will result in amino acid patterns entering the small intestine similar to those of rumen bacteria. Diets containing large amounts of ingredients such as corn will result in different amino acid patterns compared to rumen bacteria entering the small intestine since their protein is resistant to rumen degradation. Using the example of corn, the amino acid pattern will be high in sulfur amino acids and relatively low in lysine. With this in mind, the results of amino acid studies reported in the literature must be examined closely. In general, it appears that several amino acids are often co-limiting in ruminants (Merchen and Titgemeyer, 1992). This suggests that the amino acid pattern normally available to ruminants is close to the requirement, with few being in excess. This appears to be confirmed by the values for rumen bacteria and the requirements listed in Table 5.8.

Table 5.9. Amino acid requirements of fish (expressed as percent of lysine). (Data based on Choi *et al.*, 1992.)

Amino acid	Chinook salmon	Rainbow and lake trout	Carp	Catfish
Lysine	100.0	100.0	100.0	100.0
Methionine	80.0	75.0	54.4	46.0
Tryptophan	10.0	25.0	14.0	10.0
Threonine	44.0		68.4	40.0
Leucine	78.0	100.0	59.7	70.0
Valine	64.0	73.8	63.2	60.0
Isoleucine	44.0	57.5	40.4	52.0
Phenylalanine	102.0		114.0	100.0
Histidine	36.0		36.8	30.0
Arginine	120.0	111.3	73.7	86.0

Fish

The natural diet of fish is high in protein and they do not utilize carbohydrate well. Some protein therefore is metabolized for energy and hence their protein requirement is higher than animals such as ruminants, pigs and poultry. Despite this, fish, together with the other animals reviewed in this chapter, require a specific dietary pattern of amino acids for optimum growth. This pattern varies according to species, mainly in respect of methionine and arginine (Table 5.9). The patterns for chinook salmon and rainbow or lake trout, and for carp and catfish are similar. The similarity within these two groups may be explained by their similarity of habitat and natural diet. Salmon and trout are closely related species and live in cold water with high oxygen levels. They are surface and mid-depth feeders. Carp and catfish have adapted to living in warm water and are bottom feeders. From the literature it appears that more work is required in the area of amino acid requirements of trout in particular.

References

ARC (Agricultural Research Council) (1975) *The Nutrient Requirements of Farm Livestock. No. 1. Poultry*. ARC, London.

ARC (Agricultural Research Council) (1981) *The Nutrient Requirements of Pigs*. Commonwealth Agricultural Bureaux, Farnham Royal, Slough.

Batterham, E.S. (1974) The effect of frequency of feeding on the utilization of free lysine by growing pigs. *British Journal of Nutrition* 31, 237–245.

Batterham, E.S. and O'Neill, G.H. (1978) Effect of frequency of feeding on the response by growing pigs to supplements of free lysine. *British Journal of Nutrition* 39(2), 265–276.

Batterham, E.S., Anderson, L.M., Baigent, D.R. and White, E. (1990) Utilisation of ileal digestible amino acids by growing pigs: effect of dietary lysine concentration on efficiency of lysine retention. *British Journal of Nutrition* 31, 237.

Boomgardt, J. and Baker D.H. (1971) Tryptophan requirement of growing chicks as affected by dietary protein level. *Journal of Animal Science* 33, 595–599.

Boomgardt, J. and Baker, D.H. (1973) The lysine requirement of growing chicks fed sesame meal-gelatin diets at three protein levels. *Poultry Science* 52, 586–591.

Choi, Y.J., Ha, J.K. and Han, I.K. (1992) L-Lysine, *the Amino Acid for Animal Nutrition*. Miwon Food Co. Ltd.

Cole, D.J.A. (1978) Amino acid nutrition of the pig. In: Haresign, W. and Lewis, D. (eds) *Recent Advances in Animal Production*. Butterworths, London, pp. 59–72.

Cole, D.J.A. (1993) Interaction between energy and amino acid balance. In: *Proceedings of the 2nd International Feed Production Conference*. Casa Editrice Mattioli, Firenza, pp. 209–228.

Donkin, S.S., Varga, G.A., Sweeney, T.F. and Muller, L.D. (1989) Rumen protected methionine and lysine: effects on animal performance, milk yield and physiological measures. *Journal of Dairy Science* 72, 1484–1495.

FAO (Food and Agriculture Organization) (1970) *Amino Acid Content of Foods and Biological Data on Proteins*. FAO, Rome.

Fisher, C. (1980) Protein deposition in poultry. In: Buttery, P.J. and Lindsay, D.B. (eds) *Protein Deposition in Animals*. Butterworths, London, pp. 251–270.

Fuller, M.F., Livingstone, R.M., Baird, B.A. and Atkinson, T. (1979) The optimal amino acid supplementation of barley for growing pigs. 1. Response of nitrogen metabolism to progressive supplementation. *British Journal of Nutrition* 41, 321–331.

Han, Y.M. and Baker, D.H. (1991) Lysine requirement of fast-growing and slow-growing broiler chicks. *Poultry Science* 70, 2180–2114.

Merchen, N.R. and Titgemeyer, E.C. (1992) Manipulation of amino acid supply to the growing ruminant. *Journal of Animal Science* 70(10), 3238–3247.

Morris, T.R., Al-Azzawi, K., Gous, R.M. and Simpson, G.L. (1987) Effects of protein concentration on responses to dietary lysine by chicks. *British Poultry Science* 28(2), 185–195.

NRC (National Research Council) (1984) *Nutrient Requirements of Poultry*. National Academic Press, Washington, D.C.

Ørskov, E.R. (1992) *Protein Nutrition in Ruminants*. Academic Press, London.

Richardson, C.R. and Hatfield, E.E. (1978) The limiting amino acids in growing cattle. *Journal of Animal Science* 46, 740–751.

Robbins, K.R. (1987) Threonine requirement of the broiler chick as affected by protein level and source. *Poultry Science* 66(9), 1531–1534.

Storm, E. and Ørskov, E.R. (1984) The nutritive value of rumen micro-organisms in ruminants. 4. The limiting amino acids of microbial protein in growing sheep determined by a new approach. *British Journal of Nutrition* 52, 613–620.

Tanksley, T.D. Jr and Knabe, D.A. (1980) Ileal digestibilities of amino acids in pig feed and use in formulating diets. In: Haresign, W. and Cole, D.J.A. (eds) *Recent Advances in Animal Nutrition*. Butterworths, London, pp. 75–95.

Wang, T.L. and Fuller, M.F. (1990) The effect of the plane of nutrition on the optimum dietary amino acid pattern for growing pigs. *Animal Production* 50, 155–164.

Wiseman, J., Jagger, S., Cole, D.J.A. and Haresign, W. (1991) The digestion and utilization of amino acids of heat treated fish meal by growing/finishing pigs. *Animal Production* 53, 215–225.

Yen, H.T., Cole, D.J.A. and Lewis, D. (1986a) Amino acid requirements of growing pigs. 7. The response of pigs from 25 to 55 kg live weight to dietary ideal protein. *Animal Production* 43, 141–154.

Yen, H.T., Cole, D.J.A. and Lewis, D. (1986b) Amino acid requirements of growing pigs. 8. The response of pigs from 50 to 90 kg live weight to dietary ideal protein. *Animal Production* 43, 155–165.

Ileal Digestibilities of Amino Acids in Feedstuffs for Pigs

6

E.S. BATTERHAM*

NSW Agriculture, Pig Research Unit, Wollongbar
Agricultural Institute, Wollongbar, NSW 2477, Australia.

Introduction

Considerable progress has been made in recent years to develop feeding standards based on the ileal digestibility of amino acids. Ileal digestibility has the advantage over faecal digestibility in that the values more closely reflect digestibility to the point where absorption of amino acids is completed and they avoid interference of microorganisms on the metabolism of amino acids in the hind gut.

The majority of research within this field has concentrated on the development of techniques for assessing ileal digestibility. A number of comprehensive reviews have been written on these techniques, including Sauer and Ozimek (1986), Sauer *et al.* (1989), Low (1990) and Lenis (1992). Less research has concentrated on the applicability of ileal digestibility values for dietary formulations. This review will concentrate on this aspect of ileal digestibility.

Techniques for Assessing Ileal Digestibility

Intact ileal sampling (or serial slaughter)

This is the simplest method for determining ileal digestibility. The technique involves feeding the pig with the experimental diet for 5–7 days. The pig is then held under a surgical plane of anaesthesia whilst the last 1.5 m of the terminal ileum is removed. The pig is then euthanased. The use of anaesthesia avoids the shedding of the mucosal cells which may occur

---

* Deceased

with electrical stunning (Badawy, 1964). A marker (chromic oxide) is used in the feed to enable digestibility to be calculated. Comparisons with T-piece cannulation have shown the results to be similar (Moughan and Smith, 1987). The technique is also more acceptable than cannulation techniques from an ethical point of view.

It is also possible to sample faeces from the rectum at the same time that ileal samples are being collected to enable digestible energy determinations of the diet to be undertaken. However, van Barneveld *et al.* (1991a) reported that spot collections of faeces resulted in an underestimation of digestible energy compared to total collections. It appears that it is more difficult to collect a representative sample of faeces using a grab sample compared to a single sample of ileal contents.

The advantages of the intact ileal dissection technique are: (i) it is relatively quick to conduct (an experiment can be completed within 7 days); (ii) there is no disruption of the intestinal wall; (iii) it is more acceptable on animal ethics grounds; and (iv) it is possible to use the technique for determining the apparent ileal digestibility of diets at the completion of growth experiments (by sampling following slaughter). The disadvantages of the technique are: (i) only one sample of ileal digesta is collected; (ii) it is more difficult to collect adequate samples from highly digestible diets; (iii) it may not be possible to get adequate samples of faeces for digestible energy studies; and (iv) it utilizes considerably more pigs than cannulation experiments.

Simple 'T'-piece cannula

This technique involves the use of a simple 'T'-piece cannula which is inserted into the terminal ileum. Digesta is collected from the cannula after the pigs have been on the diets 5–10 days. Normally, the pigs are fed twice daily and samples are collected over an 8-hour period following feeding. Again, a marker is used to enable digestibility to be calculated. It is also possible to collect samples of faeces prior to collecting the ileal samples. The one group of cannulated pigs can be used to assess the ileal digestibility of a number of diets, with diets reallocated on either a latin-square or randomized design after each collection.

The advantages of this technique are: (i) the surgery is relatively simple; (ii) collections can be made over a longer period of time; (iii) faecal sampling can also be undertaken; and (iv) it is possible to utilize one set of pigs for a number of collections. The disadvantages are: (i) cannulation techniques are less acceptable on animal ethics grounds; (ii) leakage occurs around the base of the cannula which causes discomfort to the pig and eventually results in the termination of the collections; and (iii) considerable labour and expertise are required to minimize irritation from this leakage.

Postvalvular T-caecum cannula (PVTC)

With this technique the caecum is removed and replaced by a large T-cannula (Van Leeuwen *et al.*, 1988). When the cannula is closed the digesta flows from the terminal ileum to the colon. When the cannula is open, over-pressure from the colon forces the ileal-caecal valve into the cannula and an almost complete collection of ileal contents is possible.

Advantages of this technique include: (i) no interference with the intestinal wall, ileal-caecal valve, or colon; (ii) the ability to replace cannulas as the pig develops; (iii) the almost complete collection of digesta; (iv) the ability to recover nylon bags used in digestibility studies.

Re-entrant cannula

With these techniques, a re-entrant cannula is inserted into either the terminal ileum, or either side of the ileal caecal valve or posterior to it (Sauer *et al.*, 1989; Low, 1990). The advantage of this technique is that a total collection of digesta is possible. A major disadvantage is that the small intestine is severed leading to possible blockage of the cannula with coarsely ground or fibrous materials, particularly with a re-entrant cannula inserted in the terminal ileum. These techniques appear to offer no practical advantage over simple 'T'-piece cannulas.

Ileal-rectal shunt

With this technique the terminal ileum is anastomosed to the rectum, thereby allowing the digesta to bypass the large intestine. There are a number of variations to this technique, mainly related to the function of the large intestine (Sauer *et al.*, 1989). As the large intestine is by-passed it is necessary to compensate the pig for mineral losses (Sauer *et al.*, 1989).

The advantages of the technique are that it allows complete collection of ileal digesta over the growth span of the pig and that it is suitable for routine collections. The assumption is made that the removal of the large intestine does not interfere with the digestive physiology of the pig. The disadvantages of the technique are that substantial surgery is involved and that the technique is unacceptable in many quarters on animal ethics grounds. Collections may be messy, as the pig has little control over rectal function. There is also the question of whether the pig is functioning in a normal physiological manner and, if the large intestine remains attached to the rectum, the possibility exists that the ileal contents may be contaminated by material from the large intestine.

In essence, there are a number of techniques for determining ileal digestibilities, which vary in the degree of surgery, equipment and skill required to obtain satisfactory collections. Further work is required to determine if there are definite physiological differences in the results obtained using the various techniques. It is probable that the choice of technique will depend on the individual preferences of research workers or centres, animal ethics committees and financial considerations.

Apparent or True Digestibility?

There is considerable interest in what form the ileal digestibility measurement should be – apparent or true digestibility.

Apparent digestibility

Apparent digestibility measures both the digestibility of amino acids in the feed and those contributed from endogenous sources within the animal. These included undigested amino acids from digestive secretions and from cells sloughed from the lining of the stomach and small intestine during the passage of digesta.

Apparent digestibility values are affected by the level of crude protein (CP) in the test diet. With a low-protein diet, the amino acids from endogenous sources form a higher proportion of the total amino acids reaching the terminal ileum. As the CP level in the diet increases, the proportion from endogenous sources decreases, and the apparent digestibility of the diet increases. Thus, apparent digestibility is underestimated if it is determined in diets low in protein. For this reason, Sauer et al. (1989) recommended that diets should contain at least 150–160 g CP kg^{-1}.

True digestibility

True digestibility includes a correction for endogenous secretions. As a consequence, true digestibility values should be unaffected by the CP content of the diet. In the past, correction for the endogenous contribution was made using a protein-free diet or feeding graded levels of the test protein and extrapolating back to zero. These techniques gave reasonably similar results (Low, 1982). More recent methods involve: (i) the use of ^{15}N to label the amino acids in the feed or the pig so that the undigested amino acids can be distinguished from the unabsorbed endogenous secretions (Low, 1990; Huisman et al., 1992); and (ii) the use of homoarginine to estimate lysine digestibility (Hagemeister and Erbersdobler, 1985). The principle of this technique is that the lysine in a feed source is converted to homoarginine by a guanidination process. Homoarginine is then absorbed and is immediately reconverted to lysine so that the pig does not suffer a lysine deficiency. Any homoarginine remaining in the terminal ileum is thus of dietary origin and is a reflection of the true digestibility of lysine. Both the ^{15}N dilution and homoarginine techniques give higher estimates of true digestibility than the traditional methods (Low, 1982, 1990) and should be preferred. However, the higher cost of these procedures restricts their widespread application. It is interesting to note that the newer techniques tend to indicate that true digestibility is very high in most feeds. Therefore, variation in true digestibility becomes less of a problem.

Relationship between true and apparent ileal digestibility

The theoretical relationship between true and apparent ileal digestibility is shown in Fig. 6.1. It is thought that apparent ileal digestibility values increase until the CP content of the diet is about 140–160 g kg^{-1}. From there the difference in apparent and true digestibility reflects the endogenous cost of digestion for that protein. This will vary between protein sources, depending on the degree of difficulty in hydrolysing the particular protein source.

In terms of experimental procedures, determination of apparent digestibility is the simplest, as it requires no corrections. However, it is necessary to be aware of factors which may influence these values. This applies particularly to the protein level of the diet. The simplest way to determine the apparent ileal digestibility of amino acids in a feed source is to have the test feed as the only source of amino acids in the diet. This involves feeding single-cereal diets or protein concentrates in combination with non-protein energy sources. With single-cereal diets, the resultant CP levels may be less than 100 g kg^{-1}. This is well below the 150–160 g kg^{-1} dietary crude protein level recommended by Sauer *et al.* (1989) for

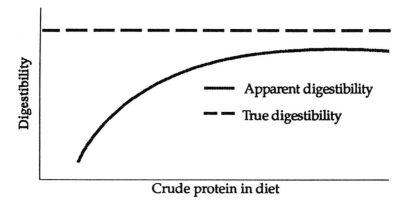

Fig. 6.1. Effect of dietary crude protein level on the relationship between apparent and true ileal digestibility.

minimizing the effects of endogenous protein losses. Thus apparent ileal
digestibility values may be underestimated by this approach. With most
protein concentrates, there should be little problem in achieving reasonably
high dietary CP levels. However, with grain legumes, low-protein test diets
can occur as grain legumes contain around 200–220 g CP kg^{-1} and, unlike
cereals, inclusion levels may need to be restricted to avoid antinutritional
factors.

Effect of dietary protein level on apparent ileal digestibility values

Investigations into the effect of CP level on apparent ileal digestibility
values have given conflicting results. For example, Imbeah *et al.* (1988)
determined the apparent ileal digestibility of amino acids in single-grain
or protein concentrate diets, as well as in diets formulated from the grain
and protein sources. The predicted ileal digestibility of amino acids in the
formulated diets agreed well with determined values. This indicated that
it was possible to determine relevant apparent ileal digestibility values,
despite the crude protein levels being less than 150–160 g kg^{-1} (only
105 g kg^{-1} in the barley diet). In contrast, Van Barneveld, Norton and
Batterham (unpublished data) found that the apparent ileal digestibility
of lysine in field peas was markedly influenced by the protein level of the
test diet. When the test diet contained 84 g CP kg^{-1} (as a result of a
400 g kg^{-1} inclusion level of field peas in a sugar-based diet) a value
of 0.75 ileal digestibility of lysine was recorded. When the apparent
digestibility was re-determined with higher inclusion levels of the field peas
and resultant crude protein levels in the diets of 105–165 g kg^{-1}, a higher
value of 0.92 apparent ileal digestibility was obtained. This indicates
that the original determination considerably underestimated apparent
ileal digestibility and, for that meal, the diets needed to contain at least
105 g CP kg^{-1}. This aspect of the effect of crude protein intake on the
values for apparent ileal digestibility needs to be clarified.

Predicting true digestibility

True digestibility can also be predicted from apparent digestibility using
various prediction equations which attempt to estimate endogenous flow
(e.g. Taverner *et al.*, 1981). These equations are normally based on dry
matter or fibre intakes. However, these techniques do not necessarily
correct for the differences in apparent digestibility as affected by low-
protein diets. As mentioned above, Van Barneveld *et al.* (unpublished
data) found that the apparent ileal digestibility in field peas was affected
by the protein level in the test diets, with estimates of 0.75 from an
84 g kg^{-1} crude protein diet, and 0.92 from higher-protein diets.
However, when the true digestibility of lysine was predicted from the 0.75

apparent value, using a series of prediction equations from the literature, estimates of true digestibility varied from 0.80 to 0.87. These values were well below the 0.92 apparent ileal digestibility value that was obtained with the higher-protein diets. This indicates that the prediction equations developed do not adequately account for differences in crude protein level in test diets. It would seem essential, when predicting true digestibility from apparent digestibility, to ensure that the apparent values have been determined in diets adequate in protein.

Apparent or true digestibility for formulating diets?

The major question is: which values have the greatest application in dietary formulations? It would seem that apparent ileal digestibility values should be used in the ordinary formulation of diets, as these values account for the endogenous cost of digestion of a feed source. This is a real cost which should be debited against the individual feed source. However, in computer simulation models of the pig, where the model predicts the endogenous cost of digestion, then true digestibility values are more applicable. The use of apparent digestibility values in such computer models would lead to a double correction for endogenous cost.

Use of Ileal Digestible Amino Acid Values in Dietary Formulations

Whilst there are a number of technical difficulties to resolve regarding apparent and true digestibility values, the main issue of concern is the overall usefulness of ileal digestibility values in dietary formulation. It had commonly been assumed that, by determining digestibility in the ileum, one was measuring amino acid availability. In other words, if an amino acid was digested and absorbed prior to reaching the terminal ileum, then that amino acid was in a form suitable for utilization. This assumption is not necessarily correct. By its very nature, a digestibility experiment measures indigestibility rather than digestibility. It is not possible to measure digestibility of amino acids as digestion and absorption occur simultaneously. With amino acids, it has been known for many years that various derivatives of lysine can be absorbed in a form(s) that cannot be used for protein synthesis (Hurrell *et al.*, 1976). However, it has always been assumed that these compounds were only of a very minor nature.

A number of different types of studies have been undertaken to determine the usefulness of ileal digestibility values and these studies have given conflicting results. Comparisons have been made between ileal and faecal values. Some of these studies have shown little or no advantage of ileal over faecal digestibility (e.g. Wiseman *et al.*, 1991). Despite this, ileal digestibility values should be the preferred system for measuring amino acid digestibility, as one can never be sure that faecal values have not been modified by microbial activity. Other studies have involved formulating diets on an ileal digestible amino acid basis, or comparing ileal values with availability values. Some of these studies question the basic assumption that ileal digestibility reflects availability.

Growth and metabolism studies

Experiments evaluating ileal determinations in growth or metabolism studies have given conflicting results. Fuller et al. (1981) reported that there was no advantage in formulating diets on an apparent ileal digestible amino acid basis when N excretion was used as the criterion of response. Tanksley and Knabe (1984) showed that diets formulated on an ileal digestible basis were superior to those formulated on a total amino acid basis, when meat- and bone-meal or cottonseed meal were used to replace half the soyabean meal in diets for growing pigs. However, with the latter, performance of pigs given the cottonseed diet did not quite equal that of pigs given the soya control. Moughan and Smith (1985) reported that apparent ileal amino acid digestibilities were a reliable measure of the degree of amino acid digestion and absorption for a diet containing barley and a mixture of protein concentrates, compared to an enzymatically hydrolyzed casein plus free amino acid control. On the other hand, Moughan et al. (1991) found that values for ileal digestible lysine in a barley/meat-and-bone meal/fish meal diet overestimated lysine availability. Similarly, Bellaver and Easter (1989) reported that diets formulated on a true ileal digestible basis, but incorporating alternative proteins to soyabean meal, did not support similar pig performance. Wiseman et al. (1991) also reported that ileal digestible amino acids in heat-processed fish meal were not fully available to pigs.

Overall, the majority of reports have indicated that formulation of diets on an ileal digestible basis does not necessarily lead to equivalent pig response.

Ileal and availability values

A number of studies have compared ileal digestibility with availability. In this regard, availability is defined as the proportion of an amino acid in a feed that is in a form suitable for digestion, absorption and protein synthesis. Thus, availability differs from ileal digestibility in that it includes a measure of utilization. Accordingly, it can only be determined by techniques which measure amino acid utilization. Availability is normally assessed using growth assays (commonly called slope-ratio assays), where the response of pigs to diets containing graded levels of, say, lysine from a protein concentrate is compared to that of pigs given graded levels of free lysine (Batterham, 1992).

Batterham *et al.* (1984) reported that the availability of lysine in a lupinseed meal (0.37) was considerably lower than the ileal digestibility (0.86). The reason for this 49% difference was not established (Batterham *et al.*, 1986). Batterham *et al.* (1990b) reported that, in meals of high quality (soyabean meal), ileal digestibility and lysine availability were similar (0-89, 0.90 respectively). However, in meals of poor quality (cottonseed), ileal digestibility (approx. 0.60–0.70) overestimated lysine availability (0.27–0.30). From this, Batterham *et al.* (1990b) suggested that, in heat-damaged meals, ileal digestibility overestimated lysine availability.

In contrast to the above studies, Leibholz (1986) reported slightly higher values for the ileal digestibility of lysine, compared to lysine availability, in a number of proteins for weaner pigs. However, in this work, most of the data failed the statistical requirement of the slope-ratio analysis and it is probable that the responses to the protein concentrates were influenced by factors other than the available lysine contents.

Leibholz (1985a,b) also reported estimates of 90% of the ileal digestible lysine and 100% of the ileal digestible methionine being recovered in the empty bodies of weaner pigs given a range of diets and suggested that this indicated that ileal digestibility reflected availability for these two amino acids. However, Leibholz (1985a,b) did not determine the actual concentrations of amino acids in the weaner pigs, but used estimated values from previous studies. The calculated recoveries are substantially greater than one would normally anticipate for lysine (0.72, Batterham *et al.*, 1990a) and particularly methionine (0.72, Chung and Baker, 1992; 0.45, Batterham *et al.*, 1993) and, as the actual concentrations were not determined, these estimates need to be treated with caution.

Similar values for true digestibilities and availabilities of lysine in soyabean meal (0.84, 0.82) and sunflower meals (0.84, 0.82) were also reported by Green and Kiener (1989). They concluded that it was difficult to make meaningful comparisons as the availability estimates had large standard errors (0.12 and 0.18 respectively).

Overall, these studies indicate that ileal digestibility values for amino acids may be preferable to total and faecal amino acid values but that they may overestimate availability in a number of protein concentrates.

Utilization of Ileal Digestible Amino Acids

Part of the variability in studies to evaluate the usefulness of the ileal digestibility assay appears due to problems in methodology. For example, some of the studies have involved feeding pigs diets containing a mixture of free and protein-bound amino acids and it was possible that the free amino acids were not fully utilized as frequent or ad lib feeding was not followed. Moreover, additional controls have not always been used to verify that the test amino acids were in fact limiting in the diets.

In order to provide additional information on the usefulness of ileal digestibility values in formulating diets and to determine the utilization of ileal digestible amino acids, a series of experiments was undertaken at Wollongbar. The aim was to examine the usefulness of the ileal digestibility assay for each individual amino acid. Diets were formulated containing a test protein in a sucrose base so that the test amino acid was from one source only. The diets were supplemented with other free amino acids to ensure that the test amino acid was limiting. Additional diets were formulated to verify that the test amino acid was in fact limiting. The pigs were fed frequently to ensure full utilization of free and protein-bound amino acids. At the completion of the experiment the pigs were slaughtered and the retention of the test amino acid determined.

Initially, the ileal digestibility of amino acids in three protein sources was determined – cottonseed meal, representing a meal of low lysine availability (0.30–0.40); meat-and-bone meal, representing a meal of medium availability (0.70); and soyabean meal, representing a meal of high availability (0.90) (Table 6.1).

Lysine

For the first experiment, diets were formulated to similar levels of ileal digestible lysine, with the three protein sources the only sources of lysine in the sucrose-based diets. As outlined above, the diets were supplemented with free amino acids to ensure a surplus of 30%, relative to lysine, and an additional three diets were supplemented with free lysine to verify that lysine was in fact the limiting amino acid in the original three diets. The diets were offered restrictively to growing pigs over the 20–45 kg growth phase. At the completion of the experiment the pigs were slaughtered and the lysine content in the empty bodies determined.

Growth, feed conversion and protein deposition were all superior in pigs given the diet containing soyabean meal (Table 6.2). The retention of ileal digestible lysine in growing pigs given cottonseed meal was only 0.36 compared to 0.75 for that from soyabean meal.

These results indicate that a considerable portion of the ileal digestible lysine in cottonseed meal was apparently absorbed in a form that was inefficiently utilized. As such, the ileal digestibility assay overestimates lysine availability in heat-damaged meals.

Table 6.1. Apparent ileal digestibility of the major essential amino acids in cottonseed, meat-and-bone and soyabean meals. (From Batterham *et al.*, 1990a.)

	Cottonseed	Meat-and-bone	Soyabean	SEM
Lysine	0.74	0.78	0.89	0.022
Threonine	0.76	0.72	0.85	0.024
Methionine	0.79	0.86	0.91	0.020
Tryptophan	0.81	0.65	0.90	0.018

Table 6.2. Growth responses and retention of ileal digestible lysine by growing pigs given diets formulated to 0.36 g ileal digestible lysine MJ^{-1} DE. (From Batterham *et al.*, 1990a.)

	Diets*			
Response	1 Cot	2 Meat	3 Soya	SEM
Gain (g d^{-1})	377	492	541	11.5
Food conversion ratio	3.5	2.6	2.3	0.07
Protein deposited (g d^{-1})	38	66	77	1.6
Lysine retained : ileal digestible lysine intake	0.36	0.60	0.75	0.012

* In Tables 6.2 to 6.6, Cot = cottonseed meal; Meat = meat-and-bone meal; Soya = soyabean meal.

Threonine and methionine

Additional experiments were conducted with threonine or methionine as the limiting amino acid. The results for threonine (Table 6.3) were similar to those for lysine. Growth response, protein deposition and retention of ileal digestible threonine were all lower in the pigs given cottonseed meal relative to soyabean meal. For methionine (Table 6.4), growth and protein deposition were also lower in pigs given cottonseed meal compared to soyabean meal, but there was little difference in methionine retentions. Also, the overall retention of ileal digestible methionine in pigs given soyabean meal (0.45) was much lower than for lysine (0.75).

These results indicate that ileal digestible threonine and methionine are also apparently absorbed in a form(s) that is (are) inefficiently utilized.

Table 6.3. Growth responses and retention of ileal digestible threonine by growing pigs given diets formulated to 0.22 g ileal digestible threonine MJ^{-1} DE. (From Beech *et al.*, 1991.)

| Response | Diets | | | SEM |
	1 Cot	2 Meat	3 Soya	
Gain (g d^{-1})	417	452	524	9.6
Food conversion ratio	3.2	2.8	2.4	0.04
Protein deposited (g d^{-1})	47	62	75	2.3
Threonine retained : ileal digestible threonine intake	0.44	0.59	0.64	0.017

Table 6.4. Growth responses and retention of ileal digestible methionine by growing pigs given diets formulated to 0.09 g ileal digestible methionine MJ^{-1} DE. (From Batterham *et al.*, 1993.)

| Response | Diets | | | SEM |
	1 Cot	2 Meat	3 Soya	
Gain (g d^{-1})	411	441	496	17.4
Protein deposited (g d^{-1})	47	57	61	2.3
Methionine retained : ileal digestible methionine intake	0.38	0.45	0.45	0.013

Tryptophan

With tryptophan, it was not possible to supply all the tryptophan from meat-and-bone meal, as excessive calcium would have resulted. Accordingly, the tryptophan was supplied from a combination of meat-and-bone meal and free tryptophan. The results of this experiment were unusual, as the meat-and-bone meal plus free tryptophan diet stimulated the greatest pig response (Table 6.5). This tends to suggest that the total values for tryptophan underestimated the amount of tryptophan in the meals.

Leucine, valine, isoleucine and phenylalanine

New samples of cottonseed and soyabean meals were used for the utilization studies with the remaining essential amino acids. Surprisingly, with

Table 6.5. Growth responses and retention of ileal digestible tryptophan by growing pigs given diets formulated to 0.065 g ileal digestible tryptophan MJ^{-1} DE. (From Batterham *et al.*, 1994.)

	Diets			
Response	1 Cot	2 Meat + Trp*	3 Soya	SEM
Gain $(g\,d^{-1})$	393	531	437	27.7
Protein deposited $(g\,d^{-1})$	54	75	63	3.5
Tryptophan retained : ileal digestible tryptophan intake	0.46	0.45	0.38	0.010

* Trp = tryptophan.

leucine, there were no differences in the growth performance or estimated retention of ileal digestible leucine (Table 6.6). This indicates that ileal digestible leucine is useful for formulating diets. Similar results were recorded with isoleucine and valine.

These results indicate that the branched-chain amino acids (valine, isoleucine and leucine), which are metabolized predominantly in the muscle rather than the liver, are less sensitive to processing damage, and the ileal digestibility assay for these amino acids reflects availability. However, for phenylalanine, which is metabolized in the liver, differences in utilization were recorded, with pigs given the cottonseed meal growing slower than those given soyabean meal. However, the difference in utilization did not appear to be as great as for the other amino acids.

The overall results of these studies are summarized in Table 6.7. For the four major amino acids, lysine, threonine, methionine and tryptophan, and for phenylalanine, which are all metabolized in the liver, heat damage during processing apparently causes changes to occur which have little effect on ileal digestibility, but result in a considerable portion of these amino acids apparently being absorbed in a form that is inefficiently utilized. Availability values are therefore needed for these amino acids in heat-damaged proteins. On the other hand, the branched-chain amino acids, valine, isoleucine and leucine, all appear to be unaffected by processing conditions, and the ileal digestibility assay appears to reflect availability.

Effect of Heat on the Utilization of Ileal Digestible Lysine

Additional experiments were undertaken at Wollongbar to verify that heat was the factor responsible for the inefficient utilization of ileal digestible

Table 6.6. Growth responses and retention of ileal digestible leucine by growing pigs given diets formulated to 0.4 g ileal digestible leucine MJ^{-1} DE.

| | Diets | | | |
| | 1 | 2 | 3 | |
Response	Cot	Lupin	Soya	SEM
Gain (g d^{-1})	536	548	550	13.7
Protein deposited (g d^{-1})	78	77	78	2.5
Estimated leucine retention : ileal digestible leucine intake	0.68	0.66	0.67	0.016

Table 6.7. Ability of ileal digestible amino acid values to predict availability in heat-damaged proteins.

Overestimates availability[*]	Predicts availability
Lysine	Valine
Threonine	Isoleucine
Methionine	Leucine
Tryptophan	
Phenylalanine	

[*] Listed in order of degree of sensitivity.

lysine, and to determine the overall relationship between the application of heat and ileal digestibility, availability and utilization of lysine in heat-processed field peas.

The results have indicated that initially, heat improves the ileal digestibility of lysine in field peas, and then gradually reduces it (Van Barneveld *et al.*, 1991b). The ileal digestibility of lysine improved from 0.75 in raw peas to 0.82 in peas heated to 110°C, then gradually declined to 0.78 (135°), 0.75 (150°) and to 0.62 in peas heated to 160°. In contrast, the utilization of the ileal digestible lysine was reduced at all levels of heat inputs (Table 6.8).

The results of Van Barneveld *et al.* (1991b) have considerable implications. They indicate that mild heat can actually improve ileal digestibility at the same time that the utilization of the ileal digestible lysine is depressed. Thus it is possible in all protein concentrates which undergo some form of heat processing, that damage could be occurring to the lysine

Table 6.8. Growth responses and estimated lysine retention of growing pigs given diets containing heat-treated field peas and formulated to 0.36 g ileal digestible lysine MJ^{-1} DE. (From Van Barneveld *et al.*, 1991b.)

Response	Treatment					
	Raw	110°	135°	150°	165°	SEM
Gain (g d^{-1})	498	482	477	450	314	12.9
Feed conversion ratio	2.6	2.7	2.8	2.9	4.5	0.12
Estimated lysine retained : ileal digestible lysine intake	0.72	0.65	0.64	0.58	0.42	0.021

and that this damage would be undetected by the ileal digestible assay. There is an urgent need for additional research to determine the effects of processing conditions on the utilization of all protein concentrates, as the conditions used in commercial processing plants probably exceed that of the lower heat treatments used by Van Barneveld *et al.* (1991b).

Relationship between Techniques for Describing Amino Acids

It is evident that crude protein and total amino acid values are only useful in describing the potential contribution of nitrogen and amino acids in a feed. Faecal digestible amino acids make some correction for loss of digestibility of amino acids. However, this system reflects the combined effects of amino acid digestibility in the small intestine and microbial activity in the hind gut. For this reason it is not appropriate for assessing digestibility. Ileal digestible amino acids take into account losses in amino acid digestibility. For non-heat-damaged meals, ileal digestibility reflects availability. In heat-damaged meals, it overestimates availability for the amino acids metabolized in the liver. Availability values reflect the proportions of amino acids in feeds which are usable by pigs. They are essential in accounting for the major amino acids in heat-damaged meals. It is the ultimate system which should be applied to all amino acids, but is the most difficult to assess.

Which Values to Use?

The most appropriate system to use depends on available feed sources and the method of formulating diets:

1. Cereals – ileal digestibility values appear the most appropriate at present, as they account for losses in digestibility. The assumption is made that ileal digestibility reflects availability. This may not be correct and this aspect requires further research.
2. Proteins that undergo minimal damage during processing – for such proteins, e.g. soyabean meal, the ileal digestibility assay is the most appropriate at present for estimating the availability of all amino acids.
3. Proteins that undergo processing damage – for these proteins, such as cottonseed meal, sunflower meal, meat-and-bone meal, batch-dried blood meal, etc., reference values for lysine availability are recommended. Corrections can also be made for the estimated availability of threonine, methionine and tryptophan, based on modifications to the lysine value. For the branched-chain amino acids, ileal digestibility values seem appropriate.
4. Diets formulated by linear programming – if diets are being formulated to least-cost using linear programming, then apparent ileal digestibility values are the most appropriate, as they take into account the endogenous cost of digestion.
5. Diets formulated by computer simulation models – if diets are being formulated by computer simulation models, then true digestibility values are the most appropriate, as the model should correct for the endogenous cost of digestion.

In the future, as the mechanisms of heat damage are better understood and computer simulation models become more sophisticated, then all of the above parameters will be utilized in dietary formulations. Total amino acid values will be needed to define the potential amino acid intake, ileal digestible values to define digestibility and faecal load, and availability values to define the proportion of usable and non-usable amino acids which are absorbed and metabolized.

As concern for the environment increases, greater attention is being paid to formulating diets which maximize the incorporation of amino acids into lean tissue and minimize the excretion of nitrogen in the urine and faeces. This is best achieved using information on the above three aspects of amino acids.

Conclusions

Despite considerable research into developing techniques for measuring the ileal digestibility of amino acids, there is still considerable uncertainty surrounding the applicability of the values in dietary formulations. The choice of technique for determining ileal digestibility appears to be more an individual preference of the researcher or institute. More uncertainty exists in the methodology for determining apparent or true digestibilities. There is a need to determine the minimum level of dietary amino acids in the test diets for determining the apparent ileal digestibility of amino acids. It is probable that many of the literature values for apparent ileal digestibilities, in cereals and grain legumes especially, may be underestimated due to low crude protein levels in the test diets. It is also apparent that corrections for true digestibility, based on feeding protein-free diets or regressing to zero, appear to underestimate true digestibility. In this regard, values based on ^{15}N-dilution or homoarginine are needed. However, apparent ileal values appear to be most applicable in the formulation of diets by linear programming, whilst true values are needed in computer simulation models of the pig.

Heat applied to a protein concentrate can actually improve the ileal digestibility of lysine, whilst at the same time inducing changes which result in a decrease in the utilization of the ileal digestible lysine. As such, ileal digestible values for lysine overestimate availability in heat-damaged meals. Threonine, methionine, tryptophan and phenylalanine appear similarly affected, whilst the branched-chain amino acids, valine, isoleucine and leucine, appear unaffected. Thus for cereals and non-heat-damaged protein concentrates, ileal digestible values appear suitable for dietary formulations. In heat-damaged meals, only values for valine, isoleucine and leucine are useful in dietary formulations. For the remainder, availability values are needed.

References

Badawy, A.M. (1964) Changes in the protein and non-protein nitrogen in the digesta of sheep. In: Munro, H.N. (ed.) *The Role of the Gastrointestinal Tract in Protein Metabolism*. Blackwell, Oxford, pp. 175–185.

Batterham, E.S. (1992) Availability and utilization of amino acids for growing pigs. *Nutrition Research Reviews* 5, 1–18.

Batterham, E.S., Murison, R.D. and Andersen, L.M. (1984) Availability of lysine in vegetable protein concentrates as determined by the slope-ratio assay with growing pigs and rats and by chemical techniques. *British Journal of Nutrition* 51, 85–99.

Batterham, E.S., Andersen, L.M., Lowe, R.F. and Darnell, R.E. (1986) Nutritional value of lupin (*Lupinus albus*)-seed meal for growing pigs: availability of lysine, effect of autoclaving and net energy content. *British Journal of Nutrition* 56, 645–659.

Batterham, E.S., Andersen, L.M., Baigent, D.R., Beech, S.A. and Elliott, R. (1990a) Utilization of ileal digestible amino acids by pigs: lysine. *British Journal of Nutrition* 64, 679–690.

Batterham, E.S., Andersen, L.M., Baigent, D.R., Darnell, R.E. and Taverner, M.R. (1990b) A comparison of the availability and ileal digestibility of lysine in cottonseed and soya-bean meals for grower/finisher pigs. *British Journal of Nutrition* 64, 663–677.

Batterham, E.S., Andersen, L.M. and Baigent, D.R. (1993) Utilization of ileal digestible amino acids by growing pigs: methionine. *British Journal of Nutrition* 70, 711–720.

Batterham, E.S., Andersen, L.M. and Baigent, D.R. (1994) Utilization of ileal digestible amino acids by growing pigs: tryptophan. *British Journal of Nutrition* 71, 345–360.

Beech, S.A., Batterham, E.S. and Elliott, R. (1991) Utilization of ileal digestible amino acids by growing pigs: threonine. *British Journal of Nutrition* 65, 381–390.

Bellaver, C. and Easter, R.A. (1989) Performance of pigs fed diets formulated on the basis of amino acid digestibility. *Journal of Animal Science* 67(1), 241A.

Chung, T.K. and Baker, D.H. (1992) Efficiency of dietary methionine utilization by young pigs. *The Journal of Nutrition* 122, 1862–1869.

Fuller, M.F., Baird, B.A., Cadenhead, A. and Aitken, R. (1981) An assessment of amino acid digestibility at the terminal ileum as a measure of the nutritive value of proteins for pigs. *Animal Production* 32, 396.

Green, S. and Kiener, T. (1989) Digestibilities of nitrogen and amino acids in soya-bean, sunflower, meat and rapeseed meals measured with pigs and poultry. *Animal Production* 48, 157–179.

Hagemeister, H. and Erbersdobler, H. (1985) Chemical labelling of dietary protein by transformation of lysine to homoarginine: a new technique to follow intestinal digestion and absorption. *Proceedings of the Nutrition Society* 44, 133A.

Huisman, J., Heinz, T.H., van der Poel, A.F.B., van Leeuwen, P., Souffrant, W.B. and Verstegen, M.W.A. (1992) True protein digestibility and amounts of endogenous protein measured with the ^{15}N-dilution technique in piglets fed on peas (*Pisum sativum*) and common beans (*Phaseolus vulgaris*). *British Journal of Nutrition* 68, 101–110.

Hurrell, R.F., Carpenter, K.J., Sinclair, W.J., Otterburn, M.S. and Asquith, R.S. (1976) Mechanisms of heat damage in proteins 7. The significance of lysine-containing isopeptides and of lanthionine in heated proteins. *British Journal of Nutrition* 35, 383–395.

Imbeah, M., Sauer, W.C. and Mosenthin, R. (1988) The prediction of the digestible amino acid supply in barley-soybean meal or canola meal diets and pancreatic enzyme secretion in pigs. *Journal of Animal Science* 66, 1409–1417.

Leibholz, J. (1985a) An evaluation of total and digestible lysine as a predictor of lysine availability in protein concentrates for young pigs. *British Journal of Nutrition* 53, 615–624.

Leibholz, J. (1985b) The digestion of protein in young pigs and the utilization of dietary methionine. *British Journal of Nutrition* 53, 137–147.

Leibholz, J. (1986) The utilization of lysine by young pigs from nine protein concentrates compared with free lysine in young pigs fed *ad lib*. *British Journal of Nutrition* 55, 179–186.

Lenis, N.P. (1992) Digestible amino acids for pigs: assessment of requirements on ileal digestible basis. *Pig News and Information* 13(1) 31N–39N.

Low, A.G. (1982) Digestibility and availability of amino acids from feedstuffs for pigs: a review. *Livestock Production Science* 9, 511–520.

Low, A.G. (1990) Protein evaluation in pigs and poultry. In: Wiseman, J. and Cole, D.J.A. (eds) *Feedstuff Evaluation*. Butterworths, London, pp. 91–114.

Moughan, P.J. and Smith, W.C. (1985) Determination and assessment of apparent ileal amino acid digestibility coefficients for the growing pig. *New Zealand Journal of Agricultural Research* 28, 365–370.

Moughan, P.J. and Smith, W.C. (1987) A note on the effect of cannulation of the terminal ileum of the growing pig on the apparent ileal digestibility of amino acids in ground barley. *Animal Production* 44, 319–321.

Moughan, P.J., Smith, W.C., Pearson, G. and James, K.A.C. (1991) Assessment of apparent ileal lysine digestibility for use in diet formulation for the growing pig. *Animal Feed Science and Technology* 34, 95–109.

Sauer, W.C. and Ozimek, L. (1986) Digestibility of amino acids in swine: Results and their practical applications. A Review. *Livestock Production Science* 15, 367–388.

Sauer, W.C. Dugan, M., de Lange, K., Imbeah, M. and Mosenthin, R. (1989) Considerations in methodology for the determination of amino acid digestibilities in feedstuffs for pigs. In: Friedman, M. (ed.) *Absorption and Utilization of Amino Acids, Vol. III*. CRC Press, Boca Raton, Florida, pp. 217–230.

Tanksley, T.D. Jr. and Knabe, D.A. (1984) Ileal digestibilities of amino acids in pig feeds and their use in formulating diets. In: Haresign W. and Cole, D.J.A. (eds) *Recent Advances in Animal Nutrition – 1984*. Butterworths, London, pp. 75–95.

Taverner, M.R., Hume, I.D. and Farrell, D.J. (1981) Availability to pigs of amino acids in cereal grains. 1. Endogenous levels of amino acids in ileal digesta and faeces of pigs given cereal diets. *British Journal of Nutrition* 46, 149–158.

Van Barneveld, R.J., Batterham, E.S. and Norton, B.W. (1991a) Comparison of faecal sampling techniques on the determination of the digestibility of energy and nitrogen in raw and heat-treated peas. In: Batterham, E.S. (ed.) *Manipulating Pig Production III*. Australasian Pig Science Association, Attwood, p. 188.

Van Barneveld, R.J., Batterham, E.S. and Norton, B.W. (1991b) Utilization of ileal digestible lysine from heat-treated field peas by growing pigs. In: Batterham, E.S. (ed.) *Manipulating Pig Production III*. Australasian Pig Science Association, Attwood, p. 184.

Van Leeuwen, P., Huisman, J., Verstegen, M.W.A., Baak, M., van Kleef, D.J., van Weerden, E.J. and den Hartog, L.A. (1988) A new technique for the collection of ileal chyme. In: Buraczewska, L., Buraczewski, S., Pastuzewska, B. and Zebrowska, T. (eds) *Digestive Physiology in the Pig*. Polish Academy of Sciences, Jablonna, pp. 289–296.

Wiseman, J., Jagger, S., Cole, D.J.A. and Haresign, W. (1991) The digestion and utilization of amino acids of heat-treated fish meal by growing/finishing pigs. *Animal Production* 53, 215–225.

Modelling Amino Acid Absorption and Metabolism in the Growing Pig

7

P.J. MOUGHAN

Department of Animal Science, Massey University,
Private Bag, Palmerston North, New Zealand.

Introduction

Biological systems often involve complex interactions among numerous factors, such that it is difficult for the unaided human mind to integrate these processes and gain insight to the system itself. Yet, if the consequences of manipulating aspects of a system are to be fully appreciated then it is critical that the system *per se* be understood. The key to understanding systems is the science of simulation or modelling.

Mammalian growth is a complex biological system, the understanding of which can be enhanced by the construction of computerized mathematical models. Models allow hypotheses on the growth process to be developed and evaluated. Moreover, suitably validated models of animal growth may be used commercially to predict the biological and economic consequences of changes in the production system. A number of deterministic biological models simulating growth processes in the pig have been developed (Whittemore and Fawcett, 1974, 1976; Whittemore, 1983; Moughan and Smith, 1984; Black *et al.*, 1986; Moughan *et al.*, 1987; Pomar *et al.*, 1991) and are now used widely for research and in commercial practice. The general development of pig growth models has been reviewed by Moughan and Verstegen (1988).

It is critical, in any biological pig growth model, to accurately simulate amino acid flow and thus be able to predict daily net body protein deposition and the supply of net energy from degraded amino acids. Implicit in this is the need to simulate the ingestion, digestion, absorption and metabolism of amino acids. It is of note that the simulated supply of amino acids available for protein synthesis has a disproportionate effect on the accuracy of prediction of body growth rate since in most models ash and water retentions are derived as a function of body protein deposition. The absorption and metabolism of amino acids in mammals is complex and highly integrated with continuous flux within and between body cells. It is useful, however, and inherently necessary when constructing a model of metabolism, to view amino acid metabolism as several discrete physiological processes (Table 7.1). The aim of the present contribution is to discuss the latter processes in the context of modelling. Important metabolic controls and quantification of the processes are highlighted. The discussion here will be restricted to the growing pig, though the principles established apply equally well to mature, reproducing and lactating animals.

Modelling the Components of Dietary Amino Acid Utilization

Pre-uptake

The ingestion of dietary amino acids

The dietary amino acids ingested by the growing pig constitute the basic drive for protein metabolism. The ingested amount of a given amino acid, in turn, is a function of the quantity of food ingested and the amino acid composition of the food. When applying models, the usual approach for calculating amino acid intakes from daily food intake has been to use tabulated values for the amino acid contents of the dietary ingredients. Care should be exercised when specifying the model's matrix of ingredient nutrient compositions. There is often considerable variation in amino acid

Table 7.1. Processes involved in the utilization of dietary amino acids by the growing pig.

Pre-uptake
Ingestion of dietary amino acids
Absorption of amino acids from unprocessed feedstuffs
Absorption of amino acids from processed feedstuffs

Post-absorptive metabolism
Use of amino acids for body protein maintenance
Use of amino acids for synthesis of non-protein nitrogen compounds
Transamination
Inevitable amino acid catabolism
Preferential catabolism of amino acids for energy supply
Amino acid imbalance
Catabolism of amino acids supplied above the amount required to meet the maximum rate of body protein retention

composition within an ingredient type which may be partly ascribed to differences (e.g. cultivar of cereal, rendering process) among feeds within that type. This variation can be minimized by carefully defining ingredient classes and sub-classes. Furthermore, amino acid compositional data entered into the model's matrix should be subjected to careful evaluation. Preferably data should only be used from internationally accredited laboratories, using standard validated analytical procedures. For some amino acids it is important to ensure that amino acid yields during analysis have been corrected for possible amino acid oxidation or incomplete release during hydrolysis. This can constitute a major source of error (Finley, 1985). For example, and in our own laboratory (Garnett, 1985; Rowan *et al.*, 1992), the yield of isoleucine has been shown to increase by up to 20–40% when hydrolysis time is extended from 24 to 96 hours. For lysine, a comparable increase of 9 to 12% has been observed. The latter data highlight the potential inaccuracies which may occur in the simulation of amino acid uptake and metabolism, simply due to consistent biases in describing the nutrient contents of the food. Such error may be considerable and can only be minimized by attention to detail in specifying the ingredient nutrient compositions.

A further problem in describing the amino acid composition of feedstuffs is found with processed material. Ingredients used to manufacture compounded pig diets have often been subjected to physical, biological or chemical treatments while being processed. During processing, particularly that involving heat, and during subsequent storage of the material, protein-bound amino acids can react with reducing sugars, fats and their oxidation products, polyphenols or chemical additives such as alkali. The amino acids lysine, methionine, cysteine and tryptophan are particularly reactive. Lysine, which is commonly the first-limiting amino acid in pig diets, is most susceptible to damage because of the ready involvement of its epsilon-NH_2 group in intra-and inter-molecular crosslinking and in reaction with other compounds. The reaction between lysine and reducing sugars (the Maillard reaction), which occurs under mild processing conditions and during storage, is well-documented. In the advanced stages of the Maillard reaction (brown pigment formation) the amino acids will be completely destroyed, and will not be recoverable following acid hydrolysis during amino acid analysis. However, in the early stages of the reaction, which occur under the normal conditions of food processing, the deoxy-ketosyl derivative (Amadori compound) formed is hydrolysed back to lysine in the presence of strong acids. Thus heat-processing and unfavourable storage conditions can reduce dietary lysine in foods to a greater extent than is reflected by conventional amino acid analysis. The amino acid intake of the pig is commonly predicted in models from the amino acid contents of foods obtained after amino acid analysis. For foods which have undergone the early Maillard reaction, however, this will lead to overestimation of the intake of lysine. Some of the determined lysine intake will be deoxyketosyl lysine, which although being partly absorbed is of no nutritional value to the pig (Hurrell and Carpenter, 1981). Recent estimates (Gall, Moughan and Wilson, unpublished) using mildly-heated model proteins indicate that the degree of overestimation of lysine may be as high as 20 to 30%.

For heat-treated proteins, Carpenter's fluoro-dinitrobenzene (FDNB) lysine assay should be used, rather than conventional amino acid analysis, to generate estimates of chemically available lysine for inclusion in a model's matrix of ingredients. Ileal digestible lysine is not expected to be a good indicator of the level of chemically available lysine (Moughan, 1991). Validated assays comparable to FDNB lysine are required for application to other amino acids. For mixed diets used in practice, the amount of damaged (i.e. structurally altered) amino acids is probably relatively low. In situations where this is not the case, however, an adequate description of the amounts of chemically available amino acids will be important if growth of the pig is to be accurately simulated.

Absorption of amino acids from unprocessed feedstuffs

An important step in modelling protein metabolism in the pig is the prediction of amino acid absorption from the digestive tract and entry to the portal blood. Although this model component is simple conceptually, it is an area where considerable predictive error can be introduced.

The amount of a dietary amino acid absorbed from the digestive tract in pigs has often been predicted during growth simulation by application of an apparent faecal digestibility coefficient entered as a model input (Whittemore, 1983). For most amino acids, however, it seems that faecal values overestimate actual digestibility. The quantitative importance of the microbial metabolism of protein in the large intestine of the pig is now well recognized (McNeil, 1988). Amino acids entering the hind gut from the small intestine may be acted upon by the microflora, leading usually to a net disappearance though sometimes a net appearance of amino acids between the ileum and rectum (Rérat, 1981). It also seems unlikely that amino acids are absorbed to any significant extent in the large intestine of the pig (Darragh *et al.*, 1994). The extent of digestibility overestimation varies with the amino acid, the type of dietary protein and the influence of other dietary components. Lenis (1983) has surveyed the world literature from 1964 to 1982 for some 35 foodstuffs. For threonine and tryptophan, the mean overestimations of apparent digestibility by the faecal method (in comparison with ileal values) were 10 and 11 percentage units, respectively. The ileal-faecal differences tended to be smaller for lysine. The faecal method overestimated (mean overestimation = 5.6% units) lysine digestibility for 11 foods and underestimated it (mean underestimation =

4.3% units) in ten further foods. Faecal values appear to often considerably underestimate the actual digestibility of methionine, though the opposite is found for cysteine.

The inability of the faecal method of analysis to account for the effect of hind-gut metabolism may explain the oft-reported poor correlations between pig growth performance and faecal estimates of amino acid uptake (Crampton and Bell, 1946; Lawrence, 1967; Cole *et al.*, 1970). Measurement of amino acid digestibility at the end of the ileum is now generally recognized as a more acceptable approach (Low, 1980; Rérat, 1981; Darcy-Vrillon *et al.*, 1985; Sauer and Ozimek, 1986) and a significant body of literature on the ileal digestibility of amino acids in foods for pigs is accumulating. A potential criticism of the ileal measure is that there may be interference from a population of microorganisms present in the upper digestive tract (Horvath *et al.*, 1958; Williams Smith, 1965). Amino acids may be catabolized or synthesized or incorporated into microbial protein. Nevertheless, ileal digestibility coefficients have been shown to be sensitive in detecting small differences in protein digestibility (van Weerden *et al.*, 1985; Sauer and Ozimek, 1986) and several studies (Tanksley and Knabe, 1980; Low *et al.*, 1982; Moughan and Smith, 1985) have demonstrated that ileal values are reasonably accurate in describing the extent of uptake of amino acids from the gut, at least for feedstuffs which have not sustained damage to their protein during processing.

Because significant quantities of endogenous amino acids are present at the terminal ileum of the pig, true coefficients (corrected for endogenous flow) of amino acid digestibility should be more accurate indicators of absorption than their apparent counterparts. The protein-free method of determining endogenous flows, however, appears to underestimate the actual endogenous loss of amino acids (Darragh *et al.*, 1990; de Lange *et al.*, 1990; Moughan and Rutherfurd, 1990). Further, and given that the regression technique generates similar values to those obtained after feeding pigs a protein-free diet (Leibholz and Mollah, 1988), use of this method may also lead to error. A new approach for accurately determining ileal endogenous amino acid loss which is applicable to the correction of ileal flows for ingredients not containing fibre and anti-nutritional factors, such as meat meals, fish meals, milk powders, etc., has been proposed and developed (Moughan *et al.*, 1990). Further research into methods for determining endogenous excretion, having a wider application, is required. The use of true estimates of ileal amino acid absorption will enhance the accuracy of model predictions of absorbed amino acids. When apparent estimates of digestibility are used, the endogenous amino acid loss from the gut is reflected in the digestibility coefficient, and this must be recognized when simulating the body amino acid losses at maintenance. Apparent estimates of digestibility may lead to errors, however, as the endogenous amino acid losses determined under experimental conditions where single ingredients are fed to the animal, are unlikely to be additive for the case whereby a mixed diet is given.

The comments made above with respect to defining ingredient amino acid composition also apply to the description of amino acid digestibility. Ileal digestibility data used for the simulation of growth should be obtained using a standard and acceptable procedure and care should be given to defining ingredient classes and sub-classes. In summary, the overestimation of digestibility for most amino acids due to using faecal values may to some extent counteract underestimation from using apparent rather than true coefficients of ileal digestibility. Overall, however, it is likely that a variable degree of error will remain if faecal rather than ileal values are used in simulation. True ileal amino acid digestibility coefficients offer the best practical means of predicting the amounts of amino acids entering the portal blood of the pig, at least for feedstuffs whereby amino acids have not been significantly damaged during processing or storage.

Absorption of amino acids from processed feedstuffs

For feedstuffs which have been subjected to processing and with consequent damage to constituent amino acids, the ileal digestibility assay is not expected to accurately indicate the absorption of all amino acids (Moughan, 1991; Moughan *et al.*, 1991). This is well exemplified for lysine. During amino acid analysis with strong hydrochloric acid, early Maillard compounds are known to partially revert to lysine. Such reversion does not occur, however, in the pig's alimentary canal during gastric digestion. Consequently, estimates of dietary and ileal digesta lysine will be in error leading to biased ileal digestibility coefficients. Although, and at least for lysine, structurally unaltered molecules can be accurately determined chemically (e.g. FDNB lysine assay), there is evidence (Hurrell and Carpenter, 1981) that the unaltered or chemically available molecules may not be fully absorbed from the damaged proteins.

The absorption (measured at terminal ileum) of reactive lysine has been determined in a recent study (Gall, Moughan and Wilson, unpublished) with the growing pig (Table 7.2). A casein-glucose mixture was heated to produce early Maillard compounds, and the amount of epsilon-*n*-deoxy-fructosyl-lysine (blocked lysine) and lysine regenerated after acid hydrolysis in the resulting material was calculated from the determined level of furosine. The amount of unaltered or reactive lysine was found by difference between the total lysine (acid hydrolysis) and regenerated lysine. The FDNB method allowed accurate assessment of the amount of chemically reactive lysine, which was grossly overestimated by conventional amino acid analysis (acid-hydrolysed lysine), but the reactive lysine was not

Table 7.2. Amounts of acid-hydrolysed lysine, FDNB lysine, reactive lysine and absorbed reactive lysine in a heated casein-glucose mixture.

	Acid-hydrolysed [1]	FDNB [2]	Reactive [3]	Absorbed reactive [4]
Lysine (g 100 g^{-1})	2.60	1.91	1.98	1.40

[1] After conventional amino acid analysis.
[2] FDNB = fluoro-dinitrobenzene.
[3] Lysine units remaining chemically reactive after heating, determined from furosine levels.
[4] Reactive lysine absorbed by the end of the small intestine.

completely absorbed. An incomplete absorption of reactive (guanidinated) lysine by growing pigs was also reported by Schmitz (1988), and by Desrosiers *et al.*, (1989) using an *in vitro* technique.

There is need for an assay, with particular application to processed feeds, which allows measurement of the levels of chemically available amino acids which are able to be released from the protein and absorbed. In the absence of such an assay, the uptake of amino acids by the pig from processed feeds is best modelled based on chemically available amino acid intakes (e.g. FDNB lysine) corrected for the true ileal digestibility of dietary total nitrogen. In situations where highly processed feedstuffs are being used in formulating pig diets, the conventional approach to modelling amino acid uptake may lead to error.

Post-absorptive metabolism

Use of amino acids for body protein maintenance

Amino acids are continually lost from the pig's body and the need for replacement of these amino acids constitutes the maintenance amino acid requirement. It is generally assumed that the amino acid requirement for resynthesis of protein will be met before the synthesis of new body protein proceeds. Usually, when modelling pig growth, maintenance amino acid requirements have been described as a simple function of body weight or protein mass. This approach, however, gives no insight into the physiological processes contributing to the maintenance requirement. It is also often assumed that amino acid requirements for maintenance are relatively insignificant in relation to total requirements. However, at low rates of body protein deposition (30–90 g d^{-1}), maintenance gives rise to 20–30% of the total amino acid requirement, while at higher levels of deposition (145 g d^{-1}) the maintenance requirement comprises at least 10%, and probably more, of the total amino acid requirement (Moughan, 1989). A greater accuracy in simulating amino acid metabolism may ensue from a more detailed description of the maintenance processes.

Amino acids as precursors for the synthesis of compounds other than proteins

A number of non-protein nitrogen-containing compounds, essential for normal functioning of the animal, are synthesized from absorbed amino acids. The amino acids are irreversibly lost from the body amino acid pool and are thus unavailable for body protein synthesis. It is often considered (Reeds, 1988) that the use of amino acids for these purposes is quantitatively low and insignificant overall. However, for some amino acids, for example methionine, this pathway of amino acid loss may not be negligible and should be considered in the construction of simulation models. Models simulating protein metabolism in poultry have included description of the synthesis of non-protein compounds from amino acids.

Amino acid transamination

Where the dietary non-essential amino acid component is limiting for protein growth and maintenance, essential amino acids may be used to synthesize dietary non-essentials, and this process has a potentially important impact on overall protein metabolism. In normal practice, however, and where conventional feed ingredients are used, this situation is rarely encountered. An early model developed to score mixed diets in terms of their overall protein quality (Moughan and Smith, 1984) included a description of the transamination process within a linear programming framework. Depending upon the intended end-use of a model, description of transamination may be important. As models become more mechanistic and the biochemical transactions of each of the individual body amino acids is described, there may be a greater need to directly describe transamination.

Inevitable amino acid catabolism

Inevitable amino acid catabolism refers to the degradation of the absorbed first-limiting amino acid, which occurs even though non-protein energy is supplied well in excess of the animal's metabolic needs. Currently, there is debate in pig growth modelling as to the quantitative significance of this process. It has often been assumed when modelling amino acid metabolism that the absorbed amount of the first-limiting amino acid remaining after maintenance costs have been met, and under the condition of adequate non-protein energy supply, is fully available for protein synthesis. However, even at low levels of intake of the first-limiting amino acid and with high levels of ATP generation from glucose and fatty acids, there is undoubtedly some inevitable degradation. The critical question concerns the magnitude of this base rate of catabolism and if and how the rate of inevitable catabolism is influenced by increasing amino acid uptake from low levels to those supporting maximal body protein retention.

Data on the rate of inevitable amino acid catabolism in the growing pig are sparse. Mean estimates in the literature range from the negligible to 40% of the absorbed amount of the first-limiting amino acid, and there is limited information on the pattern of change of inevitable catabolism with change in the amount of absorbed amino acid. Recently, Heger and Frydrych (1985) proposed that inevitable catabolism increases at a disproportionate rate with increasing amino acid absorption, with a curvilinear response being marked around the maximum level of protein retention. The study of Heger and Frydrych (1985) with the growing rat is of particular interest because inevitable catabolism was determined over a wide range of dietary amino acid intakes and there was control of a number of important experimental variables. Male rats were fed semisynthetic diets containing crystalline amino acids, and relationships between nitrogen balance and amino acid intakes were determined. The amino acid of interest was included (0–120% of NRC requirement) in iso-nitrogenous diets which supplied adequate amounts of all other amino acids and a high level of energy in the form of non-amino compounds. From information on nitrogen retention and the amino acid composition of body protein, the net retentions of body amino acids were determined. Third-degree polynomials were fitted to the data and the intake of each amino acid corresponding to the maximum response was considered to be the optimum requirement, while the maintenance requirement was estimated from the point of intersection of the regression curve with the zero balance point. The data of Heger and Frydrych (1985) allow calculation of the amounts of first-limiting amino acid inevitably catabolized, if it is assumed that the crystalline amino acids were completely absorbed and chemically available, and that there was minimal catabolism of amino acids for the express purpose of energy supply. Estimates of the inevitable catabolism of lysine calculated from the data presented by Heger and Frydrych (1985) are given in Table 7.3. The data for lysine were representative of those found for the other essential amino acids, except methionine, and demonstrate that the inevitable catabolism of amino acids may be considerable, particularly at levels of amino acid absorption supporting near maximal body protein deposition. The study of Batterham *et al.*, (1990), in which lysine retention in growing pigs was determined, gives some support to the findings of Heger and Frydrych (1985).

The way in which these data (Table 7.3) are interpreted, however, depends upon how maximal body nitrogen retention is defined and the form of the mathematical functions fitted to the raw data. An alternative interpretation is that inevitable catabolism is relatively constant with

Table 7.3. Amounts of lysine inevitably catabolized by the growing rat. (Data from Heger and Frydrych, 1985.)

	Lysine absorbed (mg d^{-1})							
	10	20	30	40	50	60	70	80*
Lysine catabolized (mg d^{-1})	0.85	1.82	3.06	4.92	7.80	12.06	18.20	26.48
% Absorbed	8.5	9.1	10.2	12.3	15.6	20.1	26.0	33.1

* Amount of absorbed lysine corresponding to the maximum rate of body protein retention.

increasing amino acid absorption (below the level of absorption supporting maximal protein deposition) at 10–12% of the absorbed amount, with catabolism only increasing once maximal protein deposition has occurred. Implicit to the problem of describing the data on inevitable catabolism is whether the real response in an individual animal is rectilinear or curvilinear.

In a review on the process of inevitable catabolism, Heger and Frydrych (1989) concluded that even for well-balanced proteins, a portion of the deficient amino acid is subject to oxidation. The loss increases with increasing intake of protein from amounts allowing only maintenance and minimum growth to those supporting growth. These authors viewed this catabolism as 'an inevitable consequence of the operation of mechanisms controlling the degradation of amino acids in the body' and an 'unavoidable tax which the animal must pay for the ability to respond quickly to quantitative and qualitative changes in protein supply'. Millward and Rivers (1988) and Millward (1989), have introduced the concept of an 'anabolic drive' to explain the process of inevitable catabolism of amino acids which is seen as beneficial to the organism. They suggest that there are advantages to the animal in consuming amounts of essential amino acids in excess of the amount required to match identifiable needs, since before their oxidation these amino acids exert a transitory regulatory influence (the anabolic drive) on growth and maintenance.

Work on 'inevitable catabolism' to date has emphasized the determination of gross estimates of the efficiency of utilization of amino acids. It may be more informative, however, to use more precise methods (e.g. radioactive tracers) to define the process. It is also necessary that studies be conducted to unbiasedly define the shape of the response of inevitable amino acid catabolism to amino acid intake.

Preferential catabolism of amino acids for energy supply

The term 'preferential' rather than 'inevitable' catabolism is used to distinguish the catabolism of amino acids for the express purpose of energy supply (ATP generation). Preferential catabolism will occur in metabolic states whereby the supply of ATP from non-protein compounds is low in relation to the animal's needs. In modelling, it becomes necessary to quantitate this process; Whittemore (1983) laid the foundation for this by proposing a biological basis for protein–energy interaction, in terms of a physiological requirement for a certain minimal amount of body lipid.

If, following the 'classic' approach to energy partitioning, residual energy (residual ME = ME − maintenance energy − total energy cost of depositing protein) is insufficient to meet the minimum lipid deposition requirement, then deamination of amino acids is triggered and amino acids are degraded to supply energy. With well-balanced diets it should be possible to avoid 'preferential' catabolism of amino acids. However, this potential source of loss of absorbed dietary amino acids may become of greater quantitative significance as the potential lean growth rate of pigs is increased by genetic means and producers adopt diets formulated to minimize the deposition of body fat during growth.

The minimum lipid constraint is specified in terms of a minimum ratio between daily lipid and protein deposition rates and is assumed to be affected by sex and genotype of the pig (Whittemore, 1983) and is probably also influenced by age and liveweight. Expression of the minimum lipid rule in terms of whole-body lipid content rather than daily lipid gain (Whittemore and Gibson, 1983; Moughan and Smith, 1984; Moughan *et al.*, 1987) appears more realistic and provides for a description of body lipid catabolism.

Black *et al.* (1986) predict the amount of body protein which may be deposited daily by applying empirical relationships between protein gain and metabolizable energy intake. Different relationships are found for pigs of different liveweight and there will be different slopes of the functions for pigs of varying genotype. If the essential lipid hypothesis is accepted, the slopes (energy-dependent phase of growth) of the functions can be explained in terms of the minimum Ld:Pd (lipid deposition to protein deposition) ratio. The essential lipid approach in growth modelling has the advantage of being mechanistic. The theory has not been fully and adequately tested empirically, however, and more work is required in this area. The factors that may influence minimum Ld:Pd need to be examined. Recently, some important new information on the effects of dietary energy intake, liveweight and nutritional history on energy partitioning in the pig has been published (de Greef, 1992). In particular, the hypothesis of a constant minimal Ld:Pd ratio at energy intakes below those supporting maximal protein deposition needs to be critically re-examined.

Amino acid imbalance

Once the tissue uptake of amino acids for maintenance purposes and the inevitable catabolism of amino acids have been modelled, it is possible to predict the array of absorbed amino acids available for new protein synthesis. The latter array of amino acids can be compared with the pattern of amino acids required for new protein synthesis in the 'typical' body cell and the fraction of imbalanced amino acids (destined for catabolism) determined. The required pattern of amino acids is assumed to be the amino acid composition of whole-body protein.

This direct approach allows mechanistic description of amino acid utilization for new protein synthesis. In the past modellers have often relied upon empirical estimates of ideal amino acid balance for calculating imbalanced amino acids, but this approach is subject to error. As pointed out by Black and Davies (1991) there is no one ideal amino acid pattern for pig growth, but rather the ideal pattern varies with a number of animal and dietary factors.

It does not appear that breed, sex or age greatly influence the whole-body amino acid composition of the growing pig.

Catabolism of amino acids supplied above the amount required to meet the maximum rate of body protein retention

The cell has a finite capacity for protein synthesis and is not able to store amino acids as such, for later use. If, after a meal, the uptake of balanced amino acids exceeds the animal's capacity for protein synthesis, the surplus amino acids will be deaminated and the carbon skeletons eventually degraded.

An intrinsic upper limit to whole body protein retention (Pd_{max}) in growing pigs, which is influenced by genotype, sex and age has been demonstrated empirically in numerous studies. It has been more difficult, however, to demonstrate the existence of Pd_{max} in pigs below 20 kg liveweight fed cereal-based diets, because appetite rather than Pd_{max} often becomes the limiting factor. Hodge (1974), however, fed 10 kg (liveweight) pigs a protein-adequate liquid milk diet and found no change in nitrogen retention when energy intake was raised from four to five times maintenance. The specification of Pd_{max} is an integral part of any biological pig growth model.

Whittemore (1983) has reviewed studies in which protein deposition (Pd) could be considered close to Pd_{max} and gives values for Pd_{max} ranging from 90 to 175 g d^{-1}. Differences in Pd_{max} between sexes and genetic strains of pigs were also noted. The study of Campbell (1985) indicates that for very well-bred boars Pd_{max} may exceed 187 g d^{-1} and estimates as high as 200 g d^{-1} have been reported.

In addition to having mean estimates for Pd_{max} for pigs growing over specified liveweight ranges it is important in growth modelling to be able to describe the relationship between Pd_{max} and liveweight. Several workers have derived relationships between P_d and liveweight and it is generally accepted that P_d increases rapidly in early life, reaches a plateau during the grower/finisher stages and then decreases towards zero at maturity. There is little argument that after the point of decline in P_d, P_d will be regulated by Pd_{max} and measurement of P_d should equate with Pd_{max}. The relationship between P_d and liveweight before the point of decline, however, does not necessarily describe the relation between Pd_{max} and liveweight. At a given liveweight it is often difficult to establish that Pd_{max} has actually been reached. This has led workers to adopt a more pragmatic approach to estimating Pd_{max}. Measurements of P_d at different liveweights for pigs fed high-quality diets ad lib and kept under optimum environmental conditions are accepted as values of Pd_{max} (Tullis, 1981; Black *et al.*, 1986; Siebrits *et al.*, 1986). The conclusions drawn by these workers vary, due in part, no doubt, to differences between the breeds and strains studied but also to the method of curve-fitting and the experimental conditions adopted.

The functions derived by Tullis (1981) and Black *et al.* (1986) demonstrate that Pd_{max} declines towards zero at maturity from around 100 to 110 kg liveweight, whereas the curves of Siebrits *et al.* (1986) show an earlier decline in Pd_{max} from around 60 kg liveweight for obese gilts, 75 kg liveweight for lean boars and gilts, and 100 kg liveweight for obese boars. Taking the results from these three studies together, it appears that determined Pd_{max} is broadly constant between 45 and 90 kg liveweight. The data from these experiments and particularly observations relating to the early growth stages (birth to 100 kg liveweight) must be interpreted with caution, however, because measured P_d ('determined' Pd_{max}) may be lower than actual Pd_{max}. By using a different diet or by altering some environmental condition, it is possible that P_d at a given liveweight may be increased. Although time-consuming and expensive a better approach to determining the relationship between Pd_{max} and liveweight would be to attempt to establish the attainment of Pd_{max} at selected liveweights using the procedure outlined by Campbell *et al.* (1983).

Some animal growth modellers have described maximal protein growth as a trajectory towards a mature protein mass (Gompertz equation). Once again such an approach has the advantage of explaining the growth process mechanistically. An interesting recent development in the description of protein growth and its control is the work of Pomar *et al.* (1991), who have taken a further step in mechanistic description by modelling potential protein accretion on the basis of body DNA content, which is a function of DNA content in the body at maturity.

Simulation of Amino Acid Flow in the 50 kg Liveweight Growing Pig – Application of a Biological Model

An important application of a model of amino acid metabolism is to provide a deeper quantitative understanding of the inherent metabolic and physiological processes. A model can be used to quantify the significance of different aspects of the growth process.

Having broadly described the processes in protein metabolism, a mechanistic model, describing amino acid flow in the 50 kg liveweight pig and embodying the concepts discussed in this chapter, is now applied to give an overall appreciation of amino acid flow. The deterministic model was based around that described by Moughan (1989) except that in the presently described model daily body protein deposition was predicted rather than given as a model input and daily food intake was a model input, with dietary energy partitioning being simulated. The model allows for specification of total and chemically available dietary amino acid intakes and predicts absorbed amino acids based on true ileal digestibility coefficients. It describes cutaneous amino acid loss as a function of metabolic body weight and endogenous gut amino acid losses as a function of dry matter intake. The model includes a weighting factor for endogenous amino acid flow to allow expression of the effect of elevated amounts of anti-nutritional factors or dietary fibre. The fractional rate of whole body protein synthesis is given as a function of the mean daily protein deposition rate (P_d) over the 3 days of growth preceding the day of simulation (assumed in this exercise to equal P_d on the day of simulation). The loss of protein nitrogen in the urine at maintenance is a set proportion of whole body protein synthesis and amino acids are assumed to be catabolized in proportion to their occurrence in body protein. Some amino acids (e.g. lysine) are assumed to be retained in the cell following protein breakdown and their catabolisms are discounted. The rates of inevitable catabolism are described as curvilinear functions of the amounts of amino acids absorbed in relation to the potential amino acid depositions (based on the genetic upper limit to protein retention, Pd_{max}). The model predicts the amount of each amino acid available for growth (after maintenance and inevitable catabolism costs have been met) and the pattern of amino acids available for growth is compared with body protein amino acid composition to identify the first-limiting amino acid and to determine the imbalanced amino acids. In the model, if balanced protein available for growth is greater than Pd_{max}, then excess amino acids are catabolized.

Net energy yields from amino acid catabolism are predicted, and with the non-protein digested energy, give metabolizable energy (ME). The daily ME is partitioned, ultimately to daily protein and lipid, facing a

Table 7.4. Ingredient composition of a commercial barley-based diet[1] formulated for growing pigs.

Ingredient	Composition (g 100 g^{-1} air-dry weight)
Barley	73.25
Peas	16.50
Meat-and-bone meal	5.00
Fish meal	5.00
Vitamins, minerals	0.25

[1] Crude protein = 17.8 g 100 g^{-1}; apparent digestible energy = 13.26 MJ kg^{-1}; total lysine = 0.92 g 100 g^{-1}.

minimal lipid : protein deposition ratio to model the process of preferential catabolism.

The simulated flow of lysine in the 50 kg liveweight pig, given a commercial grower diet (Table 7.4), is given in Table 7.5. The simulation data allow consideration of the modelled effects of level of food intake and Pd_{max}. In the model, feed intake does not influence the urinary loss of nitrogen at maintenance but Pd_{max} has a small effect. Inevitable catabolic nitrogen loss in the urine increases with increasing food intake, generally declines at a given food intake with increasing Pd_{max}, but remains constant at high food intakes. The loss of urine nitrogen from catabolism due to excess amino acid supply is quantitatively significant at high food intakes, with the opposite being true for preferential catabolic loss. Cutaneous amino acid loss is unaffected by feeding level or Pd_{max} and gut loss is influenced by feed intake but not Pd_{max}. The loss of lysine due to imbalance was zero or minimal for the examples shown in Table 7.5 as lysine was generally the first-limiting dietary amino acid.

The data in Table 7.5 demonstrate that, particularly at higher food intakes, the process of inevitable catabolism may have an important effect on the utilization of the first-limiting amino acid. Absorption and endogenous gut loss are also of importance, with body protein turnover being of lesser significance and cutaneous loss of only minor significance. Preferential catabolism may contribute relatively significantly to amino acid loss in situations where metabolizable energy limits protein deposition. Similarly, excess amino acid supply can make a major contribution to inefficiency. The relative importance of dietary amino acid imbalance to protein utilization by the pig is shown more clearly in Table 7.6. For well-formulated commercial diets the loss of amino acids due to imbalance is of relatively minor significance.

Table 7.5. Predicted (simulation model) utilization of dietary lysine by the 50 kg liveweight growing pig, at three feeding levels [1] and three maximal rates of body protein deposition (Pd_{max}).

Feeding level (g d^{-1})	Pd_{max}	Diet intake (g d^{-1})		Losses (g d^{-1})									Total lysine	Deposition (g d^{-1})	
		Total lysine	Available lysine	Unabsorbed available lysine	Urine						Cutaneous	Gut endogenous		Protein	Lipid
					Protein turnover	Inevitable catabolism	Imbalance	Excess supply	Preferential catabolism	Total					
1505	100	13.8	13.1	1.8	0.7	3.2	0	0	1.0	4.9	0.08	1.6	4.78	72	72
	130	13.8	13.1	1.8	0.8	2.5	0.2	0	1.4	4.9	0.08	1.6	4.83	73	73
	160	13.8	13.1	1.8	0.8	2.1	0	0	1.8	4.7	0.08	1.6	4.91	74	74
2069	100	18.9	17.9	2.3	0.7	4.7	0	1.6	0	7.0	0.08	1.9	6.63	100	162
	130	18.9	17.9	2.3	0.8	4.7	0	0	0	5.5	0.08	1.9	8.20	124	145
	160	18.9	17.9	2.3	0.8	3.9	0.1	0	0	4.8	0.08	1.9	8.84	133	138
2633	100	24.1	22.9	3.1	0.7	6.0	0	4.4	0	11.1	0.08	2.1	6.63	100	277
	130	24.1	22.9	3.1	0.8	6.0	0	2.3	0	9.1	0.08	2.1	8.62	130	255
	160	24.1	22.9	3.1	0.8	6.0	0	0.3	0	7.1	0.08	2.1	10.61	160	233

[1] Correspond to 8, 11 and 14% metabolic liveweight, kg$^{0.75}$.

Table 7.6. Predicted (simulation model) utilization of dietary crude protein by the 50 kg liveweight growing pig, at three feeding levels and three maximal rates of body protein deposition (Pd_{max}).

	Feeding level (g d^{-1})*								
	1505			2069			2633		
$Pd_{max} =$	100	130	160	100	130	160	100	130	160
Available protein intake (g d^{-1})	268	268	268	368	368	368	468	468	468
Absorbed available protein intake (g d^{-1})	229	229	229	315	315	315	401	401	401
Body protein synthesis (g d^{-1})	516	566	616	516	566	616	516	566	616
Protein losses (g d^{-1})									
Cutaneous loss	1.8	1.8	1.8	1.8	1.8	1.8	1.8	1.8	1.8
Urine loss at maintenance	21	23	25	21	23	25	21	23	25
Gut loss at maintenance	34	34	34	47	47	47	59	59	59
Inevitable catabolism	69	66	56	95	95	93	121	121	121
Urine loss (imbalance)	17	11	12	26	25	16	31	30	30
Urine loss (excess supply)	0	0	0	25	0	0	66	35	4
Urine loss (preferential catabolism)	15	21	28	0	0	0	0	0	0
Protein deposition (g d^{-1})	72	73	74	100	124	133	100	130	160

* Levels correspond to 8, 11 and 14% of metabolic liveweight, kg$^{0.75}$.

The influence that the different processes have on overall protein metabolism will, of course, vary with the type of diet (Porter *et al.*, 1994), age and weight of the animal. Nevertheless, application of the above simplified model gives a general view of amino acid dynamics and allows a ranking of the importance of the respective processes.

Conclusion

The present discussion has focused on the modelling of amino acid metabolism in the pig at a relatively high level of aggregation whereby the control of protein metabolism is viewed at a tissue level and the principles of growth are established with regard to the integrative functioning of the whole animal. At least when the objective for model construction is to allow prediction of protein deposition and tissue growth during production, greatest progress in developing models is likely to be made by refining description of the principles governing growth of the whole animal as an integrated system. Models at the tissue level describe growth relatively grossly, but have the distinct advantage of including well-defined and readily testable theories of growth control.

Pig growth models that are clearly conceived and that have been logically constructed from sound biological principles and that consistently give reasonable predictive accuracy, are now available. Such computerized models have numerous applications in commercial practice. Perhaps the most important application has been their use in designing economically optimal feeding regimes for growing pigs. The static empirical approach to estimating amino acid requirements has been replaced by a dynamic approach based on the principles of production economics.

References

Batterham, E.S., Andersen, L.M., Baigent, D.R. and White, E. (1990) Utilization of ileal digestible amino acids by growing pigs: Effect of dietary lysine concentration on efficiency of lysine retention. *British Journal of Nutrition* 64, 81–94.

Black, J.L. and Davies, G.T. (1991) Ideal protein – its variable composition. In: Batterham, E.S. (ed.) *Manipulating Pig Production III.* Australasian Pig Science Association, Attwood, Victoria, p. 111 (abstract).

Black, J.L., Campbell, R.G, Williams, I.H., James, K.J. and Davies, G.T. (1986) Simulation of energy and amino acid utilization in the pig. *Research and Development in Agriculture* 3, 121–145.

Campbell, R.G. (1985) Effects of sex and genotype on energy and protein metabolism in the pig. In: Cumming, R.B. (ed.) *Recent Advances in Animal Nutrition in Australia.* University of New England, Armidale.

Campbell, R.G., Taverner, M.R. and Curic, D.M. (1983) The influence of feeding level from 20 to 45 kg liveweight on the performance and body composition of female and entire-male pigs. *Animal Production* 36, 193–199.

Cole, D.J.A., Dean, G.W. and Luscombe, J.R. (1970) Single cereal diets for bacon pigs. 2. The effects of methods of storage and preparation of barley on performance and carcass quality. *Animal Production* 12, 1–6.

Crampton, E.W. and Bell, J.M. (1946) The effect of fineness of grinding on the utilisation of oats by market hogs. *Journal of Animal Science* 5, 200–210.

Darcy-Vrillon, B., Souffrant, W.B., Laplace, J.P., Rerat, A., Gebhardt, G., Vaugelade, P. and Jung, J. (1985) Digestion of proteins in the pig: Ileal and faecal digestibilities and absorption coefficients of amino acids. In: Just, A., Jorgensen, H. and Fernandez, J.A. (eds) *Digestive Physiology in the Pig*. National Institute of Animal Science, Copenhagen, pp. 326–328.

Darragh, A.J., Moughan, P.J. and Smith, W.C. (1990) The effect of amino acid and peptide alimentation on the determination of endogenous amino acid flow at the terminal ileum of the rat. *Journal of the Science of Food and Agriculture* 51, 47–56.

Darragh, A.J., Cranwell, P.D. and Moughan, P.J. (1994) Absorption of lysine and methionine from the proximal colon of the piglet. *British Journal of Nutrition* 71, 739–752.

de Greef, K.H. (1992) Prediction of production – Nutrition induced tissue partitioning in growing pigs. PhD Thesis, Wageningen Agricultural University.

de Lange, C.F.M., Souffrant, W.B. and Sauer, W.C. (1990) Real ileal protein and amino acid digestibilities in feedstuffs for growing pigs as determined with the ^{15}N-isotope dilution technique. *Journal of Animal Science* 68, 409–418.

Desrosiers, T., Savoire, L., Bergeron, G. and Parent, G. (1989) *Journal of Agricultural and Food Chemistry* 37, 1385–1391.

Finley, J.W. (1985) Reducing variability in amino acid analysis. In: Finley, J.W. and Hopkins, D.T. (eds) *Digestibility and amino acid availability in cereals and oilseeds*. American Association of Cereal Chemists, St Paul, pp. 15–30.

Garnett, N.K. (1985) The influence of amino acid hydrolysis times on the determination of amino acid requirements for poultry. In: Ideris, A. and Hamid, R. (eds) *Proceedings of the Regional Seminar on Future Developments in the Poultry Industry*. World Poultry Science Association, Kuala Lumpur, pp. 14–17.

Heger, J. and Frydrych, Z. (1985) Efficiency of utilization of essential amino acids in growing rats at different levels of intake. *British Journal of Nutrition* 54, 499–508.

Heger, J. and Frydrych, Z. (1989) Efficiency of utilization of amino acids. In: Friedman, M. (ed.) *Absorption and Utilization of Amino Acids, Vol. 1*. CRC Press, Boca Raton, Florida, pp. 31–56.

Hodge, R.W. (1974) Efficiency of food conversion and body composition of the pre-ruminant lamb and the young pig. *British Journal of Nutrition* 32, 113–126.

Horvath, D.J., Seeley, H.W., Warner, R.G. and Loosli, J.K. (1958) Microflora of intestinal contents and faeces of pigs fed different diets including pigs showing parakeratosis. *Journal of Animal Science* 17, 714–722.

Hurrell, R.F. and Carpenter, K.J. (1981) The estimation of available lysine in foodstuffs after Maillard reactions. *Progress in Food and Nutritional Science* 5, 159–176.

Lawrence, T.L.J. (1967) High level cereal diets for the growing/finishing pig. 1. The effect of cereal preparation and water level on the performance of pigs fed diets containing high levels of wheat. *Journal of Agricultural Science* 68, 269–274.

Leibholz, J. and Mollah, Y. (1988) Digestibility of threonine from protein concentrates for growing pigs. 1. The flow of endogenous amino acids at the terminal ileum of growing pigs. *Australian Journal of Agricultural Research* 39, 713–719.

Lenis, (1983) Faecal amino acid digestibility in feedstuffs for pigs. In: Arnal, M., Pion, R. and Bonin, D. (eds) *Metabolisme et Nutrition Azotes*. INRA, Paris, pp. 385–389.

Low, A.G. (1980) Nutrient absorption in pigs. *Journal of the Science of Food and Agriculture* 31, 1087–1130.

Low, A.G., Partridge, I.G., Keal, H.D. and Jones, A.R. (1982) A comparison of methods *in vitro* and *in vivo* of measuring amino acid digestibility in foodstuffs as predictors of pig growth and carcass composition. *Animal Production* 34, 403.

McNeil, N.I. (1988) Nutritional implications of human and mammalian large intestinal function. *World Review of Nutrition and Dietetics* 56, 1–42.

Millward, D.J. (1989) The endocrine response to dietary protein: the anabolic drive on growth. In: Barth, C.A. and Schlimme, E. (eds) *Milk Proteins*. Steinkopff, Darmstadt, pp. 49–61.

Millward, D.J. and Rivers, J.P.W. (1988) The nutritional role of indispensable amino acids and the metabolic basis for their requirements. *European Journal of Clinical Nutrition* 42, 367–393.

Moughan, P.J. (1989) Simulation of the daily partitioning of lysine in the 50 kg liveweight pig – A factorial approach to estimating amino acid requirements for growth and maintenance. *Research and Development in Agriculture* 6, 7–14.

Moughan, P.J. (1991) Towards an improved utilization of dietary amino acids by the growing pig. In: Haresign, W. and Cole, D.J.A. (eds) *Recent Advances in Animal Nutrition, 1991*. Butterworths, Heinemann, Oxford, pp. 45–64.

Moughan, P.J. and Rutherfurd, S.M. (1990) Endogenous flow of total lysine and other amino acids at the distal ileum of the protein- or peptide-fed rat: The chemical labelling of gelatin protein by transformation of lysine to homoarginine. *Journal of the Science of Food and Agriculture* 52, 179–192.

Moughan, P.J. and Smith, W.C. (1984) Prediction of dietary protein quality based on a model of the digestion and metabolism of nitrogen in the growing pig. *New Zealand Journal of Agricultural Research* 27, 501–507.

Moughan, P.J. and Smith, W.C. (1985) Determination and assessment of apparent ileal amino acid digestibility coefficients for the growing pig. *New Zealand Journal of Agricultural Research* 28, 365–370.

Moughan, P.J. and Verstegen, M.W.A. (1988) The modelling of growth in the pig. *Netherlands Journal of Agricultural Science*, 36, 145–166.

Moughan, P.J., Smith, W.C. and Pearson, G. (1987) Description and validation of a model simulating growth in the pig (20–90 kg liveweight). *New Zealand Journal of Agricultural Research* 30, 481–489.

Moughan, P.J., Darragh, A.J., Smith, W.C. and Butts, C.A. (1990) Perchloric and trichloroacetic acids as precipitants of protein in endogenous ileal digesta from the rat. *Journal of the Science of Food and Agriculture* 52, 13–21.

Moughan, P.J., Smith, W.C., Pearson, G. and James, K.A.C. (1991) Assessment of apparent ileal lysine digestibility for use in diet formulation for the growing pig. *Animal Feed Science and Technology* 34, 95–109.

Pomar, C., Harris, L. and Minvielle, F. (1991) Computer simulation model of swine production systems. 1. Modeling the growth of young pigs. *Journal of Animal Science* 69, 1468–1488.

Porter, A.A., Cortamira, N.O., Seve, B., Salter, D.N. and Morgan, L.M. (1994) The effects of energy source and tryptophan on the rate of protein synthesis and on hormones of the entero-insular axis in the piglet. *British Journal of Nutrition* 71, 661–674.

Reeds, P.J. (1988) Nitrogen metabolism and protein requirements. In: Blaxter, K. and Macdonald, I. (eds) *Comparative Nutrition*. John Libbey, London, pp. 55–72.

Rérat, A. (1981) Digestion and absorption of nutrients in the pig. *World Review of Nutrition and Dietetics* 37, 239–287.

Rowan, A.M., Moughan, P.J. and Wilson, M.N. (1992) The effect of hydrolysis time on determination of the amino acid composition of diet, ileal digesta and faeces samples and on the determination of dietary amino acid digestibility coefficients. *Journal of Agricultural and Food Chemistry* 40, 981–985.

Sauer, W.C. and Ozimek, L. (1986) Digestibility of amino acids in swine: Results and their practical application – A review. *Livestock Production Science* 15, 367–388.

Schmitz, M. (1988) Möglichkeiten und grenzen der Homoargininmarkierungsmethode zur messung der proteinverdaulichkeit beim Schwein. PhD Thesis, Christian-Albrechts-Universität, Kiel.

Siebrits, F.K., Kemm, E.H., Ras, M.N. and Barnes, P.M. (1986) Protein deposition in pigs as influenced by sex, type and livemass. 1. The pattern and composition of protein deposition. *South African Journal of Animal Science* 16, 23–27.

Tanksley, T.D. and Knabe, D.A. (1980) Availability of amino acids for swine. In: *Proceedings 1980 Georgia Nutrition Conference for the Feed Industry*. University of Georgia, Atlanta, pp. 157–168.

Tullis, J.B. (1981) Protein growth in pigs. PhD Thesis, University of Edinburgh.

van Weerden, E.J., Huisman, J., van Leeuwen, P. and Slump, P. (1985) The sensitivity of the ileal digestibility method as compared to the faecal digestibility method. In: Just, A., Jorgensen, H. and Fernandez, J.A. (eds) *Digestive Physiology in the Pig*. National Institute of Animal Science, Copenhagen, pp. 392–395.

Whittemore, C.T. (1983) Development of recommended energy and protein allowances for growing pigs. *Agricultural Systems* 11, 159–186.

Whittemore, C.T. and Fawcett, R.H. (1974) Model responses of the growing pig to the dietary intake of energy and protein. *Animal Production* 19, 221–231.

Whittemore, C.T. and Fawcett, R.H. (1976) Theoretical aspects of a flexible model to simulate protein and lipid growth in pigs. *Animal Production* 22, 87–96.

Whittemore, C.T. and Gibson, A. (1983) A growth model for pigs designed to include recent concepts of nutrient requirement. *Animal Production* 36, 516.

Williams Smith, H. (1965) Observations on the flora of alimentary tracts of animals and factors affecting its composition. *Journal of Pathology and Bacteriology* 89, 95–122.

Amino Acid Requirements for Maintenance, Body Protein Accretion and Reproduction in Pigs

8

M.F. FULLER

The Rowett Research Institute, Greenburn Road,
Bucksburn, Aberdeen, AB2 9SB, UK.

Introduction

Essential and non-essential amino acids

Essential amino acids are so called because the animal in question does not possess the pathways for their synthesis and they must therefore be supplied in the diet. In contrast, the non-essential amino acids can be synthesized by the animal from simpler precursors, for example by addition of an amino group to a TCA cycle intermediate. As the word 'non-essential' may suggest physiological rather than dietary inessentiality the term 'dispensable' is often considered preferable. The amino acids that are essential for the pig are the same nine that are essential for humans, the rat and indeed most mammals (Table 8.1).

As in other species, tyrosine and cysteine are semi-essential (or conditionally dispensable) as pigs have the enzymes for the hydroxylation of phenylalanine to form tyrosine and for the more complex trans-sulfuration pathway by which cysteine is formed from methionine and serine. In most species, though not in cats, cysteine can be further metabolized to taurine. Though not a constituent of protein, taurine is therefore also a conditionally dispensable amino acid for pigs.

Other amino acids, normally considered dispensable, may be required for young animals to achieve their maximum growth rate. Arginine, for example, is synthesized at a rate inadequate for the needs of growing pigs, allowing only some 60% of maximum growth rate to be achieved (Mertz *et al.*, 1952); for maximum performance, therefore, at least 40% of the arginine requirement must be supplied in the diet. More recent studies have led to the suggestion that the rate of proline synthesis in neonatal

Table 8.1. Classification of essential (or indispensable), semi-essential (or conditionally dispensable) and non-essential (or dispensable) amino acids for pigs.

Category	Amino acid
Essential	Threonine
	Methionine
	Valine
	Leucine ⎫
	Isoleucine ⎬ Branched-chain amino acids
	⎭
	Lysine
	Phenylalanine
	Tryptophan
	Histidine
Semi-essential	Cyst(e)ine
	Taurine
	Tyrosine
	Arginine
Non-essential	Glutamic acid, glutamine
	Glycine, serine, proline
	Aspartic acid, asparagine
	Alanine

pigs may also be too slow to meet their needs (Ball *et al.*, 1986), although Chung and Baker (1993) have recently shown that, by the time pigs reach 5 kg, they no longer require a dietary source of proline.

Before reviewing the quantitative estimates of amino acid requirements that have been made, and attempting some form of conspectus, the underlying metabolism of amino acids is first reviewed, which draws attention to some of the problems of measurement and particularly of interpretation which account for disparities amongst published estimates of requirements.

The Physiological Basis of Amino Acid Needs

Amino acids are required for a number of different functions in the body (Table 8.2). Some of these functions, although physiologically essential, require only minuscule quantities of amino acids and contribute negligibly to total requirements; others, which may be physiologically no more vital, constitute major components of requirements. The animal's dietary amino acid requirement, at any particular point in its life, is the sum of these components. Those needs which persist throughout life, regardless of

Table 8.2. Major components of amino acid requirements of pigs.

Replacement of obligatory losses via:
  Urinary excretion
  Irreversible modification
  Irreducible oxidation
  Losses from epithelia

Synthesis of non-protein substances

Tissue protein accretion (growth and gestation)

Milk protein secretion

whether the animal is growing or reproducing or not, are collectively considered as the maintenance requirement. It is important to realize that 'maintenance' is not a physiological reality but a conceptual convenience. The processes which collectively give rise to a maintenance requirement or a requirement for a productive process are parts of an integrated metabolism in which there is continuous tissue protein synthesis and breakdown, and continuous exchange of amino acids, keto acids and other metabolites between tissues and organs, and it is the net outcome of these exchanges which determines the amino acid requirements. In terms of quantitative needs for amino acids, protein synthesis is foremost amongst these processes and it is the difference between protein synthesis and protein breakdown which determines the rate of body protein accretion. 'Maintenance' simply describes the state in which these rates are equal. In reality, all aspects of metabolism are changing continuously in response to feeding, exercise and other stimuli. Only over relatively long periods, such as 24 hours, do the average rates of such processes approach constancy.

Maintenance requirements

Maintenance can be defined in various ways. In relation to amino acid requirements it is normally defined as nitrogen equilibrium, the state in which nitrogen intake exactly equals the sum of nitrogen losses so that the nitrogen content of the body remains constant. In mature individuals this may also mean that body weight is constant; however, immature animals kept at constant weight by undernutrition retain their impetus to grow and body protein accretion continues whilst body fat is lost (Close and Mount, 1978; Fuller *et al.*, 1976). Since few pigs are kept until they are truly mature it is likely that, in most pigs, nitrogen equilibrium is accompanied by a loss of weight and of body fat.

To maintain nitrogen equilibrium amino acids must be supplied at a rate equal to the rate at which they are lost through metabolism, secretion or excretion from the body. These obligatory losses arise through many processes, of which the following are quantitatively the most important.

Urinary excretion

Except in certain disorders of amino acid metabolism urinary excretion of free amino acids accounts for only a very small component of essential amino acid requirements in humans (Leverton et al., 1959), and the same seems to be true of the pig (Southern and Baker, 1984) though comparable data with protein-free diets are not available. The exceptions are amongst the modified amino acids described below. Small quantities of amino acids are also excreted in proteins in the urine but again account for only a minor proportion of obligatory losses.

Irreversible modification

Most amino acids released in protein turnover are available for reutilization but some which are modified in metabolism are not, and are excreted in larger quantities. Some histidine residues in contractile proteins are methylated, and significant quantities of histidine are excreted as 1-methyl histidine and 3-methyl histidine, and in the form of the β-alanyl dipeptides of these amino acids, anserine and balenine, though much greater quantities of these are retained in the tissues. In the post-translational processing of collagen certain lysine residues are modified by methylation and hydroxylation. Losses of amino acids through these irreversible modifications, although much greater than those affecting unmodified amino acids, are, nevertheless, a small component of total obligatory losses.

Synthesis of non-protein substances

Several of the essential amino acids are indispensable in the synthesis of physiologically important substances such as hormones, neurotransmitters and other metabolically important products (Table 8.3). The amino acids which have such additional roles are lysine, methionine (and cysteine), phenylalanine (mainly via tyrosine), histidine and tryptophan.

Minimum rates of amino acid oxidation

In addition to the losses of amino acids through excretion, there is a continuous loss of amino acids through oxidation in the tissues. Even though, under conditions of dietary deprivation, amino acids are well conserved, the activities of the amino acid degrading enzymes are not completely suppressed. By giving animals diets specifically devoid of an individual essential amino acid but otherwise nutritionally adequate the minimum

Table 8.3. Non-protein products of essential amino acids.

Process	Amino acid	Product
Formation of conditionally dispensable amino acids	Methionine	Cysteine - - - > taurine
	Phenylalanine	Tyrosine
Replacement of irreversibly modified amino acids	Histidine	1-methylhistidine, 3-methylhistidine
	Lysine	Hydroxylysine, methyllysine
Synthesis of non-protein products	Phenylalanine	Tyrosine - - - > melanin, thyroxine, dopamine, catecholamines
	Lysine	Carnitine
	Tryptophan	Serotonin, melatonin
	Histidine	Histamine, carnosine, anserine, balenine
	Methionine	Cysteine - - - > glutathione, polyamines

rate of oxidation can be assessed. Such estimates have been made for several amino acids (Table 8.4) but not all.

Losses from epithelia

Losses of amino acids from the epithelia occur both as free amino acids and proteins. Because the pig is a non-sweating species, losses through perspiration are expected to be small. In particular, skin and hair are shed and replaced continuously. Estimates of the nitrogen lost in skin and hair by pigs averaging 55 kg gave values in the range $0.25–0.5 \, g \, d^{-1}$ (Fuller and Boyne, 1971), depending on feeding and environmental temperature. By extrapolation, Fuller (1991) estimated that if the food were restricted to such an extent as to stop growth, these losses would be reduced to around $0.1 \, g \, N \, d^{-1}$ or $5 \, mg \, N \, kg^{-0.75} \, d^{-1}$. This is only of the order of 2% of the total N requirement for maintenance. However, bearing in mind the exceptionally high cystine content of keratins, the loss of cystine could be around $3 \, mg \, kg^{-0.75} \, d^{-1}$. However, this estimate involves extensive extrapolation; there do not appear to be any more direct estimates.

In contrast to the external epithelium, secretion into the digestive tract accounts for a large flow of protein, some secreted into the lumen and some lost from the mucosa as enterocytes are extruded from the villi. Much of this is subject to proteolysis and a large proportion of the amino acids is reabsorbed. However, some of the nitrogenous secretions into the digestive tract, the glycoconjugates in particular, are, by their nature, highly resistant to mammalian digestive enzymes. The amino acid composition of gut mucins suggests that these may account for substantial losses of threonine, serine and proline.

Table 8.4. Estimates of minimum rates of oxidation of essential amino acids by growing pigs, given diets devoid of that amino acid.

Amino acid	Rates of oxidation (mg kg$^{-0.75}$ d^{-1})
Leucine[1]	28
Phenylalanine[1]	3
Histidine[1]	3
Threonine[2]	14

[1] Beckett *et al.* (1987) and unpublished.
[2] Ballèvre *et al.* (1991).

Estimates of minimal losses by this route have been made by measurement of the flow of amino acids into the large intestine when animals are given protein-free diets (Table 8.5). This is based on the observation that amino acids entering the large intestine, whether as protein or in free form, contribute little if at all to the animal's needs (Zebrowska, 1973, 1975; Just *et al.*, 1981; Wünsche *et al.*, 1984; Schmitz *et al.*, 1991). There is a remarkable degree of concordance amongst most of these values. Although there are higher estimates in the literature, for example when fibre is added to the diet, these represent minimum estimates. To assess how important these losses are in relation to total maintenance needs, the mean values from Table 8.5 are compared in Fig. 8.1 with estimates of the maintenance needs of comparable animals, taken from Fuller *et al.* (1989) (Table 8.6).

This comparison suggests that, for many amino acids, the rate of loss in the gastrointestinal tract constitutes a major component of the animal's maintenance need, leaving little to be accounted for by other routes. Major exceptions are threonine and methionine + cystine, which are prominent in the pattern of maintenance needs.

The relative importance of oxidation and gastrointestinal losses as components of maintenance needs can be assessed by comparing the values in Table 8.4 and Fig. 8.1. Thus, in Fig. 8.1, the total maintenance requirement for threonine, 53 mg kg$^{-0.75}$ d^{-1}, seems to consist of a component of 38 mg kg$^{-0.75}$ d^{-1} lost in the intestine, and a component of tissue oxidation, 14 mg kg$^{-0.75}$ d^{-1}. Although it is interesting that these two components sum to a value almost identical to the estimated maintenance requirement this is probably fortuitous as there are undoubtedly substantial errors in both estimates.

These losses may be increased when a normal diet is given but there is little information on the effect of amino acid intake on gastrointestinal losses. Such measurements are difficult to make because of the difficulty of distinguishing between amino acids of the diet and those of exogenous

Table 8.5. Estimates of amino acid flow (mg kg$^{-0.75}$ d^{-1}) into the large intestine of pigs given protein-free diets.

Amino acid	Source reference[*]				
	1	2	3	4	5
Threonine	37	45	38	39	32
Valine	27	35	21	31	25
Isoleucine	20	23	15	22	18
Leucine	34	38	28	39	30
Tyrosine	23	23	16	73	44
Phenylalanine	34	41	16	24	24
Histidine	12	12	18	15	12
Lysine	31	31	19	37	23
Arginine	41	23	18	37	30
Cystine	nd	nd	15	nd	nd
Methionine	9	12	8	8	6
Tryptophan	6	nd	nd	nd	nd

[*]Data of: 1, de Lange *et al.* (1989a); 2, de Lange *et al.* (1989b); 3, Fuller *et al.* (1989); 4, Furuya and Kaji (1992), 49-kg pigs; 5, Furuya and Kaji (1992), 92-kg pigs.
nd = not determined.

origin. Although approaches using isotopic labels have been used there is likely to be some recycling of label into the gut. To overcome this difficulty, de Lange *et al.* (1989b) measured amino acid flow at the terminal ileum in pigs given a protein-free diet with and without a parenteral infusion of an amino acid mixture and showed that ileal amino acid flow was actually reduced when amino acids were infused: however, the only substantial decreases were in proline and glycine.

Requirements for production

Whilst maintenance needs are a component of amino acid requirements at any stage of life, the remainder, and usually the largest part, depends upon the productive state of the animal. In growing animals maintenance accounts for only a small component of requirements, approximately one-tenth, depending on the rate of growth. Although maintenance needs increase as animals approach maturity, breeding animals, as mentioned above, are rarely fully mature. Thus, in pregnant and especially lactating

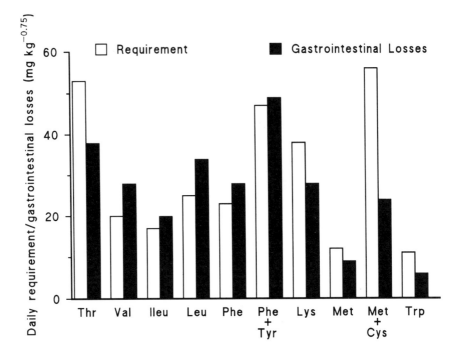

Fig. 8.1. Components of maintenance amino acid requirements of young pigs ($mg\ kg^{-0.75}\ d^{-1}$). Data on gastrointestinal amino acid losses from Table 8.5; estimates of maintenance requirements from Table 8.6.

sows, maintenance requirements are overshadowed by the needs for tissue protein accretion and for milk production.

Tissue protein accretion

In growing animals total amino acid needs are dominated by the requirements for body protein accretion, which increases from birth to a maximum at the peak of growth and then declines; this maximum and the subsequent decline vary greatly amongst pigs of different genetic stock (Fig. 8.2; Friesen *et al.*, 1994). Likewise in gestation the amino acids required for the synthesis of the proteins of the fetus and other products of conception constitute a major component of total requirements. These increase rapidly during pregnancy, typically reaching a maximum of $60–90\ g\ d^{-1}$ (Noblet *et al.*, 1985). In addition to the amino acids deposited in the products of conception, there are also increases in maternal tissue protein in pregnancy. Close *et al.* (1985) found the highest rate of nitrogen retention by pregnant sows

Table 8.6. Amino acid requirements (mg kg$^{-0.75}$ d^{-1}) for maintenance in pigs.

Amino acid	Source reference[*]	
	1	2
Threonine	53	39
Valine	20	21
Isoleucine	16	30
Leucine	23	20
Phenylalanine	18	21
Phenylalanine + tyrosine	37	nd
Lysine	36	25
Methionine	9	nd
Methionine + cystine	26	26
Tryptophan	11	5

[*] 1, for pigs of 50 kg (Fuller *et al.*, 1989); 2, for pigs of 150 kg (Baker *et al.*, 1966a,b,c; Baker and Allee, 1970).
nd = not determined.

to be in mid-gestation. This maximum rate, 0.28 g kg$^{-0.75}$ d^{-1}, is of a similar magnitude to their daily nitrogen requirement for maintenance, emphasizing that, in relation to maintenance, the demands of pregnancy are very modest compared with the needs for rapid growth. Although similar in total amount, the patterns of amino acids required for maintenance and for tissue accretion in pregnancy are very different.

Milk protein secretion

Because protein is a major constituent of milk the rate of milk protein secretion during lactation may be of the order of 400 g d^{-1}, a rate of protein production two or three times higher than that which pigs achieve as tissue protein accretion during growth.

Specification of amino acid supply

It is self-evident that animals can only use for productive purposes that fraction of their dietary amino acid supply that is absorbed *net* from the gastrointestinal tract. Also, they can only utilize that fraction of any amino acid that is chemically available, i.e. not chemically modified in ways that the animal lacks enzymes to reverse. These aspects are discussed more fully

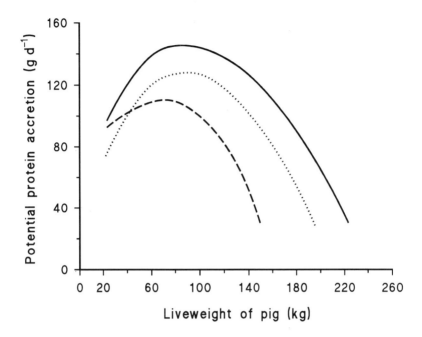

Fig. 8.2. Potential rates of protein accretion in pigs from three studies. . . . Thorbek (1975); _ _ _ _ Carr *et al.* (1977); _____ Whittemore *et al.* (1988). (Adapted from Whittemore *et al.* 1988.)

in Chapters 6, 7 and 9; for the purpose of the present discussion the important consideration is that animals' amino acid needs should be expressed in units which do not depend on the specific feedstuffs used to supply them. Equally, the value of a feedstuff as a source of amino acids to meet the animal's needs should be specified in terms of the *digestible and chemically available* amino acids that it supplies.

Digestibility and availability of amino acids

The terms 'digestibility' and 'availability' are sometimes confused and sometimes used interchangeably. As originally defined they referred to quite different attributes. Digestibility refers to that proportion of amino acid intake which is absorbed – or, strictly, that proportion which is not recovered in digesta. Availability describes the chemical integrity of the amino acid, with particular reference to those amino acids which are commonly subject to heat damage: lysine, methionine, cystine and tryptophan.

The digestibility of an amino acid cannot be accurately assessed from its faecal excretion because both synthetic and degradative metabolism by the hind-gut microflora mask any differences in real amino acid digestibility such that the amino acid composition of faecal protein is of more or less constant composition, regardless of the diet. For this reason, more precise estimates of amino acid digestibility are obtained by measurement of the amino acid flow into the large intestine, i.e at the terminal ileum, beyond which, as already discussed, little absorption of intact amino acids occurs.

Chemical availability is normally assessed by dye-binding methods (see review by Carpenter and Booth, 1973) and further discussion is beyond the scope of this chapter. For the present, however, it is sufficient to note that to specify animals' actual metabolic needs, rather than the dietary allowances which must be made to meet those needs, both the dietary amino acid supply and the animal's requirements must be described in terms of *digestible available amino acids*.

Supply of essential amino acids via intestinal microbial biosynthesis

It has conventionally been assumed that essential amino acids can only be supplied *de novo* in the diet. Although it is recognized that there is microbial activity in those sections of the gastrointestinal tract in which absorption of amino acids takes place (Jensen, 1988), the contribution of the gastrointestinal microflora to the amino acid needs of pigs has been considered negligible. This assumption has been challenged by the finding that, in animals given a source of ^{15}N, amino acids in the body become labelled (Deguchi *et al.*, 1978, 1980). It is of course to be expected that those amino acids which transaminate freely would become labelled with ^{15}N but the finding that lysine, which does not become labelled by transamination (Clark and Rittenberg, 1951), also appears in the body labelled with ^{15}N has renewed interest in the possibility that this source of essential amino acids may be more significant than previously thought. By measuring the ^{15}N- and ^{14}C-labelling of gut microbial protein as well as the total labelled lysine in the body, Torrallardona *et al.* (1993) estimated that the lysine supplied by this route may approximate to the maintenance need of the pig, a not insignificant quantity.

Quantitative Estimation of Amino Acid Requirements

The last essential amino acid to be identified was threonine (McCoy *et al.*, 1935; Meyer and Rose, 1936), opening the way for experiments to establish which of the amino acids are essential for the pig (Mertz *et al.*, 1952). These in turn were followed by numerous studies to assess the quantitative needs of pigs in practice.

Methods of estimating amino acid requirements

There have been four main approaches to estimating amino acid requirements.

Requirements from analysis of satisfactory diets

The earliest attempts to estimate amino acid requirements were made by analysing the amino acid composition of diets which were found to give satisfactory performance (Becker *et al.*, 1954). As diets with moderate excesses of amino acids may produce performance comparable to those without any excess this approach could not be expected to produce the most precise or consistent estimates. However, in the absence of more direct evidence it has acted as a useful guide (see, e.g., Rérat and Lougnon, 1968).

Requirements from tissue analysis

The second method of estimating requirements derived from the observation that proteins of high nutritional value for growing animals tended to have an amino acid composition comparable to that of the animals' bodies (Block and Bolling, 1944). This led to the concept that the amino acid requirements of a growing animal are determined primarily by the needs for body protein accretion, a notion that sounds logical, for, as Hankins and Titus (1939) wrote, 'from the standpoint of animal nutrition the growth of an animal is peculiarly characterized by the increase in the protein content of its body.' This concept was further elaborated by Mitchell (1950), who compared the amino acid requirements of the rat, then recently estimated by Rose *et al.* (1948), with its amino acid composition. Williams *et al.* (1954) made similar comparisons for the rat, chick and pig, finding generally good agreement between whole-body amino acid composition and estimates of requirements in the contemporary literature. Mitchell (1950) went on to suggest that the amino acid requirements of any animal could be assessed through a simple dose–response experiment with graded intakes of any one amino acid (he suggested lysine), the requirements for the others being derived by simple proportion, using the pattern of amino acids in tissue protein. The 'ideal protein' concept, and the derivation of 'ideal' amino acid patterns, owe a great deal to these observations (see Chapter 5).

In more recent years quantitative estimates of dietary amino acid requirements have generally been made using one of two approaches. These are referred to as the dose–response and the factorial methods. There is a fundamental difference between the two. In the factorial method the output is regarded as known and the calculation is of the input required to sustain it; in dose–response experiments the output response to a known input, or series of inputs, is assessed, usually with the aim of establishing empirically what input gives the maximum rate of production (or highest efficiency, etc.).

Requirements by empirical dose–response experiments

In dose–response experiments the requirement for a particular amino acid is deduced from the results of experiments in which animals' responses are measured to graded intakes of that amino acid when it is given in the presence of a supposedly adequate intake of every other amino acid (and of all other nutrients). The requirement is estimated as the amino acid intake which satisfies a given criterion of adequacy, such as the maintenance of nitrogen equilibrium, the attainment of maximum growth rate, minimum feed:gain ratio or other criterion deemed appropriate. An example of the results of such an experiment is given in Fig. 8.3.

Although results obtained by this approach are relatively easy to interpret and apply there are some drawbacks. First, a separate experiment must be made for each amino acid and, for the results to be comparable, must be made with similar animals. Second, an experiment on one amino acid requires that other amino acids are adequately supplied but not present in such excess as might affect the response to the limiting amino acid. To fulfil these conditions requires a preliminary estimate of the requirement for every amino acid. If these are not well made, and the conditions of the experiment are not fulfilled, the resulting estimate may be erroneous.

The third difficulty concerns the applicability of such estimates to other populations of pigs. There is a great deal of variation in performance amongst pigs, depending on their weight, sex, genotype and environment. As the results of a dose–response experiment are obtained with a particular population of pigs they can be considered appropriate only to other pigs which are similar. It may therefore be necessary to carry out further experiments of the same kind with other populations as, for example, selective breeding changes the productive potential of the animals. It is therefore questionable whether recommendations such as those made in 1981 by the Agricultural Research Council (ARC, 1981) on the basis of reviewing literature from the 1960s and '70s are relevant to the pigs being used more than a decade later.

The 'ideal protein' concept helps to overcome this difficulty. It is based on the principle that the major difference in amino acid requirements between these diverse pigs is not in the *pattern* of amino acids they require but in the *total amount of balanced protein*, the assumption being that the amino acid composition of tissue protein accreted is essentially the same, regardless of the rate at which that protein is laid down. This question is discussed later.

MODELLING RESPONSES TO AMINO ACIDS

The estimation of amino acid requirements from dose–response experiments depends crucially on their interpretation, in particular what form of response is assumed. This will be clear by considering the data shown in Fig. 8.3, from a typical simple experiment (Mitchell *et al.*, 1968). These authors interpreted their data, as have most, by fitting a broken line by the method of least squares. This is shown in Fig. 8.3b. This assumes that a pig's rate of body protein accretion, measured by nitrogen retention, increases linearly with amino acid intake, up to a certain intake – the requirement – with higher intakes producing no further response. Two assumptions of this model need to be examined. The first is that the response at suboptimal intakes is linear, i.e. that within this range of intakes, the efficiency of amino acid utilization is constant. The second is that intakes greater than the minimum requirement elicit no change in response, neither increasing nor decreasing the rate of body protein accretion.

There is an important distinction to be made between the response of an individual animal and the response of a population of animals. This theme is developed further in Chapter 11 in relation to the laying hen. In fitting a broken line response to experimental data each of the parameters which describes the broken line has a variance; statistically, when a number of individual broken line responses are pooled the resulting mean response of the population is curvilinear. However, the response of an individual may not be rectilinear, as Fuller and Garthwaite (1993) showed by giving individual pigs an ordered sequence of diets varying in protein content. Alternative models used to fit such data include quadratic and exponential functions and more complex but probably more suitable four-parameter

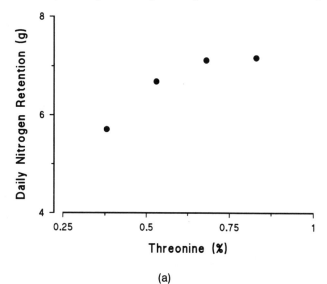

(a)

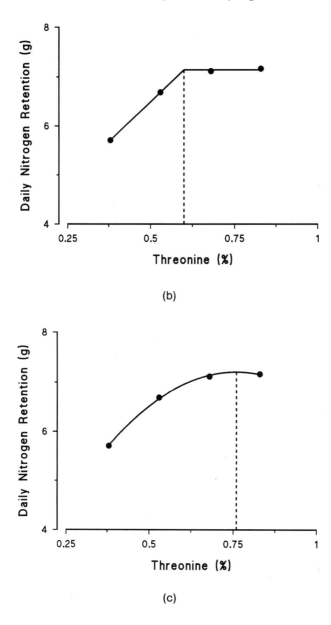

Fig. 8.3. Models for the interpretation of dose-response experiments. (a) Data from the experiments of Mitchell *et al.* (1968); (b) the same data fitted with a broken-line model; (c) the same data fitted by a simple curvilinear (quadratic) model.

equations (e.g. Mercer, 1982; Gahl *et al.*, 1991). The choice of model has a substantial effect on the estimate of requirement (Fig. 8.3b and c).

Requirements by the summation of components – the factorial approach

The factorial method treats a dietary amino acid requirement as the sum of its physiological components, some of which are regarded as constant for animals of a given weight, others as varying with the animal's productive state. The strength of the approach is that it allows requirements to be estimated for animals differing in their productive state.

For these reasons, the factorial approach, and the development of mathematical models of nutrient utilization which build on the same principles, have gained acceptance as means of estimating requirements for all the amino acids simultaneously, which do not require prior estimates of requirements and which can be applied to other populations with different performance potential. However, although there is a great deal of information on rates of protein accretion in the body and in the products of conception and on rates of milk protein secretion, one of the limitations in the application of the factorial approach is the paucity of information concerning the efficiency with which amino acids are utilized in meeting the various components of their requirements.

EFFICIENCY OF UTILIZATION
Body protein accretion, whether during growth or during gestation, is not completely efficient; in other words, not the whole quantity of dietary amino acid is retained. This is due, in part, to the obligatory losses already described. Even after these requirements have been met, however, further increments of amino acid supply are utilized with something less than 100% efficiency. For an animal to retain one gram of body protein the diet must not only supply the amino acids contained in that protein but must also cover any additional amino acid losses that are occasioned. These may arise through increased losses in the digestive tract, increased oxidation or by other routes; collectively these increased losses result in reduced efficiency of amino acid utilization. The exact value for the efficiency term is crucial to the application of the factorial method for the estimation of requirements. It can be measured either as the increment of amino acid retained (or secreted, in the case of milk protein), or from the incremental losses, expressed, in each case, per unit increment in amino acid supply.

In a series of experiments with diets each specifically limiting in one amino acid, Fuller *et al.* (1989) estimated the amino acid requirements of growing pigs for body protein accretion. These estimates are given in Table 8.8. By relating these amino acid requirements to estimated tissue amino acid contents (taken from Table 8.7) the efficiency with which absorbed

Table 8.7. Amino acid concentration (g 16 g^{-1} N) in the whole bodies of growing pigs.

Amino acid	Body weight (kg)				
	4.9[1]	7.4[2]	20[3]	45[4]	100[5]
Aspartic acid		8.1	8.22	8.23	
Threonine	3.66	3.7	3.90	3.79	3.5
Serine		4.0	4.37	4.44	
Glutamic acid		13.6	13.03	13.20	
Glycine		9.75	8.47	8.63	
Alanine		6.55	6.37	6.65	
Valine	4.50	4.7	4.83	4.18	4.9
Cystine	0.91	1.25	1.14	0.92	
Methionine	1.62	1.65	1.92	1.95	
Methionine + cystine	2.53	2.9	3.06	2.87	3.0
Isoleucine	3.23	3.5	3.50	3.20	3.9
Leucine	6.77	6.8	7.22	6.80	7.1
Tyrosine	2.89	2.75	2.67	2.66	
Phenylalanine	4.26	3.7	3.92	3.62	
Phenylalanine + tyrosine	7.15	6.45	6.59	6.28	5.6
Histidine	2.50	2.55	2.91	3.30	2.8
Ammonia			1.33	1.27	
Lysine	6.77	6.75	6.58	6.51	6.9
Arginine	6.95	6.85	6.27	6.13	
Proline		6.65	6.05	6.05[6]	
Hydroxyproline		3.2	2.69	2.69[6]	
Tryptophan			0.78	0.78[6]	1.2

[1] Zhang *et al.* (1986).
[2] Mean values from Aumaitre and Duée (1974).
[3] Mean of values from Zhang *et al.* (1986), Campbell *et al.* (1988) and Batterham *et al.* (1990).
[4] Mean values of diets 4-8, Batterham *et al.* (1990).
[5] Buraczewski (1972).
[6] Campbell *et al.* (1988).

amino acids are utilized for body protein accretion can be calculated. These estimates are also given in Table 8.8. They suggest that most essential amino acids are used with a similar efficiency, approximately 0.8; however, the value for lysine is conspicuously higher, 0.96, and that for tryptophan lower, 0.64. Unfortunately, there is little published information with pigs with which to compare these estimates. However, Batterham *et al.* (1990) reported that the efficiency of utilization of absorbed lysine by growing pigs was 0.86, a lower estimate than that suggested in Table 8.8 for lysine, but similar to that found for other amino acids. Chung and Baker (1992) reported that methionine was utilized by young pigs with an efficiency of

Table 8.8. Estimates of the amino acid requirements for tissue protein accretion in growing pigs, the amino acid composition of whole-body protein (see Table 8.7) and the efficiency of utilization, calculated from these values.

Amino acid	Requirement $(g\,g^{-1}$ protein accretion)	Whole-body amino acid composition $(g\,16\,g^{-1}\,N)$	Marginal efficiency
Threonine	4.69	3.79	0.81
Valine	5.25	4.18	0.80
Methionine + cystine	3.58	2.87	0.80
Isoleucine	4.30	3.20	0.74
Leucine	7.80	6.80	0.87
Phenylalanine + tyrosine	8.42	6.28	0.75
Lysine	6.81	6.51	0.96
Tryptophan	1.21	0.78	0.64

0.72. More complete information on this issue is available from experiments with rats. Heger and Frydrych (1985) studied the efficiency of utilization of individual essential amino acids by growing rats. They suggest maximum efficiencies of between 0.65 and 0.85 except for the sulfur amino acids, for which the value was approximately 0.55. Gahl *et al.* (1991) also reported that the efficiency of utilization by rats was similar amongst the essential amino acids. It is thus not clear whether the differences amongst amino acids in Table 8.8 are real or not. Both Heger and Frydrych (1985) and Gahl *et al.* (1991) describe efficiency, not as constant at suboptimal intakes but as diminishing continuously as optimum intakes are approached. This is, of course, simply another way of describing the diminishing response of body protein accretion to amino acid intake described in Fig. 8.3 and the accompanying text.

Comparable information on the efficiency of amino acid utilization in gestation and lactation is lacking but data on nitrogen utilization indicate a similar kind of diminishing response. The paucity of quantitative data on the efficiency of amino acid utilization and its control is one of the main obstacles to the application of the factorial approach to estimating amino acid requirements.

Amino acid requirements for maintenance, tissue protein accretion and milk protein secretion

As pointed out earlier, a pig's amino acid requirements can be considered to be made up of several components, each varying in magnitude according to the animal's physiological state at the time. In this section, quantitative estimates of each component are collated.

Maintenance

To determine the amount of each essential amino acid required to maintain nitrogen equilibrium requires a separate experiment for each amino acid in which graded levels are given over a range estimated to span the requirement. From a series of such experiments with female pigs of approximately 150 kg (Baker *et al.*, 1966a,b,c; Baker and Allee, 1970) were estimated the maintenance requirements for threonine, valine, isoleucine, phenylalanine, lysine, methionine + cystine and tryptophan. Because of the well-known tendency for nitrogen retention to be overestimated by balance methods, Baker *et al.* (1966a,b,c) suggested that maintenance needs would best be estimated as the intake required to support a positive nitrogen retention of $1 g d^{-1}$. Their estimates are given in Table 8.6. Using a similar approach, but with bladder catheters to minimize the loss of urinary nitrogen, Fuller *et al.* (1989) estimated the maintenance requirements of growing pigs (approx. 50 kg). Their estimates are also given in Table 8.6.

In making such estimates it is important to be certain that the amino acid is limiting at all intakes. This is easy to achieve with synthetic diets but more difficult when diets include protein: because the patterns of amino acids required for maintenance and for body protein accretion are so different it is possible that, as protein intake is reduced below the maintenance level the identity of the limiting amino acid may change. A protein deficient in lysine for growth may have sufficient lysine for maintenance for which it may be deficient in another amino acid, for example threonine.

Tissue protein accretion

In growing and gestating pigs the amino acids needed for the synthesis of new tissue protein constitute the largest single component of total requirements and, as discussed above, the relative proportions of amino acids required are closely related to the composition of whole-body protein. Estimates of the amino acid composition of whole-body protein of normally growing pigs are in broad agreement (Table 8.7). This apparent constancy may, however, be disturbed when growth is modified by alterations of diet. The protein mass of the body is made up of a large number of different proteins, each with its unique amino acid sequence dictated by the gene which encodes it. Different tissues contain different proteins and their amino acid composition varies greatly. During normal growth the relative proportions of different tissues follow characteristic changes which have their correlate in amino acid composition, which also changes systematically with age. In addition, specific deficiencies of individual amino acids produce small but significant changes in whole-body amino acid composition. In dietary histidine deficiency, for example, histidine is released not only by breakdown of carnosine (Easter and Baker, 1977) but also by reduction in the concentration of haemoglobin, a protein particularly rich in histidine. Recently, it has been shown that, in growing pigs fed a lysine-deficient diet, there are significant changes in whole-body amino acid composition, with reductions in the concentrations of several amino acids, including lysine, accompanied by increases in the concentrations of others (Batterham *et al.*, 1990). Similar findings have been reported by Gahl *et al.* (1992). It appears therefore that although in principle the amino acid requirements for body protein accretion are set by the amino acid composition of the tissues this pattern is not completely constant but may be somewhat modified by diet. It is not clear at present whether these changes are related specifically to individual amino acid deficiencies or are simply a consequence of slow growth, however produced.

Milk protein secretion

The amounts of protein secreted in milk by lactating sows depend on several factors, including the weight of the animal, the amount and composition of its diet and, importantly, the stage of lactation. The amino acid composition of milk, does not, however, vary greatly (see Table 8.11).

Current Recommendations

Categories of pigs

As pigs grow, mature and reproduce their daily amino acid needs change. First, the maintenance needs increase as the animal increases in size; second, needs for body protein accretion increase during early growth, reach a maximum and then decline as the pig approaches its adult size. In addition to these quantitative changes the pattern of dietary amino acid needs also alters. This reflects in part the changing relative requirements for maintenance and for body protein accretion; there may also be some alteration in the composition of the body proteins formed at different stages of growth. Thus the 'ideal protein' is expected to vary at different stages of the pig's life. These changes are continuous, suggesting that daily amino acid needs also vary continuously. To accommodate these changes in practice, diets have been designed for pigs within prescribed weight ranges or physiological states, or both. Thus, breeding sows may not be fully mature, necessitating separate recommendations according to their weight, in addition to allowances for gestation or lactation. The choice of suitable weight ranges has derived partly from considerations related to commercial practice and partly from the availability of appropriate data.

In most published recommendations pigs have been subdivided solely in terms of their body weight. Although it has long been recognized that pigs of different sex and genotype differ substantially in their capacity to grow, these differences have generally been ignored: the aim has generally been to provide guidelines for 'average' pigs. To some extent this decision was forced on reviewers from the paucity of data; with the passage of time, more experimental work has been published which allows sexes and genotypes to be distinguished. The passage of time, however, has had a negative effect in rendering obsolete some of the earlier work with animals which were much slower growing and fatter than the pigs of today. With the genetic improvements of the last decades and the development of greater control over all aspects of pig production the need for animals to be fed more precisely according to their requirements has focused attention on the need for more specific estimates of nutrient requirements for particular populations. When viewed on a worldwide scale the extreme diversity of animals and the need for separate estimates of nutrient requirements become even more evident. These considerations suggest that many separate experiments would have to be made with a range of different animals.

This difficulty provided part of the motivation for the development of the 'ideal protein' concept, by which amino acid allowances may be modified collectively to correspond to the needs of different populations.

Collation of recommendations

The recommendations made by various national authorities and other reviewers are not, of course, independent of each other; the same original work is included in many different reviews. Differences amongst them stem partly from the range of sources used (which changes with time), partly in the method of assessment, and partly in the criteria of adequacy.

To present here all the estimates that have been made for all the amino acids would be unhelpful. Rather, bearing in mind the unifying concept of an 'ideal protein', various current estimates can be compared in two aspects: (i) the requirement for a single 'specimen' amino acid; and (ii) the pattern of amino acid requirements. For the single amino acid I, like others, have chosen to use lysine, firstly because of its importance in practical pig nutrition, secondly because there are more data on lysine than on any other amino acid, and thirdly because it is quantitatively one of the larger components of amino acid requirements and provides a fairly stable baseline.

Growing pigs

Current estimates of lysine requirements from several national authorities are collated in Fig. 8.4. They accord in the general decline in lysine requirements, expressed here in relation to dietary energy, as pigs increase in weight. At any weight, however, there is a 1.5 to 2-fold variation amongst estimates from different sources. In view of the real variation that exists amongst different populations of pigs, embracing differences in breed, strain, sex and a host of environmental factors, this is probably not unreasonable.

Estimates of the 'ideal' amino acid balance for growing pigs, as suggested by the same authorities, are given in Table 8.9. The French and Australian recommendations acknowledge the derivation of their estimates from the Agricultural Research Council (ARC, 1981). Also given are the results from two series of recent experiments specifically designed to reassess these ratios (Wang and Fuller, 1989; Fuller *et al.*, 1989). They suggest

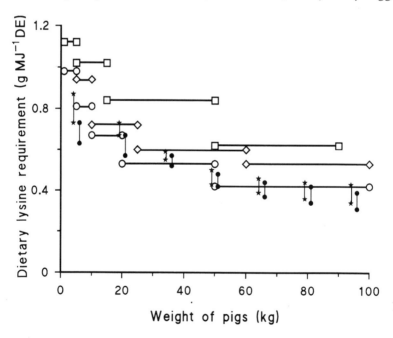

Fig. 8.4. Current estimates of the lysine requirements of growing pigs. Estimates are for Australian entire males (★) and females (●) (CSIRO, 1987); within each category the range of values is for slow- and fast-growing genotypes; NRC (1988) (○), ARC (1981) (□) and INRA (1984) (◇). Values given by the ARC for pigs of birth to 3 weeks and 3 to 8 weeks are presented here in terms of body weight.

Table 8.9. Balance of amino acids (relative to lysine = 100) recommended by various sources.

Amino acid	ARC[1]	NRC[2]	CSIRO[3]	INRA[4]	Wang and Fuller (1989)	Fuller et al. (1989)	Baker and Chung (1992)		
							5-20 kg	20-50 kg	50-100 kg
Threonine	60	64	60	60	72	75	65	67	70
Valine	70	64	70	70	75	75	68	68	68
Methionine + cystine	50	55	50	60	63	59	60	65	70
Isoleucine	55	61	54	60	60	61	60	60	60
Leucine	100	80	100	72	110	110	100	100	100
Phenylalanine + tyrosine	96	88	96	100	120	122	95	95	95
Lysine	100	100	100	100	100	100	100	100	100
Histidine	33	29	33	26	nd	nd	32	32	32
Tryptophan	15	16	14	18	18	19	18	19	20

[1] ARC (1981); [2] NRC (1988); [3] CSIRO (1987); [4] INRA (1984).
nd = not determined.

some slightly higher ratios than those originally proposed by the ARC (1981), who estimated that the lysine concentration in an 'ideal protein' should be $7.0 \, g \, 16 \, g^{-1} \, N$. Subsequent experiments have suggested lower values and this one change automatically increases the ratio of every other amino acid to lysine. However, as pointed out in Chapter 5, the values proposed by the ARC (1981) have largely been substantiated by subsequent work. Recently, Baker and Chung (1992) have suggested estimates for pigs in three weight classes: these are also given in Table 8.9.

Reproduction

Compared to the experimental data for growing pigs there is relatively little direct information on the amino acid requirements of pregnant or lactating sows and even less on boars. Most values have been derived from estimates of protein requirements, by applying the ideal protein concept, using data derived from either factorial calculations or empirical experiments. In Table 8.10 various estimates of the daily lysine requirements of pregnant and lactating sows are given, while Table 8.11 gives the suggested balance of amino acids for each of these classes of animal. Combining the information in Tables 8.10 and 8.11 allows the requirements of sows for each amino acid to be calculated. Since the daily requirements vary with the weight of the sow and with her milk yield, the average values for lysine requirements given in Table 8.10 may be varied to take account of these factors.

Table 8.10. Estimates of the daily lysine requirement ($g \, d^{-1}$) of pregnant and lactating sows.

Source	Pregnant	Lactating
ARC[1]	8-10	33
NRC[2]	8.2	31.8
CSIRO[3]	7.0	25
INRA[4]	10.0	27-33

[1] ARC (1981);
[2] NRC (1988);
[3] CSIRO (1987);
[4] INRA (1984).

Table 8.11. Patterns of amino acids proposed for pregnant and lactating sows (relative to lysine = 100).

	Pregnant			Lactating				
	NRC[1]	CSIRO[2]	INRA[3]	NRC[1]	ARC[4]	CSIRO[2]	INRA[3]	Sow's milk[5]
Threonine	70	60	85	72	70	63	76	61
Valine	74	70	108	100	70	63	70	63
Methionine + cystine	53	50	68	60	55	44	55	40
Methionine	ns	ns	ns	ns	ns	ns	ns	23
Isoleucine	70	55	85	65	70	53	70	51
Leucine	70	100	75	80	115	116	115	115
Phenylalanine + tyrosine	105	96	78	117	115	124	115	120
Phenylalanine	ns	ns	ns	ns	ns	ns	ns	56
Lysine	100	100	100	100	100	100	100	100
Histidine	35	ns	30	42	39	ns	38	45
Arginine	0	ns	0	67	67	ns	67	67
Tryptophan	21	15	18	20	19	19	20	17

[1] NRC (1988); [2] CSIRO (1987); [3] INRA (1984); [4] ARC (1981); [5] from Elliott *et al.* (1971).
ns = not specified.

References

ARC (Agricultural Research Council) (1981) *The Nutrient Requirements of Pigs.* Commonwealth Agricultural Bureaux, Farnham Royal, Slough.

Aumaitre, A. and Duée, P.H. (1974) Amino acid composition of piglet body proteins between birth and 8 weeks of age. *Annales de Zootechnie* 23, 231–236.

Baker, D.H. and Allee, G.L. (1970) Effect of dietary carbohydrate on assessment of the leucine need for maintenance of adult swine. *Journal of Nutrition* 100, 277–280.

Baker, D.H. and Chung, T.K. (1992) Ideal protein for swine and poultry. *Biokyowa Technical Review* Number 4, 16 pages.

Baker, D.H., Becker, D.E., Norton, H.W., Jensen, A.H. and Harmon, B.G. (1966a) Some qualitative amino acid needs of adult swine for maintenance. *Journal of Nutrition* 88, 382–390.

Baker, D.H., Becker, D.E., Norton, H.W., Jensen, A.H. and Harmon, B.G. (1966b) Quantitative evaluation of the threonine, isoleucine, valine and phenylalanine needs of adult swine for maintenance. *Journal of Nutrition* 88, 391–396.

Baker, D.H., Becker, D.E., Norton, H.W., Jensen, A.H. and Harmon, B.G. (1966c) Quantitative evaluation of the tryptophan, methionine and lysine needs of adult swine for maintenance. *Journal of Nutrition* 89, 441–447.

Ball, R.O., Atkinson, J.L. and Bayley, H.S. (1986) Proline as an essential amino acid for the young pig. *British Journal of Nutrition* 55, 659–668.

Ballèvre, O., Sève, B., Arnal, M., Garlick, P.J. and Fuller, M.F. (1991) Nutritional regulation of threonine metabolism in growing pigs. In: Eggum, B.O., Boison, S., Børsting, C., Danfaer, A. and Hvelplund, T. (eds) *Protein Metabolism and Nutrition.* National Institute of Animal Science, European Association for Animal Production, Publication no.59, 2, Tjele, Denmark, pp. 145–147.

Batterham, E.S., Andersen, L.M., Baigent, D.R. and White, E. (1990) Utilization of ileal digestible amino acids by growing pigs: effect of dietary lysine concentration on efficiency of lysine retention. *British Journal of Nutrition* 64, 81–94.

Becker, D.E., Ullrey, D.E. and Terrill, S.W. (1954) Protein and amino acid intakes for optimum growth rate in the young pig. *Journal of Animal Science* 13, 346–356.

Beckett, P.R., Fuller, M.F., Cadenhead, A., Rockway, J.M. and McGaw, B.A. (1987) Whole body flux and degradation of amino acids measured with ^{3}H- and ^{14}C-labels in pigs given diets deficient in histidine, phenylalanine or leucine. In: Lehmann, J. (ed.) *Protein Metabolism and Nutrition.* European Association of Animal Production (EAAP) Publication no.35. Rostock, Germany, pp. 26–27.

Block, R.J. and Bolling, D. (1944) Nutritional opportunities with amino acids. *Journal of the American Dietetic Association* 20, 69–76.

Buraczewski, S. (1972) *United Nations Economic and Social Symposium: New Developments in the Provision of Amino Acids in the Diets of Pigs and Poultry.* EC/FAO, Geneva, 23 pp.

Campbell, R.G., Taverner, M.R. and Rayner, C.J. (1988) The tissue and dietary protein and amino acid requirements of pigs from 8.0 to 20.0 kg live weight.

Animal Production 46, 283–290.

Carpenter, K.J. and Booth, V.H. (1973) Damage to lysine in food processing: its measurement and its significance. *Nutrition Abstracts and Reviews* 43, 423–451.

Carr, J.R., Boorman, K.N. and Cole, D.J.A. (1977) Nitrogen retention in the pig. *British Journal of Nutrition* 37, 143–155.

Chung, T.K. and Baker, D.H. (1992) Efficiency of dietary methionine utilization by young pigs. *Journal of Nutrition* 122, 1862–1869.

Chung, T.K. and Baker, D.H. (1993) A note on the dispensability of proline for weanling pigs. *Animal Production* 56, 407–408.

Clark, I. and Rittenberg, D. (1951) The metabolic activity of the α-hydrogen atom of lysine. *Journal of Biological Chemistry* 189, 521–528.

Close, W.H. and Mount, L.E. (1978) The effects of plane of nutrition and environmental temperature on the energy metabolism of the growing pig. 1. Heat loss and critical temperature. *British Journal of Nutrition* 40, 413–421.

Close, W.H., Noblet, J. and Heavens, R.P. (1985) Studies on the energy metabolism of pregnant sows. 2. The partition and utilization of metabolizable energy intake in pregnant and non-pregnant animals. *British Journal of Nutrition* 53, 267–279.

CSIRO (1987) *Feeding Standards for Australian Livestock. Pigs.* Commonwealth Scientific and Industrial Research Organization, East Melbourne, Australia.

Deguchi, E., Niiyama, M., Kagota, K. and Namioka, S. (1978) Incorporation of ^{15}N administered to germfree and SPF piglets as ^{15}N-urea into amino acids of hydrolyzed liver and muscle proteins. *Japanese Journal of Veterinary Research* 26, 68–73.

Deguchi, E., Niiyama, M., Kagota, K. and Namioka, S. (1980) Incorporation of nitrogen-15 from dietary [^{15}N]-diammonium citrate into amino acids of liver and muscle proteins in germfree and specific-pathogen-free neonatal pigs. *American Journal of Veterinary Research* 41, 212–214.

de Lange, C.F.M., Sauer, W.C., Mosenthin, R. and Souffrant, W.B. (1989a) The effect of feeding different protein-free diets on the recovery and amino acid composition of endogenous protein collected from the distal ileum and faeces in pigs. *Journal of Animal Science* 67, 746–754.

de Lange, C.F.M., Sauer, W.C., Mosenthin, R. and Souffrant, W.B. (1989b) The effect of protein status of the pig on the recovery and amino acid composition of endogenous protein in digesta collected from the distal ileum. *Journal of Animal Science* 67, 755–762.

Easter, R.A. and Baker, D.H. (1977) Nitrogen metabolism, tissue carnosine concentration and blood chemistry of gravid swine fed graded levels of histidine. *Journal of Nutrition* 107, 120–125.

Elliott, R.F., Vander Noot, G.W., Gilbreath, R.L. and Fisher, H. (1971) Effect of dietary protein level on composition changes in sow colostrum and milk. *Journal of Animal Science* 32, 1128–1137.

Friesen, K.G., Nelssen, J.L., Unroh, J.A., Goodband, R.D. and Tokach, M.D. (1994) Effects of the interrelationship between genotype, sex and dietary lysine on growth performance and carcass composition in finishing pigs fed to either 104 or 127 kilograms. *Journal of Animal Science* 72, 946–954.

Fuller, M.F. (1991) Present knowledge of amino acid requirements for maintenance and production: non-ruminants. In: Eggum, B.O., Boisen, S.,

Børsting, C., Danfaer, A. and Hvelplund, T. (eds) *Protein Metabolism and Nutrition*. National Institute of Animal Science, European Association for Animal Production Publication no.59, 1. Tjele, Denmark, pp. 116–126.

Fuller, M.F. and Boyne, A.W. (1971) The effects of environmental temperature on the growth and metabolism of pigs given different amounts of food. 1. Nitrogen metabolism, growth and body composition. *British Journal of Nutrition* 25, 259–272.

Fuller, M.F. and Garthwaite, P. (1993) The form of response of body protein accretion to dietary amino acid supply. *Journal of Nutrition* 123, 957–963.

Fuller, M.F., Webster, A.J.F., MacPherson, R.M. and Smith, J.S. (1976) Comparative aspects of the energy metabolism of Piétrain and Large White × Landrace pigs during growth. In: Vermorel, M. (ed.) *Energy Metabolism of Farm Animals*. European Association for Animal Production Publication no.19. G. de Bussac, Clermont-Ferrand, France, pp. 177–180.

Fuller, M.F., McWilliam, R., Wang, T.C. and Giles, L.R. (1989) The optimum dietary amino acid pattern for growing pigs. 2. Requirements for maintenance and for tissue protein accretion. *British Journal of Nutrition* 62, 255–267.

Furuya, S. and Kaji, Y. (1992) The effects of feed intake and purified cellulose on the endogenous ileal amino acid flow in growing pigs. *British Journal of Nutrition* 68, 463–472.

Gahl, M.K., Finke, M.D., Crenshaw, T.D. and Benevenga, N.J. (1991) Use of a four-parameter logistic equation to evaluate the response of growing rats to ten levels of each indispensable amino acid. *Journal of Nutrition* 121, 1720–1729.

Gahl, M.J., Crenshaw, T.D. and Benevenga, N.J. (1992) Amino acid composition in growing pigs fed graded levels of lysine. In: *Proceedings of the American Society of Animal Science*, Des Moines, Iowa, p. 66.

Hankins, O.G. and Titus, H.W. (1939) Growth, fattening, and meat production. In: US Department of Agriculture (ed.) *Food and Life, Yearbook of Agriculture*. US Printing Office, Washington, DC, pp. 450–468.

Heger, J. and Frydrych, Z. (1985) Efficiency of utilization of essential amino acids in growing rats at different levels of intake. *British Journal of Nutrition* 54, 499–508.

INRA (1984) *The Diet of Non-Ruminant Animals: Pigs, Rabbits and Poultry*. Institut National de la Recherche Agronomique, Paris.

Jensen, B.B. (1988) Effect of diet composition and virginiamycin on microbial activity in the digestive tract of pigs. In: Buraczewska, L., Buraczewska, S., Pastuzewska, B. and Zebrowska, T. (eds) *IVth International Symposium on Digestive Physiology in the Pig*. Polish Academy of Sciences, Warsaw, pp. 392–400.

Just, A., Jørgensen, H. and Fernández, J.A. (1981) The digestive capacity of the caecum-colon and the value of the nitrogen absorbed from the hind gut for protein synthesis in pigs. *British Journal of Nutrition* 46, 209–219.

Leverton, R.M., Waddill, F.S. and Skellenger, M. (1959) The urinary excretion of five essential amino acids by young women. *Journal of Nutrition* 67, 19–28.

McCoy, R.H., Meyer, C.E. and Rose, W.C. (1935) Feeding experiments with mixtures of highly purified amino acids. VIII. Isolation and identification of a new essential amino acid. *Journal of Biological Chemistry* 112, 283–302.

Mercer, L.P. (1982) The quantitative nutrient-response relationship. *Journal of Nutrition* 112, 560–566.

Mertz, E.T., Beeson, W.M. and Jackson, H.D. (1952) Classification of essential amino acids for the weanling pig. *Archives of Biochemistry and Biophysics* 38, 121–128.

Meyer, C.E. and Rose, W.C. (1936) The spatial configuration of a-amino-β-hydroxy-n-butyric acid. *Journal of Biological Chemistry* 115, 721–729.

Mitchell, H.H. (1950) Some species and age differences in amino acid requirements. In: Albanese, A.A. (ed.) *Protein and Amino Acid Requirements of Mammals*. Academic Press, New York, pp. 1–31.

Mitchell, J.R., Becker, D.E., Harmon, B.G., Norton, H.W. and Jensen, A.H. (1968) Some amino acid needs of the young pig fed a semisynthetic diet. *Journal of Animal Science* 27, 1322–1326.

Noblet, J., Close, W.H. and Heavens, R.P. (1985) Studies on the energy metabolism of the pregnant sow. 1. Uterus and mammary tissue development. *British Journal of Nutrition* 53, 251–265.

NRC (National Research Council) (1988) *Nutrient Requirements of Domestic Animals*. National Academy Press, Washington, DC.

Rérat, A. and Lougnon, J. (1968) Amino acid requirements of growing pigs. *World Review of Animal Production* 4(17), 84–95.

Rose, W.C., Oesterling, M.J. and Womack, M. (1948) Comparative growth on diets containing ten and nineteen amino acids, with further observations upon the role of glutamic and aspartic acids. *Journal of Biological Chemistry* 176, 753–762.

Schmitz, Von M., Ahrens, F., Schön, J. and Hagemeister, H. (1991) Amino acid absorption in the large intestine of cow, pig and horse and its significance for protein supply. In: Günther, K.-D. and Kirchgessner, M. (eds) *Advances in Animal Physiology and Animal Nutrition, No.22*. Verlag Paul Parey, Hamburg, pp. 67–71.

Southern, L.L. and Baker, D.H. (1984) Urinary excretion of free amino acids, cystathionine and taurine by the young pig. *Canadian Journal of Animal Science* 64, 1067–1069.

Thorbek, G. (1975) Studies on energy metabolism in growing pigs. II. Protein-and fat-gain in growing pigs fed different feed compounds. Efficiency of utilization of metabolizable energy for growth. *Beretning fra Statens Husdyrbrugs Forøg* No. 424.

Torrallardona, D., Harris, C.I., Milne, E. and Fuller, M.F. (1993) Contribution of intestinal microflora to lysine requirements in nonruminants. *Proceedings of the Nutrition Society* 52, 153A.

Wang, T.C. and Fuller, M.F. (1989) The optimum dietary amino acid pattern for growing pigs. 1. Experiments by amino acid deletion. *British Journal of Nutrition* 62, 77–89.

Whittemore, C.T., Tullis, J.B. and Emmans, G.C. (1988) Protein growth in pigs. *Animal Production* 46, 437–445.

Williams, H.H., Curtin, L.V., Abraham, J., Loosli, J.K. and Maynard, L.A. (1954) Estimation of growth requirements for amino acids by assay of the carcass. *Journal of Biological Chemistry* 208, 277–286.

Wünsche, J., Hennig, U., Meinl, M. and Bock, H.D. (1984) Amino acid absorption from the caecum of growing pigs. In: Zebrowska, T., Buraczewska, L., Buraczewska, S., Kowalczyk, J. and Pastuzewska, B. (eds) *Proceedings of the VI International Symposium on Amino Acids*. Polish Scientific Publishers, Warsaw, pp. 158–167.

Zebrowska, T. (1973) Digestion and absorption of nitrogenous compounds in the large intestine of pigs. *Roczniki Nauk Rolniczych* 95B, 85–90.

Zebrowska, T. (1975) The apparent digestibility of nitrogen and individual amino acids in the large intestine of pigs. *Roczniki Nauk Rolniczych* 97B, 117–123.

Zhang, Y., Partridge, I.G. and Mitchell, K.G. (1986) The effect of dietary energy levels and protein:energy ratio on nitrogen and energy balance, performance and carcass composition of pigs weaned at 3 weeks of age. *Animal Production* 42, 389–395.

Amino Acid Digestibility and Availability Studies with Poultry

9

J.M. McNab
Roslin Institute (Edinburgh)
Roslin, Midlothian EH25 9PS, UK.

Introduction

It is generally believed that the provision of amino acids, either free or, much more usually, in the form of protein, accounts for approximately a quarter of the cost of practical diets for poultry. However, as is the case with a number of nutrients, the economic influence of amino acids is probably much greater than this amount, because any shortfall in their dietary supply can impair productivity substantially. An important objective of animal nutritional science is to formulate diets that allow a predetermined rate of production to be achieved at least cost. In theory this implies that, as far as protein is concerned, the ideal diet should exactly satisfy the requirements of the target species for amino acids. In practice, however, this goal is probably unrealistic, partly because of variable demand by individual animals and partly because of dietary constraints imposed by the amino acid profiles of the raw materials accessible to the animal feed trade. The increasing use of synthetic amino acids in diet formulation is reducing the importance of the latter of these two limitations. Nevertheless, it is unlikely that the maximum economic return in poultry production can be achieved unless the amino acid concentrations in the diets are known to meet the requirements of the animals being fed.

In this context it is universally accepted that the contribution made by dietary protein to the nutritional needs of the animal depends not only on its amino acid composition but also on how effectively these amino acids are utilized. The advent and development of ion-exchange and high performance liquid chromatography has made the determination of most of the amino acids present in foods a relatively routine, if somewhat skilled, exercise in most laboratories (see Chapter 2). The results of such analyses are contained in regularly published databanks (Degussa, 1990; NRC, 1994; Rhône Poulenc, 1987; WPSA, 1992) which list the amino acid components of most feedstuffs used in poultry diets. However, figures such as these are only useful in predicting the maximum value of the protein. It has long been recognized that for almost all foods the contribution made by the dietary amino acids to the animal's requirements falls significantly short of this maximum. The reason for this shortfall is known to be a consequence of the fact that not all amino acids supplied by the dietary protein become available to the animal during the course of digestion and metabolism. This has led to the concept of availability.

Sibbald (1987) uses the word 'bioavailable' to describe the proportion of an ingested nutrient which can be used for normal metabolic functions, but in this review the simpler adjective 'available' is preferred to indicate the same quantity. Available amino acids are those which are actually supplied at the sites of protein synthesis. Despite many attempts to devise methods capable of measuring what proportions of the amino acids contained in the ingested protein reach these sites, quantitative data which can be applied to diet formulation are very limited, more often than not restricted to one amino acid (usually lysine) and are not universally accepted. Although few, if any, would disagree with the view (Carpenter, 1973) that responses in performance are the only direct way to test the extent to which dietary protein is broken down and absorbed into the portal blood during digestion and incorporated into body protein or metabolized for some other essential function, the inability of nutritionists to devise an effective and valid assay of wide applicability has proved a formidable barrier to progress. Little progress has been made in this area during the past 20 years and there is virtually nothing new to add to earlier reviews of the topic (Engster, 1986; McNab, 1979a,b; Papadopoulos, 1985; Sibbald, 1987; Whiteacre and Tanner, 1988). Almost invariably based on a slope-ratio growth assay described by Finney (1964), these tests are complex to design, laborious and expensive to carry out and, most seriously, lack precision (standard errors are usually in excess of 10% of the mean). Nevertheless they are the ultimate measure of an amino acid's availability and the only direct means whereby the validity of other less time-consuming methods can be tested.

At the present time about the only acknowledgement that is made to the availability of amino acids in commercial diet formulation is to increase their specifications by a small percentage, the precise amounts depending on the nature of the ingredients, the marginal cost and the judgement of the nutritionist. In the current circumstances of relatively high food costs and small profit margins in poultry production there is considerable pressure to reduce the extent of overformulation, at least of price-sensitive nutrients. The prodigal policy of overformulation is also coming under mounting criticism on environmental grounds, because amino acids in excess of the birds' requirements are excreted and ultimately serve as a source of pollution. Excretion of excess amino acids can also cause welfare problems; carcass defects such as breast blisters and hock burn are often attributable to high uric acid excretion rates caused by excesses of or poorly available protein.

It is, therefore, increasingly being recognized that, unless some means can be devised whereby the amount of each dietary amino acid becoming available to the animal can be determined, little improvement can be made to current formulation practice and the maximum economic output from flocks of poultry will remain an elusive goal.

Digestibility

The property of amino acid availability arises from two processes: digestion (protein hydrolysis and absorption of the products) and utilization (retention of absorbed amino acids). As a first step, therefore, to describing amino acids in terms of their availability, it seems worthwhile to establish the extent by which the amino acids contained in the dietary protein disappear from the gastrointestinal tract during digestion, the so-called digestibility coefficients. Although it is possible to imagine circumstances whereby an amino acid could be digested but not become available for use by the host animal, it is obvious that undigested amino acids (those appearing in the faeces) have made no contribution to the needs of the animal. Therefore, describing the proteins in feedstuffs in terms of their digestible amino acids, although perhaps not optimal, comes nearest to defining the actual amounts which become available to the animal for maintenance and production.

Digestible amino acids are generally derived from the results of balance experiments which measure the difference between their input and output. Such an approach has been used for at least 100 years and has made a considerable contribution to nutritional knowledge. It is usual to express the difference between the amounts eaten and excreted as faeces as a proportion of the amount consumed (the digestibility coefficient) as follows:

$$\text{Amino acid digestibility} = \frac{\text{Amino acid consumed} - \text{Amino acid in faeces}}{\text{Amino acid consumed}}$$

In definitions of this sort, confusion often arises over the terminology used. Strictly speaking the above term should be referred to as apparent digestibility because, of the amino acids present in the faeces, only some have arisen from undigested food residues. Part has come from the animal itself and consists of unabsorbed digestive secretions, sloughed-off gut tissue and bacteria. Sibbald (1987) distinguishes between what he calls the metabolic faecal component (secretions, abraded cells, mucus, bile) and the endogenous faecal fraction (bacteria and bacterial debris) but in this chapter both will be grouped together and referred to as endogenous faecal material. Its measurement allows true digestibility to be calculated thus:

$$\text{True amino acid digestibility} = \frac{\begin{array}{ccc} \text{Amino acid} & \text{Amino acid} & \text{Endogenous amino} \\ \text{consumed} & - \text{ in faeces} & + \text{ acid in faeces} \end{array}}{\text{Amino acid consumed}}$$

A further source of debate in measurements of digestibility in poultry is the effect exerted by the bacteria in the hind gut, an activity that could influence the amounts of both exogenous and endogenous amino acids excreted. Definitions of digestibility can take into account, at least partly, the actions of the microfloral population, either by deriving results from caecectomized birds (the caeca are generally acknowledged to be the principal site of bacterial activity) or, arguably better, by basing values on amino acid concentrations in the terminal ileum (i.e. before the bacteria exert any effect), as is commonly done with pigs (Sibbald, 1987). To relate the amino acid concentration at the ileum to that in the diet requires the addition of an indigestible marker (such as chromium sesquioxide) to the food; measurements of the marker in both the diet and ileal contents allow the apparent digestibility coefficient to be derived as follows:

$$\text{Apparent amino acid digestibility} = \frac{\dfrac{\text{Amino acid in food}}{\text{Marker in food}} - \dfrac{\text{Amino acid in ileum}}{\text{Marker in ileum}}}{\dfrac{\text{Amino acid in food}}{\text{Marker in food}}}$$

Provided a means of measuring the endogenous amino acid concentration in the ileum can be devised then derivations of true ileal digestibilities will follow. Measurements of concentrations of components of the ileum almost invariably require the birds to be killed. Although birds fitted with ileal cannulae have been used for this type of assay, cannulation requires skilful surgery, which is laborious and expensive to carry out; moreover, maintaining the cannulated birds requires considerably more than normal husbandry skills.

One final factor which complicates the determination of digestibility with poultry is the fact that birds excrete faeces and urine together and the collection of faeces requires that birds be colostomized. It has, however, become increasingly common to overlook the urinary component of excreta in assays designed to derive digestibility coefficients, the rationale being that the urinary contribution to poultry excreta is small and its amino acid concentration is unaffected by the nature of the diet. However, these assumptions should be examined critically, particularly when comparisons are frequently being drawn from data derived from ileal and excreta balance studies.

In addressing the topic of amino acid digestion by poultry and devising techniques for its measurement, all the factors outlined above have, at one time or another, been considered sufficiently important to have been taken into account. Despite this there is still no clear indication whether data derived from excreta differ significantly from measurements made on faeces, what effects the microflora have on digestibility coefficients and their significance, or whether the difficulties that have to be overcome in moving from apparent to true digestibility are worth the effort.

Methods for Determining Amino Acid Digestibilities

Although digestibility is frequently regarded as a property of a diet or feedingstuff it is really a characteristic of an animal to which the food is given. It is, for example, a matter for debate whether the extent to which the protein in a particular raw material is digested is the same across all monogastric species (Slump *et al.*, 1977; Terpstra, 1977; Skrede *et al.*, 1980; Campbell *et al.*, 1981; Parsons *et al.*, 1981; Picard *et al.*, 1985; Green and Kiener, 1989). Digestibility measurements relate to the complete diet consumed and values for ingredients (arguably of greatest interest in diet formulation) must, in most cases, be obtained by comparing results from two or more appropriate diets (substitution methods). The validity of the hypothesis that digestibility coefficients are additive across feedingstuffs is essential and, at least at this stage, little progress can be made if this assumption is not upheld, although results from some experiments suggest that amino acid digestibility may be influenced by interactions between dietary ingredients (Wallis *et al.*, 1985).

Three observations are required from a bioassay designed to determine the digestibility of amino acids: (i) the amounts of the amino acids consumed; (ii) the amounts excreted ; and (iii) a measure of the endogenous amino acid losses. When discussing methods it is useful to remember the relationships shown in Figs 9.1. and 9.2. Figure 9.1 illustrates the regression of excreted amino acid against amino acid intake. The intercept on the y axis represents the endogenous amino acid loss (i.e. the amino acid excretion at zero amino acid intake) while the slope of the line, b, gives the true digestibility of the amino acid as follows:

$$\text{True digestibility} = \text{Amino acid intake} \times (1 - b)$$

Estimates of apparent digestibility correspond in a similar way to the slopes of lines joining given amino acid balances with the origin; thus, in Fig. 9.1,

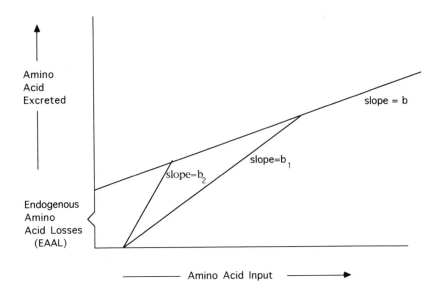

Fig. 9.1. Regression of excreta amino acid content on amino acid input.

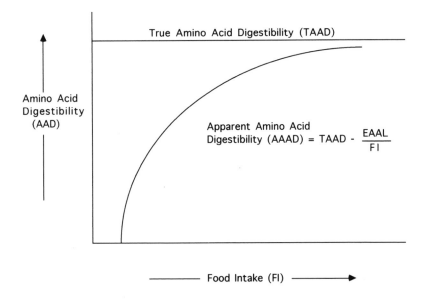

Fig. 9.2. Relationship between true and apparent amino acid digestibility and food intake as derived from Fig. 9.1 assuming that endogenous amino acid losses do not equal zero.

apparent digestibility is equal to the amino acid intake $\times$ $(1 - b_1)$, or $(1 - b_2)$, depending on the amount consumed. The derivation of the response outlined in Fig. 9.2 is obvious if a range of food intakes is envisaged. It should also be noted that if the intercept is zero then apparent and true digestibility are the same and apparent digestibility is independent of intake. Negative intercepts imply an artefact in the measurement.

There are three general types of amino acid balance experiments which have been used to derive digestibility coefficients (*in vitro* tests have also been used, but these are not pursued here). These three types are as follows:

1. Traditional assays which almost always involve preliminary feeding periods to establish equilibrium conditions within the digestive tract of the bird. Differences in carryover between the beginning and end of the period of the assay (the 'end effects') are controlled by trying to ensure that they are the same. In most cases complete diets must be fed and substitution methods used for ingredients.
2. Rapid assays, using starvation before and after giving a known aliquot of test diet to control the 'end effects' but which permit the birds free access to the diet. Again, in most cases, complete diets must be fed and substitution methods used for ingredients.
3. Rapid assays (as above) which rely on tube-feeding to place the test material directly into the birds' crops. These methods almost always avoid the need to substitute ingredients into a basal diet.

While many variations are found within these three general groups, this classification provides a convenient framework within which the many details of procedure can be discussed.

Balance Experiment Methodology

Diet presentation and the measurement of intake are the two most difficult aspects of balance experiments to control in the derivation of digestibility coefficients. If birds are given free access to the diet, a technique which still seems to have widest acceptance and is subject to the least criticism, great care must be taken to avoid food loss, to recover lost food (including any in the drinker), to avoid separation of food components, to monitor dry matter changes and to take representative samples for analysis. In total these are difficult to control in a consistent way but specially designed systems have been described and used successfully (Terpstra and Janssen, 1975).

Such free-feeding methods are used in type 1 assays which historically form the bulk of the experiments designed to measure amino acid digestibility coefficients. Farrell (1978) proposed that the advantages of a rapid type 2 assay could be obtained by training birds to consume satisfactory intakes in 1 h after a 23 h fast. In this assay, which was originally designed to measure metabolizable energy, equal quantities of a basal diet and test ingredient were combined, although pelleting the diet was recommended to sustain food intakes across a range of test materials. Difficulty in achieving this goal has often been reported.

In type 3 assays the presentation of the material under test by tube permits food intake to be controlled to a high degree of precision and offers the means whereby some problems, such as dry matter content of the food, are no longer a variable. However, because input size is reduced, problems of sampling become more important. The only valid objections to this technique are the limits on the size of the input and, perhaps, attitudes towards the acceptability of a procedure which is frequently called 'force-feeding'. Experience in our laboratory is that, with practice, the procedure is extremely rapid (15–30 seconds per bird) and that there is little evidence of more stress beyond that involved in handling. Experience and skill must be attained, although with practice this is readily acquired by most operators.

Excreta collection is another simple task which is difficult to do well in routine experiments. When done with trays under the cages the problems include adherence to feathers; physical losses; contamination with food, feathers and scurf; fermentation and losses in collection and transfer. Sibbald (1986) lists useful precautions to be taken: for example, frequent collection (12-hourly), and continuous mechanical blowing to remove scurf are the sorts of devices which might be judged beneficial.

The only alternative to collection trays is to attach a bag to the bird for collection of droppings and, although methods for doing this have been described (Blakely, 1963; Hayes and Austic, 1982; Almeida and Baptista, 1984; Sibbald, 1986), experience in our laboratory is not encouraging as far as routine programmes are concerned. At present further experimentation is necessary before bags can be considered a realistic alternative to trays.

Droppings versus faecal collection

The difficulty involved in separating faeces from urine in poultry has meant that almost all published digestibility values are based on the amino acid recovery in droppings rather than the more technically correct faeces. It is generally assumed that the amino acid concentration of urine is low and can be ignored. Amino acids account for about 2% of the nitrogen of chicken urine and the concentrations have been reported to be independent of the nature of the diet (O'Dell et al., 1960; Teekell et al., 1968). While some of the variations reported for urinary amino acid excretion appear large when reported in percentages, the absolute changes relative to intake tend to be extremely small. Terpstra (1977) has calculated that between 86% (lysine) to 97% (isoleucine, leucine and valine) of the amino acids excreted by starved birds are derived from faeces and this finding (generally) supports the opinion that in balance studies the amino acid content of urine is low and can be ignored. An experiment by Bragg *et al.* (1969) compared results from normal and colostomized birds. These data suggested that digestibility values derived from normal birds were slightly but significantly different from those derived from colostomized birds, differences which were attributed to the colostomized birds excreting greater quantities of endogenous amino acids. Because these larger concentrations occasionally led to digestibility values greater than 100% (particularly with cysteine), they were considered to be artefacts of the modification and that normal birds gave more realistic digestibility coefficients. These findings are generally endorsed by some Japanese data (Yamazaki *et al.*, 1977; Yamazaki, 1983) which showed there was very little, if any, difference between digestibility coefficients derived with normal and colostomized birds, and that the excretion of endogenous amino acids tended to be higher from colostomized birds than from normal cockerels. This supports the hypothesis of Bragg *et al.* (1969) that surgery had affected the pattern of nitrogen excretion. A recent study (Karasawa and Maeda, 1992) has shown that the colostomy operation can alter the nitrogen economy of the bird by preventing urinary backflow and decreasing nitrogen retention.

Effects of the gastrointestinal microflora on digestibility

The effects that the microflora exert on the digestion of protein and on the values that are derived for amino acid digestibilities are further largely unresolved issues. It has been recognized for many years that incompletely digested proteins which reach the hind gut of the bird come under the influence of the intestinal bacteria. Four approaches have been adopted in attempts to unravel the consequences of this interaction: (i) the use of germ-free animals; (ii) the feeding of antibiotics; (iii) sampling ileal contents either after killing or from birds fitted with appropriate cannulae; (iv) sampling droppings from birds whose caeca have been surgically removed. Perhaps not surprisingly with four such contrasting model systems, results have often been conflicting and no clear consensus has emerged on either the digestive role of the hind-gut bacteria or their effects. Bacterial proteases may release amino acids that are absorbed and ultimately benefit the host animal or amino acids that are converted into bacterial protein and consequently lost to the host. These pathways have contrasting consequences for the birds, but result in the same observation in balance experiments, namely the excretion of fewer amino acids.

Although Salter (1973) has tabulated all the possible effects that the microflora may have on the utilization and excretion of protein and concluded that, at least under adequate dietary regimes, the intestinal bacteria have only a marginal effect on the protein economy of the host (Salter and Coates, 1971; Salter and Fulford, 1974; Salter *et al.*, 1974; Erbersdobler *et al.*, 1975; Fuller and Coates, 1983), the feeling has persisted that the disappearance of amino acids from the contents of the lower intestine as a result of the action of the microflora has not resulted in any nutritional benefit to the animal, and that amino acid digestibility coefficients derived from balance experiments using normal birds will overstate the contribution made by the dietary protein to the amino acid requirements of the bird (Nesheim, 1965; Nesheim and Carpenter, 1967; Parsons, 1985). It is also possible that the effects of the microflora may depend on whether true or apparent digestibility coefficients are being derived. For example, it has been shown that, whereas dietary protein which has resisted breakdown by the digestive enzymes in the duodenum and jejunum is also extremely resistant to degradation by the bacterial proteases, endogenously derived protein is extensively modified by the actions of the microflora (Erbersdobler and Riedel, 1972; Salter and Fulford, 1974); it has, therefore, been suggested that the hind-gut bacteria may serve an important role in recycling body nitrogen. Furthermore, if the diet affects endogenous protein (amino acid) production, any interaction with the microflora would complicate matters considerably and make results from balance experiments even more challenging to interpret usefully. Skilton *et al.* (1988) and Moughan and Rutherford (1990) have illustrated that endogenous amino acid flow at the terminal ileum of rats is underestimated when a nitrogen-free diet is fed and question the validity of this model system, which is very frequently used to derive endogenous amino acid concentrations for the derivation of true digestibility coefficients. This conclusion contrasts with results which show that when a protein-free diet is fed, the absorption of endogenous amino acids is almost complete (95%) by the time the digesta reach the ileocaecal junction (Bielorai *et al.*, 1985). However, on excretion, the digesta appear to have gained endogenous material as there were about 20% more endogenous amino acids present in the droppings than those present earlier in the gastrointestinal tract. These amino acids could explain differences between the true (89%) and apparent (84%) digestibility coefficients derived from the excreta.

In another study (Bielorai and Iosif, 1987), which compared true digestibility coefficients derived after feeding soyabean meal at different dietary concentrations, it was observed that endogenous amino acid concentrations were higher when a protein-free diet had been fed than the values derived by regressing to zero soyabean meal inclusion. This was seen in both ileal contents and excreta and supports the opinion that feeding protein-free diets produces atypically high amino acid concentrations and consequently misleadingly high true digestibility coefficients. Endogenous amino acid concentrations in excreta from starved (i.e. stressed) birds are frequently believed to be higher than those from birds fed a protein-free diet (Bielorai and Iosif, 1987) although hard evidence to support this view is relatively scant (Chae and Han, 1984).

Higher dietary apparent amino acid digestibility coefficients are almost invariably found when germ-free birds are the experimental animals (Soares *et al.*, 1971; Soares and Kifer, 1971; Kussaibati *et al.*, 1982) and Furuse *et al.* (1985) have shown that germ-free birds also retain more protein. However, in contrast to this conclusion, Nitsan and Alumot (1963) and Nesheim and Carpenter (1967) carried out experiments which suggested that substantial proteolysis takes place within the caeca of chickens fed ad lib. It was implied that the products made a contribution to the birds' needs and that the reactions compensated for digestive shortcomings in the duodenum, jejunum and ileum. Furthermore, although Payne *et al.* (1968) reported that apparent amino acid digestibilities were lower in caecectomized than in intact birds, subsequent reports which appear to be based on the same data (Payne, 1968; Payne *et al.*, 1971) claimed that there were no differences across the bird types. Similar confusing reports have been made by Raharjo and Farrell (1984a,b) and others (Johns *et al.*, 1986a,b). For example, in a study concerned with the digestion of heat-damaged meat-and-bone meals, it was concluded that both the true and apparent amino acid digestibility coefficients derived from balance experiments depended on the bird model, values from caecectomized birds being lower than those from intact birds. More importantly, however, decreases in digestibility attributable to heat treatment were not consistent across the different types of bird. In a subsequent study (Johns *et al.*, 1987), in apparent contradiction to this conclusion, it was found that either bird model could be used to predict the available lysine contents of heat-damaged meat-and-bone meals, as determined by a growth assay. It was, therefore, recommended that, on the grounds of efficiency and cost-effectiveness, the tube-fed adult cockerel was the preferred model for the derivation of amino acid digestibility coefficients. This view is endorsed by Green *et al.* (1987a), who showed that the true digestibilities of the amino acids in maize, wheat and barley did not differ between intact and caecectomized adult cockerels. However, in a similar study (Green *et al.*, 1987b) with soyabean, sunflower and groundnut meals, it was shown that the true digestibilities of threonine and glycine were higher and that of lysine lower when intact birds were compared to their caecectomized counterparts. These differences were attributed to the deamination of threonine and glycine and the synthesis of lysine within the caeca. Parsons (1984) had also found caecectomized laying hens excreted more threonine, serine and isoleucine than intact control birds.

Caecectomized cockerels excrete more endogenous amino acids than intact birds, as judged by the output from starved cockerels (Kessler *et al.*, 1981; Parsons, 1984, 1985) and those fed glucose (McNab, 1990). However, in contrast to these data, Green *et al.* (1987a) reported that the differences in endogenous amino acid output from intact and caecectomized birds fed on a protein-free diet were no greater than those between birds of the same type. Although caecectomized birds invariably excreted numerically higher amounts of amino acids than intact birds, only threonine excretion was significantly greater. It is sometimes suggested that increasing fibre intake will increase endogenous amino acid loss but neither higher dietary cellulose nor maize husks had any effect on the amount of amino acids excreted (Sibbald, 1980; Muztar and Slinger, 1980; Parsons, 1984; Sibbald and Wolynetz, 1985; Green, 1988). However, when the carbohydrate source consisted of a mixture of uncooked potato starch $(400\,\mathrm{g\,kg^{-1}})$ and pectin $(100\,\mathrm{g\,kg^{-1}})$ excretion of amino acids increased in both types of birds, the response being more pronounced in intact birds (Parsons, 1984). It had been established earlier that poultry excreta contains much fewer microbial cells than human or pig faeces and that only between 20 and 30% of the constituent amino acids in poultry droppings are bacterial in origin (Parsons *et al.*, 1982a,b). In evaluating data from experiments of this sort it should be remembered that, although it is generally agreed that much microbial activity in chickens resides in the caeca (McNab, 1973), there are microorganisms present throughout the alimentary tract (Ratcliffe, 1991) and removal of the caeca does not necessarily eliminate all bacteria. In addition, whether another part of the gastrointestinal tract assumes a caecal role after their removal is not known.

At present it is difficult to disagree with Sibbald's (1987) view that studies designed to elucidate the effects of the microflora on digestibility coefficients are confusing and further more definitive experiments are required to resolve their role in the degradation of both exogenous and endogenous protein. Although Ratcliffe (1991) stresses the importance of the gut microorganisms in the digestion of protein, the prevailing opinion appears to favour the use of caecectomized birds, particularly if the protein source is of low digestibility.

However, digestibility values based on the droppings from normal intact birds has the decided advantage of simplicity over those taken from caecectomized birds or of values based on ileal concentrations. It encourages assays to be carried out on larger numbers of birds and this can lead to increases in the precision and usefulness of the data. Historically, the majority of published data have been derived from measurements made on droppings from intact birds and, with few exceptions, these values have seemed intuitively reasonable (Bragg *et al.*, 1969; Burgos *et al.*, 1973, 1974; Rostagno *et al.*, 1973; Elwell and Soares, 1975; Nelson *et al.*, 1975, 1982; Nwokolo *et al.*, 1976; Hvidsten and Bjørnstad, 1978; Keulder, 1978; Kirby *et al.*, 1978; Nitsan *et al.*, 1981; Nordheim and Coon, 1984).

True or apparent digestibility coefficients

Because apparent digestibilities of amino acids derived from balance experiments depend on the amount of food eaten (Fig. 9.1), care must be taken to ensure that, when comparisons are made across foodstuffs, near constant food intakes are maintained. Otherwise a systematic bias may inadvertently be incurred. For this reason it is probably preferable to express values in terms of true digestibility coefficients which are independent of food intake (Sibbald, 1979a,b), although how the endogenous amino acid contributions are determined is still a matter for debate. This problem is undoubtedly behind the claim that correcting apparent digestibility for endogenous losses can introduce artefacts and mask important differences between feedingstuffs (Muztar and Slinger, 1981). Because some believe there are no reliable methods for measuring endogenous secretion in a given set of dietary circumstances a system based on apparent digestibility is a better practical basis on which to evaluate protein quality (Van Es and Rérat, 1980). However, because endogenous excretion of amino acids by poultry generally represents a relatively small proportion of the total amino acids excreted after feeding, the uncertainty associated with these values has been exaggerated and has less impact than endogenous energy losses exert on the derivation of true metabolizable energy values.

Conclusions

Although certain issues still remain to be resolved on the most valid techniques to derive amino acid digestibility coefficients, there are good grounds for optimism. More detailed studies are required to determine precisely what factors affect endogenous losses and whether the use of caecectomized birds results in the derivation of significantly different and more meaningful coefficients. The age of the bird is another factor which may have to be taken into consideration. Although in our laboratory no substantial or consistent differences have been observed between the extent of digestion of the amino acids in ten feedingstuffs by 2-week-old broilers and 36-week-old cockerels (McNab, 1991), results from elsewhere suggest that, at least with broilers aged between 3 and 8 weeks of age, apparent digestibility of protein decreased as the birds got older (Håkansson and Eriksson, 1974; Fonolla *et al.*, 1981; Zuprizal *et al.*, 1992). Whether digestibility coefficients derived with chickens are applicable to ducks and turkeys is also not known although one experiment suggested there is little difference between the digestion of protein by chickens and muscovy ducklings (Mohamed *et al.*, 1986).

I believe the prospects for poultry nutrition are exciting. The introduction and development of rapid bioassays based on tube-feeding allows all raw materials to be studied directly and values are no longer subject to either the vagaries of food intake or the uncertainties of extrapolation. Their cheapness and speed will allow many more experiments to be carried out with a consequent increase in the amount of nutritional information available (Johnson, 1992).

References

Almeida, J.A. and Baptista, E.S. (1984) A new approach to the quantitative collection of excreta from birds in a true metabolizable energy bioassay. *Poultry Science* 63, 2501–2503.
Bielorai, R. and Iosif, B. (1987) Amino acid absorption and endogenous amino acids in the lower ileum and excreta of chicks. *Journal of Nutrition* 117, 1459–1462.
Bielorai, R., Iosif, B. and Neumark, H. (1985) Nitrogen absorption and endogenous nitrogen along the intestinal tract of chicks. *Journal of Nutrition* 115, 568–572.
Blakely, R.M. (1963) Note on an apparatus for the collection of turkey feces. *Canadian Journal of Animal Science* 43, 386–388.
Bragg, D.B., Ivy, C.A. and Stephenson, E.L. (1969) Method for determining amino acid availability of feeds. *Poultry Science* 48, 2135–2137.
Burgos, A., Caviness, C.E., Floyd, J.I. and Stephenson, E.L. (1973) Comparison of amino acid content and availability of different soybean varieties in the broiler chick. *Poultry Science* 52, 1822–1827.
Burgos, A., Floyd, J.I. and Stephenson, E.L. (1974) The amino acid content and availability of different samples of poultry by-product meal, and feather meal. *Poultry Science* 53, 198–203.
Campbell, L.D., Eggum, B.O. and Jacobsen, I. (1981) Biological value, amino acid availability and true metabolizable energy of low-glucosinolate rapeseed meal (*canola*) determined with rats and/or roosters. *Nutrition Reports International* 1981, 791–797.
Carpenter, K.J. (1973) The use of ileal content analysis to assess the digestibility of amino acids. In: Porter, J.W.G. and Rolls, B.A. (eds) *Proteins in Human Nutrition*. Academic Press, London, pp. 343–347.
Chae, B.J. and Han, I.K. (1984) Studies on the determination of amino acid bioavailability for broiler-type chicks. 1. Bioavailable amino acid values of corn, sorghum, soybean meal and fish meal as measured by two different bioassays. *Korean Journal of Animal Science* 26, 372–399.
Degussa (1990) *The Amino Acid Composition of Feedstuffs*. Degussa, Frankfurt.
Elwell, D. and Soares, J.H. (1975) Amino acid bioavailability – a comparative evaluation of several assay techniques. *Poultry Science* 54, 78–85.
Engster, H.M. (1986) Amino acid availability in poultry feeds. *4th World Congress Animal Feeding* IX, 177–186.
Erbersdobler, H. and Riedel, G. (1972) Bestimmung der aminosäurenverdaulichkeit bei keimfrei und konventionell gehaltenen Küken. 1. Mitteilung. *Archiv für Geflügelkunde* 6, 218–222.

Erbersdobler, H., Gropp, J. and Beck, H. (1975) Utilization of proteins for growth and egg production. *Proceedings of the Nutrition Society* 34, 21–28.

Farrell, D.J. (1978) Rapid determination of metabolizable energy of foods using cockerels. *British Poultry Science* 19, 303–308.

Finney, D.J. (1964) *Statistical Method in Biological Assay*, 2nd edn. Hafner Publishing Company, New York.

Fonolla, J., Prieto, C. and Sanz, R. (1981) Influence of age on the nutrient utilization of diets for broilers. *Animal Feed Science and Technology* 6, 405–411.

Fuller, R. and Coates, M.E. (1983) Influence of the intestinal microflora on nutrition. In: Freeman, B.M. (ed.) *Physiology and Biochemistry of the Domestic Fowl*, Vol. 4. Academic Press, London, pp. 51–61.

Furuse, M., Yokota, H. and Tasaki, I. (1985) Influence of energy intake on growth and utilisation of dietary protein and energy in germ-free and conventional chicks. *British Poultry Science* 26, 389–397.

Green, S. (1988) Effect of dietary fibre and caecectomy on the excretion of endogenous amino acids from adult cockerels. *British Poultry Science* 29, 419–429.

Green, S. and Kiener, T. (1989) Digestibilities of nitrogen and amino acids in soyabean, sunflower, meat and rapeseed meals measured with pigs and poultry. *Animal Production* 48, 157–179.

Green, S., Bertrand, S.L., Duron, M.J.C. and Maillard, R. (1987a) Digestibilities of amino acids in maize, wheat and barley meals, determined with intact and caecectomised cockerels. *British Poultry Science* 28, 631–641.

Green, S., Bertrand, S.L., Duron, M.J.C. and Maillard, R. (1987b) Digestibilities of amino acids in soyabean, sunflower and groundnut meals, determined with intact and caecectomised cockerels. *British Poultry Science* 28, 643–652.

Håkansson, J. and Eriksson, S. (1974) Digestibility, nitrogen retention and consumption of metabolizable energy by chickens on feeds of low and high concentration. *Swedish Journal of Agricultural Research* 4, 195–207.

Hayes, J.P. and Austic, R.E. (1982) An easy and accurate technique for feces collection in adult roosters. *Poultry Science* 61, 2294–2295.

Hvidsten, H. and Bjørnstad, J. (1978) The digestibility of amino acids in different poultry feeds and comparison with a chick growth test of lysine availability. *Proceedings and Abstracts of XVI World's Poultry Congress* Volume X, 1720–1725.

Johns, D.C., Low, C.K., Sedcole, J.R. and James, K.A.C. (1986a) Determination of amino acid digestibility using caecectomised and intact adult cockerels. *British Poultry Science* 27, 451–461.

Johns, D.C., Low, C.K. and James, K.A.C. (1986b) Comparison of amino acid digestibility using the ileal digesta from growing chickens and cannulated adult cockerels. *British Poultry Science* 27, 679–685.

Johns, D.C., Low, C.K., Sedcole, J.R., Gurnsey, M.P. and James, K.A.C. (1987) Comparison of several *in vivo* digestibility procedures to determine lysine digestibility in poultry diets containing heat treated meat and bone meals. *British Poultry Science* 28, 397–406.

Johnson, R.J. (1992) Principles, problems and application of amino acid digestibility in poultry. *World's Poultry Science Journal* 48, 232–246.

Karasawa, Y. and Maeda, M. (1992) Effect of colostomy on the utilisation of dietary nitrogen in the fowl fed on a low protein diet. *British Poultry Science* 33, 815–820.

Kessler, J.W., Nguyen, T.H. and Thomas, O.P. (1981) The amino acid excretion values in intact and cecectomised negative control roosters used for determining metabolic plus endogenous urinary losses. *Poultry Science* 60, 1576–1577.

Keulder, H.F. (1978) The development of a standardised procedure for the determination of true digestibility of amino acids in protein sources. *Proceedings XVI World's Poultry Congress* 11, 9–18.

Kirby, L.K., Nelson, T.S., Johnson, Z. and Waldroup, P.W. (1978) Content and digestibility by chicks of the amino acids in wheat, fish meal and animal by-products. *Nutrition Reports International* 18, 591–598.

Kussaibati, R., Guillaume, J. and Leclercq, B. (1982) The effects of the gut microflora on the digestibility of starch and protein in young chicks. *Annales de Zootechnie* 31, 483–488.

McNab, J.M. (1973) The avian caeca: a review. *World's Poultry Science Journal* 29, 251–263.

McNab, J.M. (1979a) The concept of amino acid availability in farm animals. In: Haresign, W. and Lewis, D. (eds) *Recent Advances in Animal Nutrition.* Butterworths, London, pp. 1–9.

McNab, J.M. (1979b) Growth tests for the determination of available amino acids. In: Kan, C.A. and Simons, P.C.M. (eds) *Proceedings of the 2nd European Symposium of Poultry Nutrition*, Spelderholt Institute for Poultry Research, Beekbergen, The Netherlands. pp. 102–106.

McNab, J.M. (1990) Measuring availability of amino acids from digestibility experiments. In: Institute de Recerca i Tecnologia Agroalimentaries (ed.) *Proceedings of the 7th European Symposium on Poultry Nutrition*, IRTA, Barcelona, Spain. pp. 45–53.

McNab, J.M. (1991) Problems in the assessment of total and digestible amino acids in feedingstuffs. *Zootecnica International* XIV (2), 42–48.

Mohamed, K., Larbier, M. and Leclercq, B. (1986) A comparative study of the digestibility of soyabean and cottonseed meal amino acids in domestic chicks and muscovy ducklings. *Annales de Zootechnie* 35, 79–86.

Moughan, P.J. and Rutherford, S.M. (1990) Endogenous flow of total lysine and other amino acids at the distal ileum of the protein- or peptide-fed rat: the chemical labelling of gelatin protein by transformation of lysine to homoarginine. *Journal of the Science of Food and Agriculture* 52, 179–192.

Muztar, A.J. and Slinger, S.J. (1980) Effect of level of dietary fiber on nitrogen and amino acid excretion in the fasted mature rooster. *Nutrition Reports International* 22, 863–868.

Muztar, A.J. and Slinger, S.J. (1981) Relationship between body weight and amino acid excretion in fasted mature cockerels. *Poultry Science* 60, 790–794.

Nelson, T.S., Stephenson, E.L., Burgos, A., Floyd, J. and York, J.O. (1975) Effect of tannin content and dry matter digestion on energy utilization and average amino acid availability of hybrid sorghum grains. *Poultry Science* 54, 1620–1623.

Nelson, T.S., Johnson, Z.B., Kirby, L.K. and Beasley, J.N. (1982) Digestion of dry matter and amino acids and energy utilization by chicks fed molded corn containing mycotoxins. *Poultry Science* 61, 584–585.

Nesheim, M.C. (1965) Amino acid availability in processed proteins. *Proceedings of the 1965 Cornell Nutrition Conference.* pp. 112–118.

Nesheim, M.C. and Carpenter, K.J. (1967) The digestion of heat damaged protein. *British Journal of Nutrition* 21, 399–411.

Nitsan, Z. and Alumot, E. (1963) Role of the cecum in the utilization of raw soybean in chicks. *Journal of Nutrition* 80, 299–304.

Nitsan, Z., Dvorin, A. and Nir, I. (1981) Availability of amino acids from soybean, corn, and milo for goslings. *Poultry Science* 60, 2724–2725.

Nordheim, J.P. and Coon, C.N. (1984) A comparison of four methods for determining available lysine in animal protein meals. *Poultry Science* 63, 1040–1051.

NRC (National Research Council) (1994) *Nutrient Requirements of Domestic Animals. Number 1. Nutrient Requirements of Poultry*, 9th revised edn. National Academy of Sciences, Washington, D.C.

Nwokolo, E.N., Bragg, D.B. and Kitts, W.D. (1976) The availability of amino acids from palm kernel, soybean, cottonseed and rapeseed meal for the growing chick. *Poultry Science* 55, 2300–2304.

O'Dell, B.L., Woods, W.D., Laerdal, O.A., Geffay, A.M. and Savage, J.E. (1960) Distribution of the major nitrogenous compounds and amino acids in chick urine. *Poultry Science* 39, 426–432.

Papadopoulos, M.C. (1985) Estimations of amino acid digestibility and availability in feedstuffs for poultry. *World's Poultry Science Journal* 41, 64–71.

Parsons, C.M. (1984) Influence of caecectomy and source of dietary fibre or starch on excretion of endogenous amino acids by laying hens. *British Journal of Nutrition* 51, 541–548.

Parsons, C.M. (1985) Influence of caecectomy on digestibility of amino acids by roosters fed distillers' dried grains with solubles. *Journal of Agricultural Science (Cambridge)* 104, 469–472.

Parsons, C.M., Potter, L.M. and Brown, R.D. (1981) True metabolizable energy and amino acid digestibility of dehulled soybean meal. *Poultry Science* 60, 2687–2696.

Parsons, C.M., Potter, L.M., Brown, R.D., Wilkins, T.D. and Bliss, B.A. (1982a) Microbial contribution to dry matter and amino acid content of poultry excreta. *Poultry Science* 61, 925–932.

Parsons, C.M., Potter, L.M. and Brown, R.D. (1982b) Effects of dietary protein and intestinal microflora on excretion of amino acids in poultry. *Poultry Science* 61, 939–946.

Payne, W.L. (1968) Investigation of apparent amino acid digestibility as a method to determine protein quality. *Proceedings of the Maryland Nutrition Conference for Feed Manufacturers.* pp. 73–83.

Payne, W.L., Combs, G.F., Kifer, R.R. and Snyder, D.G. (1968) Investigation of protein quality – ileal recovery of amino acids. *Federation Proceedings* 27, 1199–1203.

Payne, W.L., Kifer, R.R., Snyder, D.G. and Combs, G.F. (1971) Studies of protein digestion in the chicken. 1. Investigation of apparent amino acid digestibility of fish meal protein using caecectomized adult male chickens. *Poultry Science* 50, 143–150.

Picard, M., Bertrand, S., Duron, M. and Maillard, R. (1985) *Towards a Concept of Amino Acid Digestibility in Monogastric Animals.* AEC, Commentry France.

Raharjo, Y.C. and Farrell, D.J. (1984a) Effects of caecetomy and dietary antibiotics on the digestibility of dry matter and amino acids in poultry feeds determined by excreta analysis. *Australian Journal of Experimental Agriculture* 24, 516–521.

Raharjo, Y. and Farrell, D.J. (1984b) A new biological method for determining amino acid digestibility in poultry feedstuffs using a simple cannula, and the influence of dietary fibre on endogenous amino acid output. *Animal Feed Science and Technology* 12, 29–45.

Ratcliffe, B. (1991) The role of the microflora in digestion. In: Fuller, M.F. (ed.) In Vitro *Digestion for Pigs and Poultry*. CAB International, Wallingford, pp. 19–34.

Rhône-Poulenc (1987) *Recommendations for Animal Nutrition*, 5th edn. Rhône-Poulenc, Commentry, France.

Rostagno, H.S., Rogler, J.C. and Featherston, W.R. (1973) Studies on the nutritional value of sorghum grains with varying tannin contents for chicks. 2. Amino acid digestibility studies. *Poultry Science* 52, 772–778.

Salter, D.N. (1973) The influence of gut micro-organisms on utilization of dietary protein. *Proceedings of the Nutrition Society* 32, 65–71.

Salter, D.N. and Coates, M.E. (1971) The influence of the microflora of the alimentary tract on protein digestion in the chick. *British Journal of Nutrition* 26, 55–69.

Salter, D.N. and Fulford, R.J. (1974) The influence of the gut microflora on the digestion of dietary and endogenous proteins: studies of the amino acid composition of the excreta of germ-free and conventional chicks. *British Journal of Nutrition* 32, 625–637.

Salter, D.N., Coates, M.E. and Hewitt, D. (1974) The utilization of protein and excretion of uric acid in germ-free and conventional chicks. *British Journal of Nutrition* 31, 307-318.

Sibbald, I.R. (1979a) A bioassay for available amino acids and true metabolizable energy in feedstuffs. *Poultry Science* 58, 668–675.

Sibbald, I.R. (1979b) Bioavailable amino acids and true metabolizable energy of cereal grains. *Poultry Science* 58, 934–939.

Sibbald, I.R. (1980) The effects of dietary cellulose and sand on the combined metabolic plus endogenous energy and amino acid outputs of adult cockerels. *Poultry Science* 59, 836–844.

Sibbald, I.R. (1986) The T.M.E. system of feed evaluation: methodology, feed composition data and bibliography. Technical Bulletin 1986–4E. Animal Research Centre, Research Branch, Agriculture Canada, Ottawa, Canada.

Sibbald, I.R. (1987) Estimation of bioavailable amino acids in feedingstuffs for poultry and pigs: a review with emphasis on balance experiments. *Canadian Journal of Animal Science* 67, 221–300.

Sibbald, I.R. and Wolynetz, M.S. (1985) The bioavailability of supplementary lysine and its effect on the energy and nitrogen excretion of adult cockerels fed diets diluted with cellulose. *Poultry Science* 64, 1972-1975.

Skilton, G.A., Moughan, P.J. and Smith, W.C. (1988) Determination of endogenous amino acid flow at the terminal ileum of the rat. *Journal of the Science of Food and Agriculture* 44, 227–235.

Skrede, A., Krogdahl, Å. and Austreng, E. (1980) Digestibility of amino acids in raw fish flesh and meat-and-bone meal for the chicken, fox, mink and rainbow trout. *Zeitschrift für Tierphysiologie Tierernährung und Futtermittelkunde* 43, 92–101.

Slump, P., van Beek, L., Janssen, W.M.M.A., Terpstra, K., Lenis, N.P. and Smits, B. (1977) A comparative study with pigs, poultry and rats of the amino acid digestibility of diets containing crude protein with diverging digestibilities. *Zeitschrift für Tierphysiologie Tierernährung und Futtermittelkunde* 40, 257–272.

Soares, J.H. and Kifer, R.R. (1971) Evaluation of protein quality based on residual amino acids of the ileal contents of chicks. *Poultry Science* 50, 41–46.

Soares, J.H., Miller, D., Fitz, N. and Sanders, M. (1971) Some factors affecting the biological availability of amino acids in fish protein. *Poultry Science* 50, 1134–1143.

Teekell, R.A., Richardson, C.E. and Watts, A.B. (1968) Dietary protein effects on urinary nitrogen components of the hen. *Poultry Science* 47, 1260–1266.

Terpstra, K. (1977) Determination of the digestibility of protein and amino acids in poultry feeds. *Proceedings of the Vth International Symposium on Amino Acids*. Budapest, pp. 1–8.

Terpstra, K. and Janssen, W.M.M.A. (1975) Methods for the determination of metabolizable energy and digestibility coefficients of poultry feeds. *Spelderholt Report* 101, 75.

Van Es, A.J.H. and Rérat, A. (1980) In: Oslage, H.J. and Rohr, K. (eds) *Proceedings of 3rd EAAP Symposium on Protein Metabolism and Nutrition*. European Association of Animal Production, Braun, West Germany, Vol. 3, p. 32.

Wallis, I.R., Mollah, Y. and Balnave, D. (1985) Interactions between wheat and other dietary cereals with respect to metabolisable energy and digestible amino acids. *British Poultry Science* 26, 265–274.

Whiteacre, M.E. and Tanner, H. (1988) Methods of determining the bioavailability of amino acids for poultry. In: Friedman, M. (ed.) *Absorption and Utilization of Amino Acids, Vol. III*. CRC Press, Boca Raton, pp. 129–141.

WPSA (1992) *European Amino Acid Table*. Working Group Number 2 (Nutrition) of the World Poultry Science Association, Beekbergen, The Netherlands.

Yamazaki, M. (1983) A comparison of two methods in determining amino acid availability of feed ingredients. *Japanese Journal of Zootechnical Science* 54, 729–733.

Yamazaki, M., Ando, M. and Kubota, D. (1977) Studies on the digestibility of single cell protein grown on various nutrient substrates for colostomized laying hens. *Japanese Poultry Science* 14, 232–235.

Zuprizal, Larbier, M. and Chagneau, A.M. (1992) Effect of age and sex on true digestibility of amino acids of rapeseed and soybean meals in growing broilers. *Poultry Science* 71, 1486–1492.

Responses of Growing Poultry to Amino Acids 10

J.P.F. D'MELLO

The Scottish Agricultural College, West Mains Road, Edinburgh, EH9 3JG, UK.

Introduction

In common with other vertebrate species, poultry (chickens, turkeys, ducks and geese) require a core of ten amino acids for optimum growth and food utilization (Table 10.1). This need emanates from the inability of all animals to synthesize the corresponding carbon skeleton or keto acid. These amino acids are classified as 'indispensable' or 'essential' and must be supplied in the diet in balanced proportions if growth performance is to be maximized. Those amino acids which growing poultry are able to synthesize within their tissues are termed 'dispensable' or 'non-essential' amino acids. However, the distinction between indispensable and dispensable amino acids depends upon the criteria adopted in such an evaluation. If maximum growth rate is the overriding consideration, then glycine and proline should be added to the list of indispensable amino acids for young avian species. In devising the classification shown in Table 10.1, however, the ability of growing poultry to synthesize glycine and proline has been adopted as the principal criterion. Glycine is synthesized from serine, albeit at rates insufficient to sustain maximum growth. Satisfactory growth performance can still be achieved even under conditions of complete deprivation of glycine (D'Mello, 1973a). In contrast, dietary omission of any indispensable amino acid elicits profound biochemical and morphological lesions leading to death within a relatively short time span (Ousterhout, 1960).

Growing poultry, like other farm animals, do not possess the capacity to synthesize the aromatic ring of phenylalanine and tyrosine. Only phenylalanine, however, is classified as indispensable due to the ability of all animals to synthesize tyrosine from phenylalanine. Tyrosine may,

Table 10.1. Nutritional classification of amino acids for poultry. (From D'Mello, 1979. Reproduced permission of Butterworth-Heinemann Ltd.)

Indispensable	Dispensable
Lysine	Glycine
Methionine	Cystine
Threonine	Serine
Tryptophan	Proline
Isoleucine	Alanine
Leucine	Glutamic acid
Valine	Aspartic acid
Phenylalanine	Tyrosine
Arginine	
Histidine	

therefore, be regarded as dispensable providing the diet contains adequate quantities of its precursor. Although tyrosine cannot act as a precursor of phenylalanine its presence in the diet may reduce the dietary requirement for phenylalanine. However, this sparing effect of tyrosine is limited and, consequently, a minimum quantity of phenylalanine should always be provided in the diet. For growing poultry at least 58% of the combined aromatic amino acid requirement should be supplied in the form of phenylalanine. An analogous situation exists with the sulfur-containing amino acids. At least 50% of this requirement should be provided in the form of methionine. The unique relationship between tryptophan and the B-complex vitamin, nicotinamide, represents yet another example of a sparing effect in that endogenous synthesis of the latter may be accomplished at the expense of dietary tryptophan. The sparing effect of the vitamin on tryptophan requirements has not been quantified. As regards glycine and serine, the metabolic interconversions allow either to satisfy the requirement, despite earlier reports that glycine was the more effective amino acid.

It is now widely acknowledged that, in common with other non-ruminant animals, growing poultry will not achieve their genetically determined potential if the dietary nitrogen is supplied exclusively in the form of the indispensable amino acids. Additional dietary nitrogen is required and highly effective individual sources of this non-specific nitrogen include glutamic acid, alanine and diammonium citrate. However, the most effective source of this nitrogen is a mixture of the dispensable amino acids.

Consequently, although young poultry have specific dietary requirements for the indispensable amino acids, some combination of the dispensable amino acids should also be provided to maximize growth potential. The data of Bedford and Summers (1985) illustrate the importance of maintaining a satisfactory ratio between the two groups of amino acids. When the proportion of indispensable to dispensable amino acids is fixed at 55:45, growth performance, efficiency of food utilization and total carcass protein of broilers are optimized. At the higher ratio of 65:35, the rate of deamination of the surplus indispensable amino acids is insufficient to satisfy the requirements for the synthesis of the dispensable amino acids. At the other extreme, a ratio of 35:65 provides an excess of dispensable amino acids relative to indispensable amino acids resulting in both a deficiency of the latter amino acids and an additional burden of energy expenditure for the excretion of surplus nitrogen.

This chapter reviews the methods used to determine the responses of growing poultry to the indispensable amino acids and attempts to evaluate the wide range of factors which modify these responses. Estimation of 'requirements' of growing poultry for individual amino acids will not be considered here since such information is of limited interpretive value. As Morris (1983) states: 'what the practical nutritionist needs to know is the rate at which animals in a given class, in a reasonably well-defined nutritional and environmental context, will respond to incremental inputs of a given nutrient. Armed with this information, and a knowledge of his marginal costs and the value of extra output, he can calculate an optimum dose'.

Methodology

As with other animal species, the methods used to determine the responses of growing poultry to individual amino acids fall into two major categories: empirical and factorial. The latter is often projected as the method with a future as research funds for dose–response experimentation decline. However, cognizance should be taken of the mass of empirically derived data which may serve not only as a means of evaluating responses but may also contribute elements to factorial models (see Boorman and Burgess, 1986). There is little doubt that factorial models are likely to contribute significantly to future thinking since they provide a means of assessing requirements in flexible and economic terms. It is claimed that the model approach provides a mechanism for identifying important gaps in existing knowledge. However, empirical methods have contributed significantly to current understanding of factors which genuinely affect the responses of growing poultry to amino acids.

It has been conventional to consider amino acid responses of growing poultry in relation to dietary concentrations (e.g. $g\,kg^{-1}$ diet). This appears to be a convenient and satisfactory approach for a number of purposes, including the practical formulation of diets. Another way is to express requirements as a proportion of dietary crude protein (Barbour et al., 1993). It should be recognized, however, that the use of dietary concentrations can obscure important issues when attempts are made to resolve the wide range of factors which influence the responses of growing poultry to individual amino acids. This is due to the well-known but often-ignored observation that the responses of growing poultry depend not only on dietary amino acid concentration but also on food intake. Food intake, in turn, is influenced by a number of factors including environmental temperature, dietary energy concentration and breed and species of animal. The measurement of food intake is thus crucial in any investigation of amino acid responses in ad lib fed animals and recent studies continue to emphasize this point (Noble et al., 1993; Picard et al., 1993). The expression of input in terms of daily intake of an amino acid overcomes the constraints outlined above and provides a means of discriminating between those factors which genuinely influence responses and those which exert their effects through alterations in voluntary food intake. The use of intake rather than dietary concentrations to interpret responses of poultry to the first-limiting amino acid has long been advocated by Morris (1972, 1983). It may be regarded as irrational that in a subsequent and crucial study, Morris et al. (1987) rejected this fundamental approach and resumed the use of dietary concentrations to consider the responses of broiler chicks to lysine and protein.

Empirical methods

The empirical method most commonly used to determine amino acid responses in growing poultry involves the addition of graded supplements of the amino acid under test to a basal diet deficient in that amino acid (D'Mello, 1982). These graded additions are accomplished with the crystalline form of the amino acid. A number of criteria must be fulfilled for satisfactory results. It is imperative that the basal diet is deficient enough in the amino acid under investigation and that graded doses of the pure amino acid are employed to generate a smooth and full response curve encompassing both growth-limiting doses and those which elicit maximum response. It follows that data derived from supplementation experiments involving just one or two additions or those from bioassays which lack maximum response values are unsuitable for assessing optimal economic doses. In certain cases it may be necessary to use different combinations of those amino acids involved in antagonistic relationships. The measurement of food intake is an essential prerequisite for satisfactory interpretation of data. Providing that these criteria are satisfied, the response curve may be used to estimate optimal doses for given rates of growth or food efficiency. In addition, the response curve may be used to determine estimates of slope and plateau values required as important components for interpretation by the Reading model (Fisher *et al.*, 1973; see also Chapter 11).

Fisher and Morris (1970) proposed an alternative empirical method based on the sequential dilution of a high-protein 'summit' diet with an isoenergetic protein-free mixture. This method has been described at length in Chapter 4 since the technique relies on the deliberate creation of an amino acid imbalance in the summit diet. The imbalance is accomplished by maintaining large dietary excesses of all indispensable amino acids except the one under test, which is fixed at a markedly lower level.

Neither method is without its limitations. In particular, the graded supplementation technique has been criticized on several counts (Fisher and Morris, 1970; Gous, 1980). Firstly, it is claimed that dietary amino acid balance changes with each successive dose of the limiting amino acid and that the response may be influenced by this factor. The exact manner in which changing amino acid balance might affect the response has never been amplified by the critics. Since any change in amino acid balance would affect food intake rather than the efficiency of utilization of the limiting amino acid (see Chapter 4), the first criticism may be discounted. It is also argued that at high levels of supplementation, the amino acid under test may no longer be first-limiting and that further responses to this amino acid might be prevented by the second-limiting amino acid(s). An additional criticism centres on the alleged difficulty in devising a suitable basal diet which is sufficiently deficient to allow the use of a wide range of input levels of the limiting amino acid. The final criticism relates to the cost of certain synthetic amino acids which, Gous (1980) maintains, might prevent supplementation studies with the more expensive amino acids.

It is argued that none of the disadvantages just described apply to the diet dilution technique. Indeed, Fisher and Morris (1970) claimed that this technique satisfied 'all the requirements for a successful assay' and Gous (1980, 1986) described it as an 'improved method'. These assertions imply that the diet dilution technique is capable of yielding more valid response data than the graded supplementation method but no evidence was presented to substantiate this claim.

It is instructive to recall that although the initial description of the diet dilution technique appeared over 20 years ago, the graded supplementation procedure still remains the method of choice in the vast majority of studies on the amino acid responses of growing poultry (Hewitt and Lewis, 1972; Boomgaardt and Baker, 1973; D'Mello, 1974; D'Mello and Emmans, 1975; Thomas *et al.*, 1977). It is clear that cost of amino acids has not been a major deterrent to this research where the primary costs are likely to be associated with labour and capital expenditure. The other criticisms levelled at the graded supplementation technique may also be rejected. In particular it is possible to discount the assertion that other limiting amino acids may inhibit the maximum response to the first-limiting amino acid. It is now common practice to include generous levels of protein in the diet, and by judicious supplementation it should be possible to ensure adequacy of all other amino acids. In addition, the use of combined supplements of the first- and second-limiting amino acids may be employed (D'Mello and Lewis, 1970; D'Mello and Emmans, 1975). Such an approach has yielded satisfactory responses to arginine and lysine in turkey poults (D'Mello and Emmans, 1975), with maximum growth rates comparable to those observed in groups fed a standard diet. Furthermore, in many studies a wide range of input levels of the first-limiting amino acid have been provided (Hewitt and Lewis, 1972; Boomgaardt and Baker, 1973; D'Mello and Emmans, 1975; D'Mello, 1990). Thus there appears to have been little difficulty in devising suitably deficient basal diets or in providing a satisfactory range of intakes of the limiting amino acid. The most serious criticism levelled at the graded supplementation technique relates to the variation in amino acid balance of successive diets within a supplementation series (Gous, 1980). However, it will be apparent from the account in Chapter 4 that the diet dilution technique relies on the deliberate imposition of an amino acid imbalance in the summit diet by the provision of all amino acids, except that under test, in substantial excess. This is the precise mechanism whereby the deleterious effects of amino acid imbalance have been induced (Harper, 1964; D'Mello and Lewis, 1971; D'Mello, 1990).

D'Mello (1982) stated that while it is relatively straightforward to generate response curves to a single amino acid by the diet dilution technique, considerable difficulties would emerge when responses to interacting pairs such as lysine and arginine or leucine and valine are to be determined. It was suggested that the responses to arginine or valine would be unreliable owing to the constraint to design the summit diet with large excesses of other amino acids including lysine and leucine. The adverse effects of these amino acids on the responses of growing poultry to arginine and valine respectively are well documented and will be discussed later in this chapter. Gous (1980), however, argues that this issue is readily resolved by formulating a number of dilution series each containing different concentrations of the antagonizing amino acid. The degree of interaction would be indicated by the resultant difference in slope of individual response curves across the series.

In a direct comparison of the two methods, D'Mello (1982) used data obtained by the diet dilution procedure (Gous, 1980) and by the graded supplementation technique (Boomgaardt and Baker, 1973) with respect to growth responses of broiler chicks to varying lysine intakes (Fig. 10.1). The data of Stockland *et al.* (1970), obtained with rats fed graded supplements

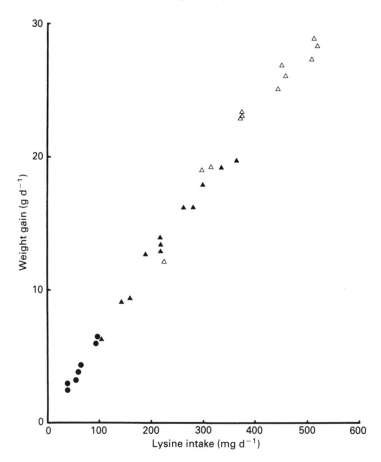

Fig. 10.1. Daily weight gain of growing rats and chicks in relation to daily lysine intake. (From D'Mello, 1982; source of data: rats (●), Stockland *et al.*, 1970; chicks (▲), Boomgaardt and Baker, 1973; chicks (△), Gous, 1980. Reproduced permission of The World's Poultry Science Association.)

of lysine, were also incorporated in this figure. It became clear that when growth increments were considered in relation to daily lysine intake both methods yielded similar results with all 'limiting points' contributing to a single response curve. In a subsequent examination of data obtained with turkey poults, D'Mello (1983) confirmed the high degree of compatibility between the two methods. If the diet dilution technique is a superior method, then the responses generated by the supplementation procedure would be displaced positively with respect to the data of Gous (1980). The agreement between the rat and chick data, irrespective of the method used to elicit these responses, is noteworthy. Such interspecies compatibility in amino acid response data is reviewed later in this chapter, but it is worth emphasizing here that these similarities were obtained with the graded supplementation technique despite wide differences in dietary amino acid balance and in the nature of the protein ingredients used. On the basis of the evidence presented in Fig. 10.1. D'Mello (1982) concluded that the lack of confidence in the graded supplementation technique and the projection of the diet dilution procedure as an improved method could not be justified since both methods yielded concordant growth responses. Subsequently, Boorman and Burgess (1986) arrived at a similar conclusion after a detailed consideration of 15 data sets derived by the supplementation technique and 11 data sets obtained by the diet dilution method in studies on lysine responses in broiler chicks.

Since the reviews of D'Mello (1982) and Boorman and Burgess (1986), a disquieting feature of the diet dilution technique has emerged with far-reaching consequences which renews doubt about the authenticity of earlier attempts to validate the technique (Gous, 1980; Gous and Morris, 1985). It is a condition of this procedure that growth responses are not confounded by the deliberate variation in dietary protein content which occurs on dilution of the summit diet. The method relies on the interpretation of responses to different rates of dilution as responses to the first-limiting amino acid and not to changing dietary protein contents. In this particular respect, Gous and Morris (1985) confirmed that the response in gain when lysine was added to diets in the dilution series 'corresponded almost exactly to the level of lysine in the diet, irrespective of protein content'. Indeed, multiple regression analysis of the data indicated that protein intake did not contribute significantly to the best-fit model. Gous and Morris (1985) proceeded to question the need to use isonitrogenous diets when determining the responses of growing poultry to an amino acid. However, in subsequent and more detailed investigations with chicks, these authors (Morris *et al.*, 1987; Abebe and Morris, 1990a,b) demonstrated that crude protein (CP) level unequivocally influences the growth response to an amino acid. The interaction between dietary protein level and lysine intake, shown in Fig. 10.2, represents a reassessment by D'Mello (1988) of the original data published by Morris *et al.* (1987). In this study, the summit diet containing 280 g CP kg^{-1}, first-limiting in lysine, was diluted with a basal diet containing 140 g CP kg^{-1} of identical amino acid balance and metabolizable energy (ME) content to yield a series of diets varying in CP content and first-limiting in lysine. Each diluted diet was supplemented with graded levels of pure lysine which, in theory, should elicit responses compatible with those obtained merely by diluting the summit diet. However, it is readily seen (Fig. 10.2) that the growth response obtained on dilution is quite distinct from that obtained on supplementation of each diluted diet with pure lysine. The displacement of response curves at each protein level and the appearance of discrete

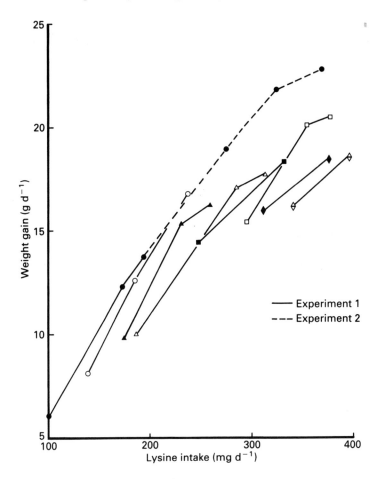

Fig. 10.2. Daily weight gain and lysine intake of chicks fed diets containing 140 (●), 160 (○), 180 (▲), 200 (△), 220 (■), 240 (□), 260 (♦) and 280 (◇) g crude protein kg^{-1}. (From D'Mello, 1988; source of data: Morris *et al.*, 1987. Reproduced permission of the World's Poultry Science Association.)

responses to supplementation with lysine question the authenticity of earlier efforts to validate the diet dilution technique. The protein effect on growth responses to amino acids also holds for tryptophan (Abebe and Morris, 1990b) and methionine (Morris *et al.*, 1992).

Discrepancies recorded between the diet dilution and the graded supplementation techniques are not just restricted to growth responses. As will be seen later in this chapter, marked differences also occur in carcass fat contents of chicks on diluted and amino acid supplemented regimes (Velu *et al.*, 1972; Seaton *et al.*, 1978; Gous and Morris, 1985; Burnham *et al.*, 1992).

Factorial methods

In view of the doubts concerning the diet dilution technique and the criticisms levelled at the supplementation method, it may well be argued that future prospects lie in the development of factorial models for the prediction of growth responses to amino acids. Such a view would imply a clear distinction between empirical and factorial approaches which is not justified in practice since factorial calculations invariably rely on empirically derived values. Thus factorial models depend upon some estimate of maintenance requirements for individual amino acids. For example, in the factorial method proposed by Hurwitz *et al.* (1978), estimates of maintenance amino acid requirements of chicks were derived from empirically derived values of Leveille *et al.* (1960) obtained with the adult rooster. Again, in the mathematical model used by Gous and Morris (1985) an estimate of the lysine requirement for maintenance was derived from empirical studies. In factorial calculations, it is common practice to use single maintenance requirement values for each amino acid. While such an approach may be justified for most amino acids, in the cases of those involved in deleterious interactions maintenance requirements may not be constant. Ousterhout (1960) showed that young chicks fed a diet devoid of lysine survived longer and lost less weight than those fed a diet lacking in arginine. Chicks fed a diet devoid of the three branched-chain amino acids survived longer and lost less weight than those fed diets devoid of either isoleucine or valine. In other studies (Okumura *et al.*, 1985) chicks were fed graded quantities of a standard diet or diets containing only half the recommended concentrations of leucine, isoleucine or valine. Chicks fed the leucine-deficient diet grew at a similar rate to those given similar quantities of the standard diet. In contrast, chicks fed either the valine- or isoleucine-deficient diet grew at a markedly poorer rate than control animals. Overall, these observations imply that amino acid antagonisms may operate in deficiency states and point to the dominant role of leucine in its interactions with its structural analogues. These observations also suggest that the maintenance requirements for valine are unlikely to remain constant and may be influenced to an appreciable degree by the relative dietary proportions of the other two branched-chain amino acids. For similar reasons, the isoleucine requirement for maintenance may also vary. In a recent study, however, Burnham and Gous (1992) discounted any such effects of branched-chain amino acid antagonisms on maintenance isoleucine requirements of adult chickens despite observing an unexpectedly low value for the efficiency of utilization of isoleucine for maintenance. It is not unreasonable to propose that the maintenance requirement for arginine may depend upon the lysine content of the diet.

Factorial models, in general, ignore the effects of interactions between amino acids. Thus, Hurwitz *et al.* (1978) specifically excluded such considerations in the development of their model. However, in a new computerized mathematical model based on a four-parameter kinetic equation, it is claimed that full account is taken of amino acid antagonisms (Muramatsu *et al.*, 1991). Assumptions are included on the basis of the experimental evidence of Austic and Scott (1975) for the lysine–arginine antagonism and Allen and Baker (1972) for the branched-chain amino acid interactions, further emphasizing the interdependence between factorial and empirical approaches.

Another problematic issue arises from a consideration of efficiency of utilization of amino acids for protein gain and for liveweight gain. Boorman and Burgess (1986) illustrate the dilemma with regard to lysine, calculating a net efficiency of utilization of lysine for protein gain as 0.89 whereas a much lower efficiency of 0.71 emerged for liveweight gain. In the cases of several other amino acids, including threonine, leucine and valine, efficiency could not be estimated from existing data and an assumed value of 0.70 was used in the factorial calculations (Boorman and Burgess, 1986).

Factors Affecting Responses to Amino Acids

The response of growing poultry to an amino acid is influenced by a wide range of factors. These include: immunological stress, environmental temperature, sex, age, species and several dietary factors (D'Mello, 1979, 1987, 1988). Critical evaluation indicates that these factors readily resolve into those which influence food intake and those which reduce the efficiency of utilization of an amino acid (D'Mello, 1979; Han and Baker, 1991). The distinction between the two categories of factors only emerges if responses are considered in relation to amino acid intake. Food intake-mediated factors may elicit differences in responses when these are plotted against dietary amino acid concentrations. However, any disparity effectively disappears on considering responses in terms of amino acid intake, as data points converge along a single response curve. On the other hand, factors reducing the efficiency of utilization of an amino acid induce differences which are sustained even when responses are viewed in the context of amino acid intake. In particular, the appearance of discrete response curves, within the range of limiting intakes, is clear evidence that efficiency of utilization has been affected, whereas differences in asymptote are not relevant in this respect.

Food intake-mediated factors

Immunological stress

Klasing and Barnes (1988) suggested reduced requirements of growing chicks for the sulfur amino acids and for lysine following immunological challenge with injected bacterial antigens. The immunogens injected included *Escherichia coli* lipopolysaccharide, *Salmonella typhimurium* lipopolysaccharide and heat-treated *Staphylococcus aureus* singly or in rotation. On the basis of the growth results, it was suggested that the lysine requirement of saline-injected control chicks was in excess of $9.5 \, \text{g kg}^{-1}$ diet, whereas immunogen-injected chicks required dietary lysine concentrations of between 7 and $9.5 \, \text{g kg}^{-1}$. However, when the growth responses are plotted against lysine intake (Fig. 10.3) it becomes apparent that the principal effect of the immunogens is to reduce food intake. Lysine utilization remains largely unaffected since much of the data, particularly those relating to lysine deficiency, appear along a single response curve. The data points for immunogen-treated chicks yield a lower plateau than that for the control group, but this is associated with factors other than lysine intake. If efficiency of utilization of dietary lysine had been affected by immunogen then any differences would have been reflected in discrete response curves over the entire range of lysine intake. Notwithstanding the evidence presented in Fig. 10.3, it should be stated that growth and food utilization efficiency may not represent the most appropriate criteria for quantifying the effects of immunological stress on amino acid requirements. There are reports that amino acid deficiency impairs immune function in broiler chicks (Tsiagbe *et al.*, 1987a,b).

Environmental temperature

March and Biely (1972) and McNaughton *et al.* (1978) have investigated the effects of environmental temperature on the growth responses of cockerels fed graded levels of lysine. The results of March and Biely (1972), shown in Fig. 10.4, indicate that when growth responses are considered in relation to dietary lysine content, two discrete curves are obtained for 20° and 31.1°C, implying a decrease in the efficiency of utilization of lysine at the higher temperature. However, if the same responses are plotted against daily intake of lysine (Fig. 10.5) it becomes apparent from the single response curve that lysine utilization has not changed. Chicks at 31.1°C merely consumed less food and higher dietary concentrations of lysine were therefore required to compensate for this reduction in appetite. The daily intake of lysine required to support a given rate of growth was similar at the two temperatures. It is likely that the data of McNaughton *et al.* (1978) would reflect a similar pattern when responses are considered in relation to lysine intake.

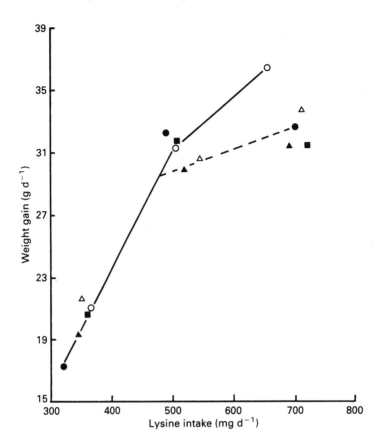

Fig. 10.3. Daily weight gain and lysine intake of chicks injected with saline (○) or with immunogens in the form of *E. coli* lipopolysaccharide (●), *S. typhimurium* lipopolysaccharide (▲), or heat-killed *S. aureus* (△). Additional groups of chicks (■) received the three immunogens in rotation. (Source of data: Klasing and Barnes, 1988.)

Sex

The responses of male and female broilers to available lysine concentrations in the diet have been published by Thomas *et al.* (1977). Viewed in terms of dietary concentrations of total lysine, two growth response curves are apparent (Fig. 10.6) implying genuine differences in the efficiency of utilization of lysine between males and females. Indeed, having arrived at this conclusion, Thomas *et al.* (1977) developed two regression equations for lysine requirements of male and female broilers. However, when the responses are considered as a function of daily lysine intake (Fig. 10.7) similarities in lysine utilization are seen with a daily intake of, say, 600 mg lysine supporting equivalent rates of growth in both males and females.

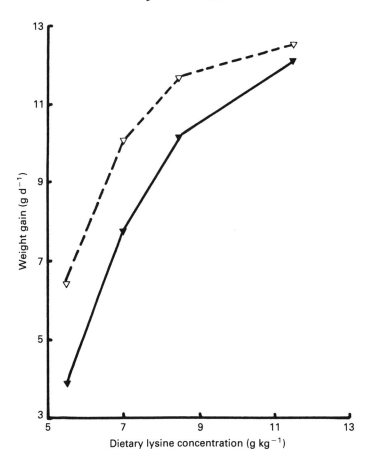

Fig. 10.4. The effects of two environmental temperatures (20°C, ▽, and 31.1°C, ▼) on daily weight gain of chicks fed graded concentrations of lysine. (From D'Mello, 1979; source of data: March and Biely, 1972. Reproduced permission of Butterworth-Heinemann Ltd.)

Age

It is consistently recorded that amino acid requirements, expressed as proportions of the diet, decrease with age (see Agricultural Research Council, 1975). In another review, D'Mello (1983) summarized published estimates of the requirements of turkeys for sulfur amino acids and these are reproduced in Table 10.2. A selection of the response data of Murillo and Jensen (1976) and of Behrends and Waibel (1980) are shown in Fig. 10.8. Expressed in these terms, these is little doubt that dietary requirements do decline with age. Of some concern, however, are the wide discrepancies

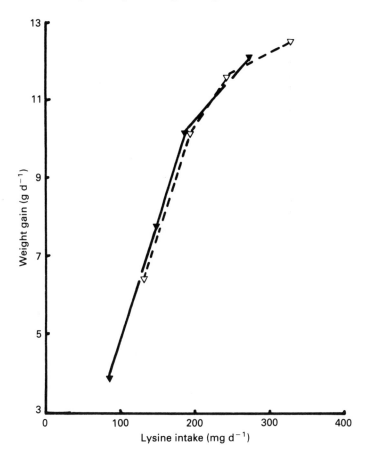

Fig. 10.5. Daily weight gain and lysine intake of chicks maintained at two environmental temperatures (20°C, ▽, and 31.1°C, ▼) (From D'Mello, 1979; source of data: March and Biely 1972. Reproduced permission of Butterworth-Heinemann Ltd.)

in recommendations offered by the different authors. Viewed in terms of dietary concentrations, the responses in Fig. 10.8 and the estimates in Table 10.2 appear irreconcilable. However, a different and more reassuring pattern emerges when growth responses from these experiments are considered in relation to daily intakes of the sulfur amino acids (Fig. 10.9). Considering the wide differences in dietary CP contents, ratios of methionine to cystine and ages of turkeys used, there appears to be a marked degree of homogeneity among the different data sets. This agreement is all the more significant in view of the widely held belief that empirical data obtained under one particular set of conditions are unlikely to have universal application. Irrespective of data source, the responses in Fig. 10.9

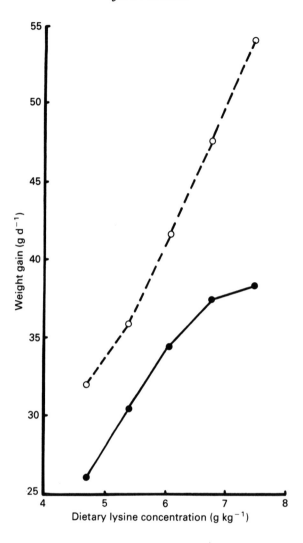

Fig. 10.6. Daily weight gain of male (○) and female (●) broiler chicks in relation to total dietary lysine concentration. (From D'Mello, 1979; source of data: Thomas *et al.*, 1977. Reproduced permission of Butterworth-Heinemann Ltd.)

suggest that turkeys growing at about $30\,\mathrm{g\,d^{-1}}$ require $500\,\mathrm{mg}$ of sulfur amino acids per day while older turkeys growing at $90\,\mathrm{g\,d^{-1}}$ require a daily intake of approximately $1700\,\mathrm{mg}$ of these amino acids. The improved agreement between the data of Behrends and Waibel (1980) and Murillo and Jensen (1976) when responses are considered in relation to methionine and cystine intake (Fig. 10.9) may be attributed to variations in food intake caused by differences in the environmental temperatures used in the two studies.

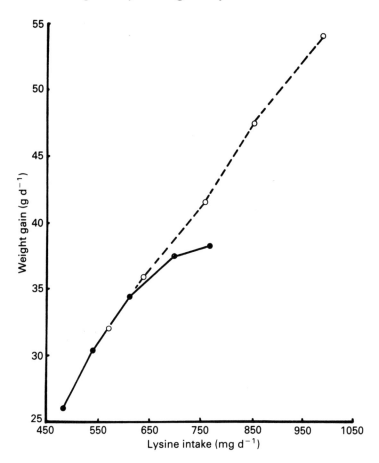

Fig. 10.7. Daily weight gain and lysine intakes of male (○) and female (●) broiler chicks. (From D'Mello, 1979; source of data: Thomas *et al.*, 1977. Reproduced permission of Butterworth-Heinemann Ltd.)

Species

The amino acid requirements of different species of growing poultry have been reviewed by the Agricultural Research Council (1975), and their recommendations are listed in Table 10.3. When expressed as dietary concentrations, differences in requirements are evident, particularly with respect to arginine. The question arises as to whether these differences reflect genuine variations in the efficiency of amino acid utilization. D'Mello (1979) addressed this issue at some length using evidence from studies specifically designed to compare amino acid utilization in young turkeys and chicks. With respect to arginine (Fig. 10.10) it was suggested that turkey poults and chicks respond similarly to different intakes of this

Table 10.2. Published estimates of the requirements of turkeys (g kg^{-1} diet) for methionine + cystine. (From D'Mello, 1983.)

Reference	Age of turkey (weeks)						
	1-3	1-4	0-4	0-7	4-8	8-12	16-20
D'Mello (1976)	8.3	-	-	-	-	-	-
Murillo and Jensen (1976)	-	-	-	-	-	8.1	-
Potter and Shelton (1976)	-	-	-	10.3	-	-	-
Potter *et al.* (1977)	-	-	-	11.0-12.0	-	-	-
Potter and Shelton (1978)	-	-	-	9.5	-	-	-
Potter and Shelton (1979)	-	-	11.0	-	10.0	-	-
Behrends and Waibel (1980)	-	9.5-10.1	-	-	-	7.0	4.3-4.8

Table 10.3. Amino acid requirements (g kg^{-1} diet) of growing poultry. (From Agricultural Research Council, 1975.)

Amino acid	Chickens		Turkeys		Ducklings	Goslings
	0-4 weeks	4-8 weeks	Poults	Growers		
Arginine	10.3	7.6	13.0	10.4	8.4	8.5
Glycine + serine	14.0	10.2	9.0	7.2	11.3	11.4
Histidine	4.8	3.6	5.0	4.0	3.9	4.0
Isoleucine	8.5	6.4	9.0	7.2	6.9	7.0
Leucine	14.7	10.7	14.0	11.2	11.8	12.0
Lysine	11.0	8.0	13.0	10.4	8.9	9.0
Methionine	4.8	3.6	5.0	3.8	3.9	4.0
Methionine + cystine	9.2	6.7	8.0	6.4	7.4	7.5
Phenylalanine	8.5	6.4	8.0	6.4	6.9	7.0
Phenylalanine + tyrosine	15.8	11.6	14.0	11.2	12.8	13.0
Threonine	7.4	5.3	9.0	7.2	5.9	6.0
Tryptophan	2.1	1.5	2.2	1.8	1.7	1.7
Valine	9.8	7.1	10.0	8.0	7.9	8.0

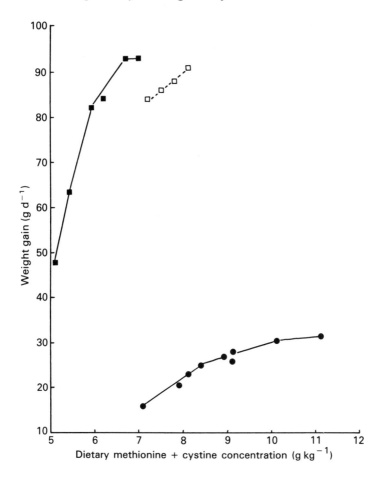

Fig. 10.8. The effects of dietary methionine + cystine concentration and age on daily weight gain in turkeys. Source of data: Behrends and Waibel (1980), 1 to 4 (●) and 8 to 12 (■) weeks of age; Murillo and Jensen (1976), 8 to 12 weeks of age (□).

amino acid. Potential growth rates in the turkey, however, are much greater and consequently responses occur to higher intakes of arginine. Food intake per unit of liveweight gain is higher in the chick than in the poult and these two factors together contribute to the much lower arginine requirement of the chick when this value is expressed in terms of dietary concentrations (D'Mello and Emmans, 1975). There is no evidence of any differences in the efficiency of arginine utilization between the two species. Any such difference would have been represented by two discrete response curves in Fig. 10.10. Interspecies similarities were later extended to isoleucine and valine responses for turkey poults and chicks (D'Mello, 1975) and in the case of methionine + cystine (D'Mello, 1976), it was observed that data

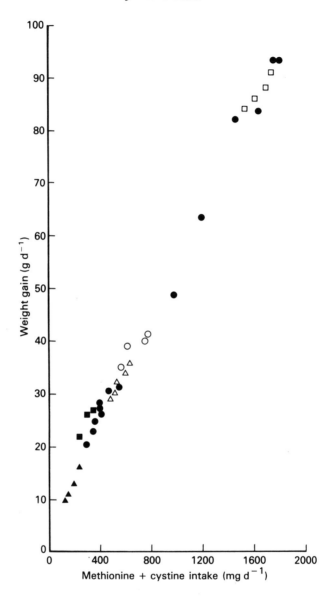

Fig. 10.9. Daily growth rates of turkeys in relation to daily intake of methionine + cystine. (From D'Mello, 1983; source of data: Behrends and Waibel, 1980 (●); D'Mello, 1976 (■); Murillo and Jensen, 1976 (□); Potter and Shelton, 1978 (△); Potter and Shelton, 1979 (▲); Potter *et al.*, 1977 (○).)

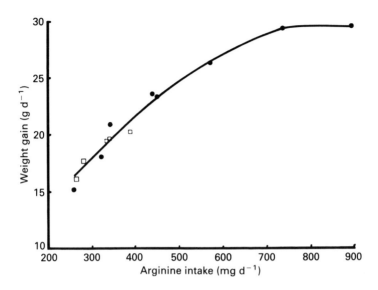

Fig. 10.10. Daily weight gain responses of turkey poults (●) and young chicks (□) in relation to daily arginine intake. (From D'Mello and Emmans, 1975; source of data: turkeys, D'Mello and Emmans, 1975; chicks, D'Mello and Lewis, 1970. Reproduced permission of British Poultry Science Ltd.)

obtained with the laboratory rat also conformed with the responses seen in avian species (Fig. 10.11). The plateaux for the various sets of data indicate the maximum growth possible in the respective experiments with the animals and diets used (Stockland *et al.*, 1973; Boomgaardt and Baker, 1973; D'Mello, 1973b, 1976).

The demonstration of strain differences in arginine utilization by chicks (Nesheim, 1968) is less readily explained and may represent an exception to the general rule. It should be recalled, however, that the strains differing in arginine requirements were selected on the basis of their responses to high-casein diets containing excess lysine. Under these conditions there is a clear difference in responses whether these are considered in relation to dietary concentrations or to daily intakes (D'Mello, 1973c). But when an arginine-deficient amino acid mixture is used in place of casein, the strain differences in responses to this deficiency largely disappear. The absence of breed differences in arginine responses of commercial broiler stocks was demonstrated by Wilburn and Fuller (1975). Their studies with Cobb and Hubbard chicks, fed high-casein diets to accentuate any potential breed differences, indicated only minor discrepancies when growth responses were considered in relation to dietary concentrations. These differences virtually disappeared when the responses were plotted against arginine intake (see D'Mello, 1979), suggesting that variations

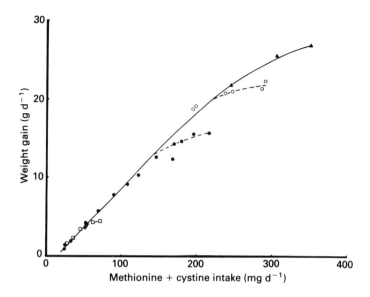

Fig. 10.11. Weight gain responses of growing rats (□), slow-growing chicks (●), fast-growing chicks (○) and turkey poults (▲) in relation to methionine and cystine intake. (From D'Mello, 1976; source of data: rats, Stockland *et al.*, 1973; slow-growing chicks, Boomgaardt and Baker, 1973; fast-growing chicks, D'Mello, 1973b; turkeys, D'Mello, 1976. Reproduced permission of British Poultry Science Ltd.)

in food intake were primarily responsible for the differences in dietary requirements.

There is currently some interest in the protein and amino acid requirements of genetically lean and fat chickens. The results of one study (Leclercq and Guy, 1991) further illustrate the importance of considering food intake. When growth responses were plotted against dietary CP content no difference was observed between the two genotypes. However, if growth responses were plotted against protein intake, the slope of the regression for the lean chickens was steeper than that for the fat genotype and the plateau was attained at a lower protein intake with the lean birds. In a subsequent study on methionine + cystine utilization, Leclercq *et al.* (1993) were unable to distinguish between the two genotypes when growth responses were considered as a function of sulfur amino acid intake, although lean chickens deposited these amino acids more efficiently in feather and body proteins. Food intake was depressed, and feather synthesis enhanced, in lean relative to fat chickens.

Dietary factors

METABOLIZABLE ENERGY CONCENTRATION
Dietary ME content is widely acknowledged to exert a dominant role in the regulation of food intake in growing poultry. Boomgaardt and Baker (1973) examined the effects of dietary ME concentration on the response of chicks to graded doses of methionine + cystine. A cursory evaluation of their results indicates three distinct growth response curves for the three dietary ME concentrations used (Fig. 10.12). However, the efficiency of utilization of these amino acids is unaffected by energy content of the diet since a single response curve is obtained on plotting weight gain against methionine + cystine intake (Fig. 10.13). It is clear that dietary ME, within the range tested, exerts its effect principally through variations in food intake and without affecting amino acid utilization (D'Mello, 1979).

AMINO ACID IMBALANCE
As discussed in Chapter 4, dietary amino acid imbalance precipitates its adverse effects by reducing food intake while efficiency of amino acid utilization remains unimpaired (Harper and Rogers, 1965). However, the results of Morris *et al.* (1987) presented in Fig. 10.2 and those of Mendonca and Jensen (1989) and Abebe and Morris (1990b) appear to challenge this universally accepted rule. The responses in Fig. 10.2 indicate that as CP content increases from 140 to 280 g kg^{-1} diet there is a marked and progressive reduction in the efficiency of utilization of the first-limiting amino acid, lysine. Both the positive displacement of the response curves and the reduction in the slope of these curves, particularly at the higher CP concentrations (260 and 280 g kg^{-1} diet) imply substantial decreases in the efficiency of utilization of lysine. It will also be apparent that the displacement of the response curves occurs even at suboptimal levels of dietary CP. Abebe and Morris (1990b) attribute these responses to an effect of 'general imbalance caused by the excess of amino acids absorbed following digestion of a high protein diet'. Such an explanation is inconsistent with the evidence and with the hypothesis of Harper and Rogers (1965). This hypothesis was originally developed on the basis of metabolic studies with rats and has not been tested exhaustively in supplementation studies using pure amino acids. Accordingly, D'Mello (1990) conducted an investigation employing crystalline amino acids to bring about changes in CP concentrations and degree of imbalance in lysine-deficient diets fed to young chicks. Three basal diets composed mainly of maize gluten meal, wheat, glucose and amino acids were each formulated to contain 5.1 g lysine kg^{-1} dry matter (DM). The first, containing 225 g CP kg^{-1} DM served as the control diet. The second and third basal diets were similar to the control apart from the inclusion of a moderately or of a severely imbalanced mixture of amino acids devoid of lysine. These mixtures were added at the expense

J.P.F. D'Mello

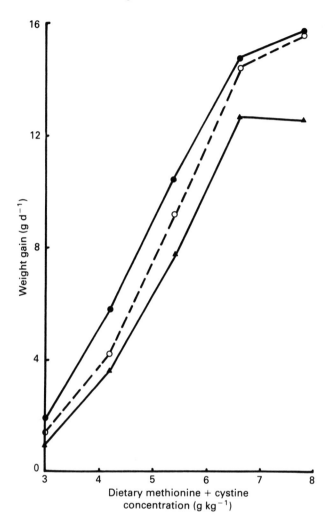

Fig. 10.12. Daily growth rates of young chicks in relation to dietary methionine + cystine and metabolizable energy concentrations. Energy levels (MJ kg^{-1}): (●) 10.9; (○) 12.6; (▲) 14.2. (From D'Mello, 1979; source of data: Boomgaardt and Baker, 1973. Reproduced permission of Butterworth-Heinemann Ltd.)

of glucose thereby increasing CP concentrations to 315 g kg^{-1} DM. Each of these three basal diets was supplemented with graded levels of lysine. The relatively poor growth performance of chicks fed the control diet was depressed further by additions of the two amino acid mixtures lacking lysine (Fig. 10.14). This follows the classic pattern established with the rat (Harper and Rogers, 1965). The severity of the adverse effects was directly proportional to the degree of imbalance in the mixtures. In all three dietary

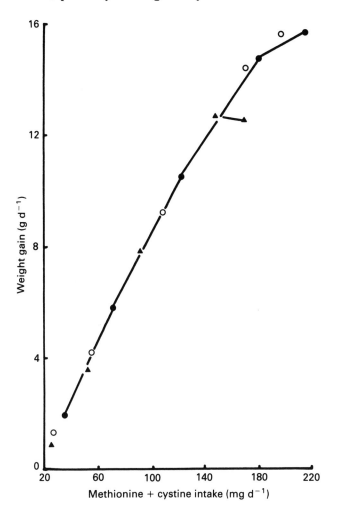

Fig. 10.13. Daily chick growth and methionine + cystine intake at three dietary concentrations of metabolizable energy (MJ kg^{-1}): (●) 10.9; (○) 12.6; (▲) 14.2. (From D'Mello, 1979; source of data: Boomgaardt and Baker, 1973. Reproduced permission of Butteworth-Heinemann Ltd.)

regimes graded growth responses occurred to lysine supplementation but chicks in the two imbalanced series failed to attain comparable growth rates to those in the control regimes at equivalent levels of lysine addition. Consequently, growth remained depressed, relative to control, in the imbalanced groups at all levels of additional lysine, the retardation being particularly marked in the severely imbalanced groups. However, a highly significant linear relationship was observed between lysine intake and

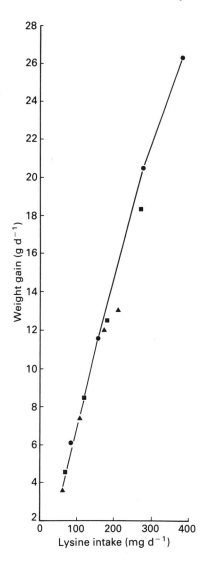

Fig. 10.14. Daily weight gain and lysine intake of chicks fed diets containing 225 g crude protein kg^{-1} dry matter (●) or similar diets supplemented with a moderately (■) or severely (▲) imbalanced mixture of amino acids lacking lysine which increased crude protein content to 315 g kg^{-1} dry matter. (Source of data: D'Mello, 1990.)

weight gain with all data points contributing to a single response curve. The appearance on a single response curve of all data points drawn from three diverse dietary regimes suggests that neither CP level nor severity of amino acid imbalance exerts any effect on lysine utilization. This finding is entirely consistent with the observations of Harper and Rogers (1965) and their hypothesis therefore remains intact. It thus appears unlikely that amino acid imbalance is a satisfactory explanation for the protein effect on lysine utilization (Fig. 10.2) and the issues raised by the data of Morris *et al.* (1987) and others (Mendonca and Jensen, 1989; Abebe and Morris, 1990a,b) remain essentially unresolved.

EFFECTS OF VITAMINS AND COCCIDIOSTATS
A number of other nutritional factors affect the growth responses of poultry to amino acids by altering food intake. D'Mello (1979) indicated that the results of Looi and Renner (1974) suggesting differences in methionine + cystine responses between chicks fed adequate and vitamin B_{12}-deficient diets could be explained in terms of variations in food intake. A similar reason may also be advanced for the proposed interaction between ionophore coccidiostats and dietary amino acids. Although Willis and Baker (1980) suggested the existence of a striking interaction between lasalocid and methionine + cystine in diets severely deficient in these amino acids, inspection of Fig. 10.15 suggests that methionine + cystine utilization is unaffected by lasalocid supplementation. The coccidiostat merely enhances food intake in chicks fed the deficient diets resulting in higher intakes of methionine + cystine with consequent improvements in growth.

Factors reducing amino acid utilization

The lysine–arginine antagonism provides a distinctive, but not unique, example of how one amino acid may reduce the efficiency of utilization of another. The results of one such study (D'Mello and Lewis, 1970) are shown in Table 10.4. A basal diet marginally deficient in arginine but adequate in lysine was employed. Addition of excess lysine to this basal diet to give dietary concentrations of 13.5, 16.0 and 18.5 g kg^{-1}, progressively reduced growth performance and enhanced the quantity of arginine required to reverse these adverse effects. Thus, at dietary lysine concentrations of 16 and 18.5 g kg^{-1} arginine requirements had increased from 10 to 14.5 and 16.0 g kg^{-1} diet respectively. These increases in arginine requirements remain prominent even after responses are considered in relation to arginine intake (Fig. 10.16). The discrete response curves and the changes in slope indicate that arginine utilization is markedly reduced by excess dietary lysine.

The response of growing chicks to the lysine–arginine interaction may be modified by at least two dietary factors. Supplementation with electrolytes, particularly potassium acetate, reduces the severity of this antagonism (O'Dell and Savage, 1966). Subsequent studies (Scott and Austic, 1978) indicated that the primary effect of dietary cations is not mediated via arginine metabolism, since renal arginase activity remained unaffected, but through increased catabolism of lysine. A second factor influencing the lysine–arginine antagonism is the presence of structural analogues of arginine in the diet. Canavanine is one such analogue

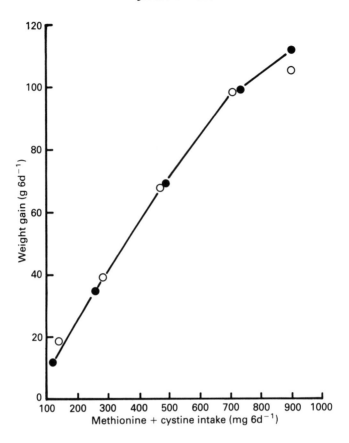

Fig. 10.15. Weight gain and methionine + cystine intake of chicks fed purified diets without (●) or with (○) lasalocid, 125 mg kg^{-1}. (From D'Mello, 1988; source of data: Willis and Baker, 1980. Reproduced permission of World's Poultry Science Association.)

Table 10.4. Effects of dietary lysine and arginine on daily weight gain (g) of chicks. (From D'Mello and Lewis, 1970.)

Dietary arginine (g kg^{-1})	Dietary lysine (g kg^{-1})			
	11.0	13.5	16.0	18.5
8.5	17.7	16.1	13.0	7.2
10.0	19.6	19.5	17.1	14.0
11.5	19.8	20.3	18.4	15.6
13.0	19.8	19.3	19.1	17.2
14.5	19.9	19.4	20.3	19.0
16.0	19.5	19.4	19.4	19.8

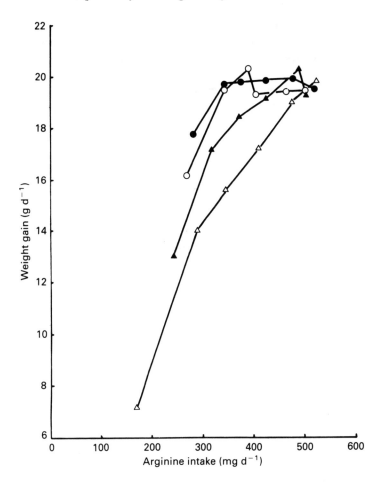

Fig. 10.16. Daily growth rates and arginine intake of chicks fed 11.0 (●), 13.5 (○), 16.0 (▲) and 18.5 (△) g lysine kg^{-1} diet. (From D'Mello, 1973c. Reproduced permission of British Poultry Science Ltd.)

occurring naturally in the seed of the legume *Canavalia ensiformis* (see D'Mello, 1991). The existence of a canavanine–arginine antagonism in chicks fed this bean has been proposed (D'Mello *et al.*, 1989) and lysine exacerbates this interaction (see also Chapter 4).

The practical significance of the lysine–arginine antagonism not only emanates from its relevance to the assessment of arginine requirements of growing poultry but also from its effect in determining the nutritive value of certain feedingstuffs. Thus Miller and Kifer (1970) noted that the nutritive value of an aged sample of fishmeal could be enhanced by arginine supplementation and impaired by lysine or methionine addition. Leslie *et al.* (1976) reported the existence of an adverse lysine:arginine ratio in rapeseed meal when this ingredient served as the sole source of dietary protein. Under these conditions, arginine supplementation improved the nutritive value of rapeseed meal whereas lysine addition precipitated severe arginine-responsive growth depressions. Arginine supplementation is also beneficial when chicks are fed diets containing single-cell protein sources such as hydrocarbon-grown yeast (D'Mello, 1973b) and methanol-grown bacteria (D'Mello, 1978). An interesting feature of the bacterial source has been the observation that chicks also respond to supplementation with electrolytes, which act by reducing mortality (Talbot, 1978). It remains to be established whether this represents another dimension of the lysine–arginine–electrolyte interactions discussed earlier.

Several studies with the young chick and the turkey poult have demonstrated clear patterns of interdependence in the metabolism of and requirements for the branched-chain amino acids, leucine, isoleucine and valine (D'Mello and Lewis, 1970; Allen and Baker, 1972; D'Mello, 1974, 1975). For example, dietary leucine exerts a profound effect on the valine requirements of the chick. Thus, for dietary leucine concentrations of 14, 24 and 34 $g\,kg^{-1}$, valine requirements are 7.7, 8.9 and 10.1 $g\,kg^{-1}$ diet respectively (D'Mello and Lewis, 1970). The leucine–isoleucine antagonism has also been described in quantitative terms. Increasing dietary leucine concentrations from 14 to 21.5 and 29 $g\,kg^{-1}$ enhances isoleucine requirements of the chick from 5.8 to 6.2 and 6.5 $g\,kg^{-1}$ diet respectively (D'Mello and Lewis, 1970). It will be noted that the leucine–isoleucine antagonism is considerably less potent than that between leucine and valine, a feature alluded to by D'Mello and Lewis (1970). Consequently, the recent observation by Burnham *et al.* (1992) that dietary leucine at 1.76 times requirement depresses chick growth without enhancing isoleucine requirements is consistent with the weak antagonism between these two amino acids. The complexity of antagonisms among the branched-chain amino acids is further illustrated by the effects of dietary isoleucine on the requirements for valine and leucine. Thus, an isoleucine concentration of 5.2 $g\,kg^{-1}$ diet permits satisfactory chick growth and efficiency of food utilization with the dietary concentrations of leucine and valine set at 9.8 and 6.3 $g\,kg^{-1}$ respectively. However, a higher isoleucine concentration of 7.6 $g\,kg^{-1}$ diet elicits increases in the requirements for leucine and valine to 11.0 and 7.5 $g\,kg^{-1}$ diet respectively (D'Mello, 1974). The extent to which the utilization of the branched-chain amino acids is affected by their antagonisms may be gauged by the growth responses of turkey poults to excess leucine (Fig. 10.17). As the leucine content is increased from 14.2 to 20.2 $g\,kg^{-1}$ diet, there is a positive displacement in the response curve to valine indicating that valine utilization is impaired by excess leucine. The lower plateau in the response curve to excess leucine suggests that isoleucine may now be the limiting factor.

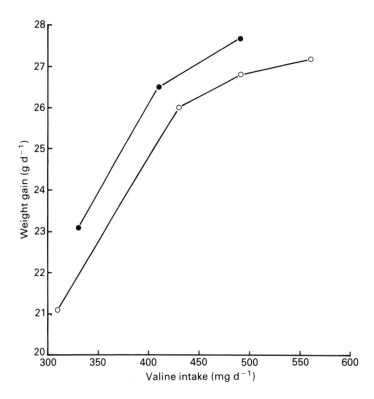

Fig. 10.17. Daily weight gain and valine intake of turkey poults fed 14.2 (●) and 20.2 (○) g leucine kg^{-1} diet. (From D'Mello, 1988; source of data: D'Mello, 1975. Reproduced permission of World's Poultry Science Association.)

Effects of Dietary Amino Acids on Carcass Composition

The effects of varying dietary concentrations of amino acids on body composition of growing poultry are imperfectly documented. However, the results of three studies (Velu *et al*, 1972; Gous and Morris, 1985; Burnham *et al.*, 1992) indicated striking effects of dietary isoleucine and lysine on fat content of 3-week-old broiler chicks. Contrasting effects were observed, depending upon the degree of deficiency of either amino acid (Figs. 10.18 and 10.19). At very low levels of isoleucine or lysine, fat content was relatively low but this increased progressively with graded dietary supplements of the amino acid. This, presumably, is a reflection of extremely low food intakes which are a characteristic feature of severe amino acid deficiency (D'Mello and Lewis, 1978). However, a point was reached for each amino acid when further dietary additions reduced carcass fat concentrations. Thus the effect of an amino acid on carcass fat content depends

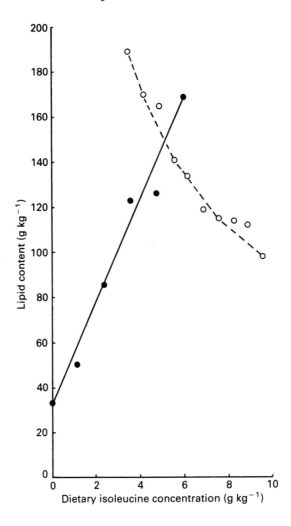

Fig. 10.18. The effects of dietary isoleucine concentration on lipid content of broiler chicks at 3 weeks of age. Experimental diets were fed from 8 days of age. Source of data: Velu *et al.*, 1972 (●); Burnham *et al.*, 1992 (○).

upon the extent of deficiency, a severe inadequacy eliciting much lower fat concentrations than diets with a moderate deficiency or with adequate levels of the amino acid. Consequently, statements indicating a general effect of amino acid deficiency in increasing carcass fat deposition (Boorman and Burgess, 1986) require qualifications as to the degree of deficiency. It may be a argued that it would be more instructive to consider lipid gain in relation to amino acid intake. However, the lack of requisite

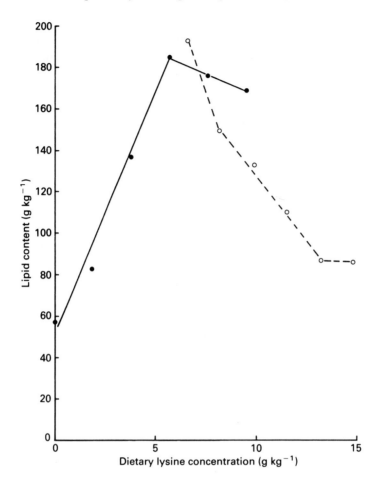

Fig. 10.19. The effects of dietary lysine concentration on lipid content of broiler chicks at 3 weeks of age. Experimental diets were fed from 7 or 8 days of age. Source of data: Velu *et al.*, 1972 (●); Gous and Morris, 1985 (○).

data in the paper of Velu *et al.* (1972) does not permit such an approach without employing untenable assumptions. In any event, it is unlikely that any manipulation of data would substantially alter the pattern of contrasting responses shown in Figs. 10.18 and 10.19.

It should be pointed out that the responses relating to severe deficiency (Velu *et al.*, 1972) were derived by the graded supplementation procedure with purified diets, whereas those referring to moderate deficiency were obtained by the diet dilution technique using diets based largely on conventional intact protein sources (Gous and Morris, 1985; Burnham *et al.*, 1992). The carcass fat data shown in Fig. 10.19 obtained by the diet dilution method are markedly different from those observed by Seaton *et al.* (1978), with the graded supplementation technique indicating no effect of dietary lysine, within the range 7.2 to 16.8 g kg^{-1}, on carcass fat composition of 3-week-old chicks despite growth increments up to 10.4 g lysine kg^{-1} diet. Thus methodological aspects should be considered in any interpretation of carcass fat responses to dietary amino acid concentrations. The axiom that broiler chickens 'seek low levels of fatness' (Gous *et al.*, 1990) requires further appraisal.

Conclusions

The responses of growing poultry to individual amino acids may be determined by empirical approaches involving the supplementation technique or the diet dilution method. Although earlier reviews indicated considerable compatibility of growth responses derived by these methods, more recent research revives the debate concerning the validity of the diet dilution technique. It is a condition of this procedure that responses are not confounded by the unavoidable variation in protein contents of diluted diets. It is now known that this condition is not satisfied since distinct and disparate responses occur to graded supplementation with the limiting amino acid at each protein level in the diluted series. Indeed, it has now been concluded that the amino acid requirements of the growing chick are 'a simple linear function of the dietary protein content' (Morris *et al.*, 1987). These results also imply reduced utilization of the first-limiting amino acid as dietary crude protein content increases. The protein effect on amino acid utilization has been attributed to amino acid imbalance. However, a recent study (D'Mello, 1990) employing the supplementation technique has shown that moderate or severe imbalance exerts no effect on the utilization of the first-limiting amino acid. Consequently, the biochemical basis of the protein effect on amino acid responses of growing chicks remains elusive. A limited analysis of data reveals that the diet dilution and graded supplementation techniques may elicit contrasting effects on carcass fat content in chicks fed varying dietary concentrations of an amino acid.

A critical review of the factors affecting amino acid responses in growing poultry indicates that immunological challenge; environmental temperature; sex, age and species of poultry; dietary energy content; dietary amino acid imbalance; and ionophore coccidiostats all exert their effects by altering voluntary food intake. However, in the case of deleterious antagonisms such as those involving the branched-chain amino acids or that between lysine and arginine there are indications of genuine changes in the efficiency of utilization of dietary amino acids. In addition there is limited evidence of genetic differences in amino acid utilization in growing chicks. The implications for other genetic models, including lean and obese lines and transgenic poultry, have yet to be established.

References

Abebe, S. and Morris, T.R. (1990a) Note on the effects of protein concentration on responses to dietary lysine by chicks. *British Poultry Science* 31, 255–260.

Abebe, S. and Morris, T.R. (1990b) Effects of protein concentration on responses to dietary tryptophan by chicks. *British Poultry Science* 31, 267–272.

Agricultural Research Council (1975) *The Nutrient Requirements of Farm Livestock: No. 1, Poultry.* Agricultural Research Council, London.

Allen, N.K. and Baker, D.H. (1972) Quantitative efficacy of dietary isoleucine and valine for chick growth as influenced by variable quantities of excess dietary leucine. *Poultry Science* 51, 1291–1298.

Austic, R.E. and Scott, R.L. (1975) Involvement of food intake in the lysine-arginine antagonism in chicks. *Journal of Nutrition* 105, 1122–1131.

Barbour, G., Latshaw, J.D. and Bishop, B. (1993) Lysine requirement of broiler chicks as affected by protein source and method of statistical evaluation. *British Poultry Science* 34, 747–756.

Bedford, M.R. and Summers, J.D. (1985) Influence of the ratio of essential to non-essential amino acids on performance and carcase composition of the broiler chick. *British Poultry Science* 26, 483–491.

Behrends, B.R. and Waibel, P.E. (1980) Methionine and cystine requirements of growing turkeys. *Poultry Science* 59, 849–859.

Boomgaardt, J. and Baker, D.H. (1973) Effect of dietary energy concentration on sulfur amino acid requirements and body composition of young chicks. *Journal of Animal Science* 36, 307–311.

Boorman, K.N. and Burgess, A.D. (1986) Responses to amino acids. In: Fisher, C. and Boorman, K.N. (eds) *Nutrient Requirements of Poultry and Nutritional Research.* Butterworths, London, pp. 99–123.

Burnham, D. and Gous, R.M. (1992) Isoleucine requirements of the chicken: requirement for maintenance. *British Poultry Science* 33, 59–69.

Burnham, D., Emmans, G.C. and Gous, R.M. (1992) Isoleucine requirements of the chicken: the effect of excess leucine and valine on the response to isoleucine. *British Poultry Science* 33, 71–87.

D'Mello, J.P.F. (1973a) Aspects of threonine and glycine metabolism in the chick (*Gallus domesticus*). *Nutrition and Metabolism* 15, 357–363.

D'Mello, J.P.F. (1973b) Amino acid supplementation of hydrocarbon-grown yeast in diets for young chicks. *Nutrition Reports International* 8, 105–109.

D'Mello, J.P.F. (1973c) Amino acid interactions in poultry nutrition. In: *Proceedings of 4th European Poultry Conference.* World's Poultry Science Association, London, pp. 331–336.

D'Mello, J.P.F. (1974) Plasma concentrations and dietary requirements of leucine, isoleucine and valine: studies with the young chick. *Journal of the Science of Food and Agriculture* 25, 187–196.

D'Mello, J.P.F. (1975) Amino acid requirements of the young turkey: leucine, isoleucine and valine. *British Poultry Science* 16, 607–615.

D'Mello, J.P.F. (1976) Requirements of the young turkey for sulphur amino acids and threonine: comparison with other species. *British Poultry Science* 17, 157–162.

D'Mello, J.P.F. (1978) Responses of young chicks to amino acid supplementation of methanol-grown dried microbial cells. *Journal of the Science of Food and Agriculture* 29, 453–460.

D'Mello, J.P.F. (1979) Factors affecting amino acid requirements of meat birds. In: Haresign, W. and Lewis, D. (eds) *Recent Advances in Animal Nutrition*. Butterworths, London, pp. 1–15.

D'Mello, J.P.F. (1982) A comparison of two empirical methods of determining amino acid requirements. *World's Poultry Science Journal* 38, 114–119.

D'Mello, J.P.F. (1983) Amino acid requirements of the turkey poult. In: Larbier, M. (ed.) *Proceedings of the 4th European Symposium on Poultry Nutrition*, World's Poultry Science Association, Tours, France, pp. 66–73.

D'Mello, J.P.F. (1987) Dietary interactions influencing amino acid utilisation. In: *Proceedings of the 6th European Symposium on Poultry Nutrition*, Königslutter, pp. RT15–RT16.

D'Mello, J.P.F. (1988) Dietary interactions influencing amino acid utilisation by poultry. *World's Poultry Science Journal* 44, 92–102.

D'Mello, J.P.F. (1990) Lysine utilisation by broiler chicks. In: *Proceedings of the VIII European Poultry Conference*, Barcelona, Spain, pp. 302–305.

D'Mello, J.P.F. (1991) Toxic amino acids. In: D'Mello, J.P.F., Duffus, C.M. and Duffus, J.H. (eds) *Toxic Substances in Crop Plants*. The Royal Society of Chemistry, Cambridge, pp. 21–48.

D'Mello, J.P.F. and Emmans, G.C. (1975) Amino acid requirements of the young turkey: lysine and arginine. *British Poultry Science* 16, 297–306.

D'Mello, J.P.F. and Lewis, D. (1970) Amino acid interactions in chick nutrition. 3. Interdependence in amino acid requirements. *British Poultry Science* 11, 367–385.

D'Mello, J.P.F. and Lewis, D. (1971) Amino acid interactions in chick nutrition. 4. Growth, food intake and plasma amino acid patterns. *British Poultry Science* 12, 345–358.

D'Mello, J.P.F. and Lewis, D. (1978) Effect of nutrient deficiencies in animals: amino acids. In: Rechcigl, M. (ed.) *CRC Handbook Series in Nutrition and Food, Section E: Nutritional Disorders, Vol. II.* CRC Press, Boca Raton, Florida, pp. 441–490.

D'Mello, J.P.F., Acamovic, T. and Walker, A.G. (1989) Nutritive value of jack beans (*Canavalia ensiformis*) (L). (DC). for young chicks: effect of amino acid supplementation. *Tropical Agriculture (Trinidad)* 66, 201–205.

Fisher, C. and Morris, T.R. (1970) The determination of the methionine requirement of laying pullets by a diet dilution technique. *British Poultry Science* 11, 67–82.

Fisher, C., Morris, T.R. and Jennings, R.C. (1973) A model for the description and prediction of the response of laying hens to amino acid intake. *British Poultry Science* 14, 469–484.

Gous, R.M. (1980) An improved method for measuring the response of broiler chickens to increasing dietary concentrations of an amino acid. In: *Proceedings of 6th European Poultry Conference*, World's Poultry Science Association, Hamburg, Vol. III, pp. 32–39.

Gous, R.M. (1986) Measurement of response in nutritional experiments. In: Fisher, C. and Boorman, K.N. (eds) *Nutrient Requirements of Poultry and Nutritional Research*. Butterworths, London, pp. 41–57.

Gous, R.M. and Morris, T.R. (1985) Evaluation of a diet dilution technique for measuring the response of broiler chickens to increasing concentrations of lysine. *British Poultry Science* 26, 147–161.

Gous, R.M., Emmans, G.C., Broadbent. L.A. and Fisher. C. (1990) Nutritional effects on the growth and fatness of broilers. *British Poultry Science* 31, 495–505.

Han, Y.M. and Baker, D.H. (1991) Lysine requirements of fast-growing and slow-growing broiler chicks. *Poultry Science* 70, 2108–2114.

Harper, A.E. (1964) Amino acid toxicities and imbalances. In: Munro, H.N. and Allison, J.B. (eds) *Mammalian Protein Metabolism, Vol. II*. Academic Press, New York, pp. 87–134.

Harper, A.E. and Rogers, Q.R. (1965) Amino acid imbalance. *Proceedings of the Nutrition Society* 24, 173–190.

Hewitt, D. and Lewis, D. (1972) The effect of dietary lysine level, restriction of food intake and sampling time on levels of amino acids in the blood plasma of chicks. *British Poultry Science* 13, 387–398.

Hurwitz, S., Sklan, D. and Bartov, I. (1978) New formal approaches to the determination of energy and amino acid requirements of chicks. *Poultry Science* 57, 197–205.

Klasing, K.C. and Barnes, D.M. (1988) Decreased amino acid requirements of growing chicks due to immunologic stress. *Journal of Nutrition* 118, 1158–1164.

Leclercq, B. and Guy, G. (1991) Further investigations on protein requirement of genetically lean and fat chickens. *British Poultry Science* 32, 789–798.

Leclercq, B., Chagneau, A.M., Cochard, T., Hamzaoui, S. and Larbier, M. (1993) Comparative utilization of sulfur amino acids by genetically lean or fat chickens. *British Poultry Science* 34, 383–391.

Leslie, A.J., Summers, J.D., Grandhi, R. and Leeson, S. (1976) Arginine-lysine relationship in rapeseed meal. *Poultry Science* 55, 631–637.

Leveille, G.A., Shapiro, R. and Fisher, H. (1960) Amino acid requirements for maintenance in the adult rooster. *Journal of Nutrition* 72, 8–15.

Looi, S.H. and Renner, R. (1974) Effect of feeding 'carbohydrate-free' diets on the chick's requirement for methionine. *Journal of Nutrition* 104, 400–404.

March, B.E. and Biely, J. (1972) The effect of energy supplied from the diet and from environment heat on the response of chicks to different levels of dietary lysine. *Poultry Science* 51, 665–668.

McNaughton, J.L., May, J.D., Reece, F.N. and Deaton, J.W. (1978) Lysine requirement of broilers as influenced by environmental temperatures. *Poultry Science* 57, 57–64.

Mendonca, C.X. and Jensen. L.S. (1989) Influence of protein concentration on the sulphur-containing amino acid requirement of broiler chickens. *British Poultry Science* 30, 889–898.

Miller, D. and Kifer, R.R. (1970) Factors affecting protein evaluation of fish meal by chick bioassay. *Poultry Science* 49, 999–1004.

Morris, T.R. (1972) Prospects for improving the efficiency of nutrient utilization. In: Freeman, B.M. and Lake, P.E. (eds) *Egg Formation and Production*. British Poultry Science Ltd, Edinburgh, pp. 139–159.

Morris, T.R. (1983) The interpretation of response data from animal feeding trials. In: Haresign, W. (ed.) *Recent Advances in Animal Nutrition*. Butterworths, London, pp. 13–23.

Morris, T.R., Al-Azzawi, K., Gous, R.M. and Jackson, G.L. (1987) Effects of protein concentration on responses to dietary lysine by chicks. *British Poultry Science* 28, 185–195.

Morris, T.R., Gous, R.M. and Abebe, S. (1992) Effects of dietary protein concentration on the response of growing chicks to methionine. *British Poultry Science* 33, 795–803.

Muramatsu, T., Ohshima, H., Goto, M., Mori, S. and Okumura, J. (1991) Growth prediction of young chicks: do equal deficiencies of essential amino acids produce equal growth responses? *British Poultry Science* 32, 139–149.

Murillo, M.G. and Jensen, L.S. (1976) Methionine requirement of developing turkeys from 8–12 weeks of age. *Poultry Science* 55, 1414–1418.

Nesheim, M.C. (1968) Genetic variation in arginine and lysine utilization. *Federation Proceedings* 27, 1210–1214.

Noble, D.O., Pickard, M.L., Dunnington, E.A., Uzu, G., Larsen, A.S. and Siegel, P.B. (1993) Food intake adjustments of chicks: short term reactions of genetic stocks to deficiencies in lysine, methionine or tryptophan. *British Poultry Science* 34, 725–735.

O'Dell, B.L. and Savage, J.E. (1966) Arginine-lysine antagonism in the chick and its relationship to dietary cations. *Journal of Nutrition* 90, 364–370.

Okumura, J., Mori, S. and Muramatsu, T. (1985) Relationship between food consumption and energy and nitrogen utilisation by chicks given varying amounts of standard and leucine-, isoleucine- and valine-deficient diets. *British Poultry Science* 26, 519–525.

Ousterhout, L.E. (1960) Survival time and biochemical changes in chicks fed diets lacking different essential amino acids. *Journal of Nutrition* 70, 226–234.

Picard, M.L., Uzu, G., Dunnington, E.A. and Siegel, P.B. (1993) Food intake adjustments of chicks: short term reactions to deficiencies in lysine, methionine and tryptophan. *British Poultry Science* 34, 737–746.

Potter, L.M. and Shelton, J.R. (1976) Protein, methionine, lysine and a fermentation residue as variables in diets of young turkeys. *Poultry Science* 55, 1535–1543.

Potter, L.M. and Shelton, J.R. (1978) Evaluation of corn fermentation solubles, menhaden fish meal, methionine and hydrolyzed feather meal in diets of young turkeys. *Poultry Science* 57, 1586–1593.

Potter, L.M. and Shelton, J.R. (1979) Methionine and protein requirements of young turkeys. *Poultry Science* 58, 609–615.

Potter, L.M., Shelton, J.R. and Pierson, E.E. (1977) Menhaden fish meal, dried fish solubles, methionine and zinc bacitracin in diets of young turkeys. *Poultry Science* 56, 1189–1200.

Scott, R.L. and Austic, R.E. (1978) Influence of dietary potassium on lysine metabolism in the chick. *Journal of Nutrition* 108, 137–144.

Seaton, K.W., Thomas, O.P., Gous, R.M. and Bossard, E.H. (1978) The effect of diet on liver glycogen and body composition in the chick. *Poultry Science* 57, 692–698.

Stockland, W.L., Meade, R.J. and Melliere, A.L. (1970) Lysine requirement of the growing rat: plasma-free lysine as a response criterion. *Journal of Nutrition* 100, 925–933.

Stockland, W.L., Meade, R.J., Wass, D.F. and Sowers, J.E. (1973) Influence of levels of methionine and cystine on the total sulfur amino acid requirement of the growing rat. *Journal of Animal Science* 36, 526–530.

Talbot, C.J. (1978) Sodium, potassium and chloride imbalance in broiler diets. *Proceedings of the Nutrition Society* 37, 53A.

Thomas, O.P., Twining, P.V. and Bossard, E.H. (1977) The available lysine requirement of 7–9 week old sexed broiler chicks. *Poultry Science* 56, 57–60.

Tsiagbe, V.K., Cook, M.E., Harper, A.E. and Sunde, M.L. (1987a) Efficacy of cysteine in replacing methionine in the immune responses of broiler chicks. *Poultry Science* 66, 1138–1146.

Tsiagbe, V.K., Cook, M.E., Harper, A.E. and Sunde, M.L. (1987b) Enhanced immune responses in broiler chicks fed methionine-supplemented diets. *Poultry Science* 66, 1147–1154.

Velu, J.G., Scott, H.M. and Baker, D.H. (1972) Body composition and nutrient utilization of chicks fed amino acid diets containing graded amounts of either isoleucine or lysine. *Journal of Nutrition* 102, 741–748.

Wilburn, D.R. and Fuller, H.L. (1975) The effect of methionine and lysine levels on the arginine requirement of the chick. *Poultry Science* 54, 248–256.

Willis, G.M. and Baker, D.H. (1980) Lasalocid-sulfur amino acid interrelationship in the chick. *Poultry Science* 59, 2538–2543.

Responses of Laying Hens to Amino Acids

11

C. FISHER

Leyden Old House, Kirknewton,
Midlothian EH27 8DQ, UK.

... the requirement is for the various amino acids.... In practice it is necessary to meet the requirements of as many of the individuals as we can economically.... The actual need is probably on the basis of certain amounts of the various amino acids per unit weight of maintenance plus definite additional quantities for productive increases such as units of growth and quantity of eggs. Meeting these minimum needs will be materially influenced by food consumption.

(Heuser, 1940)

The response of laying hens to amino acids is usually considered in relation to nutritional requirements. A requirement can be seen as a single point on a response curve which relates some measure of output (e.g. weight or value of eggs produced) to amino acid supply. The quotation above summarizes very elegantly a conceptual theory of amino acid requirements that is suitable for resolving practical nutritional questions. In general, however, this perceptive statement by Heuser was largely ignored by the research community until the 1970s and most information about response to amino acid supply is to be obtained from empirical feeding trials. Much of the discussion below can be related back to this statement of 50 years ago.

The topic is of course of practical importance. On a worldwide basis about 36 million tonnes of hens' eggs are produced (Gillin, 1992), representing some 4.5 million tonnes of egg protein. If the gross efficiency with which feed protein is utilized is about 0.30 then 15 million tonnes of high-quality feed protein are used in this production. Nutritional information which ensures efficient use of resources on this scale is obviously important.

From a scientific dimension the laying hen is of interest since it exhibits a high and continuous rate of protein production under the control of primary reproductive mechanisms. Mechanistic descriptions of amino acid utilization in laying hens have not been widely discussed in the literature but the simple models available have been successful in answering practical questions and also allow information collected on one type of bird (egg-laying hens) to be used for others (broiler breeders, turkeys, ducks, etc.).

The discussion in this chapter relates mainly to the response of egg output or mass (grams egg per hen day) to amino acid level in the diet or to amino acid intake. There is a widespread view, expressed in the literature, that increments in the average weight of eggs continue to be observed as amino acid supply increases after the rate of egg production (number of eggs) has ceased to respond. However, this issue has been examined in detail by Morris and Gous (1988) who concluded that this hypothesis is not true and that it arises from the different relative sensitivity with which the two output characteristics can be measured. The relationship between egg output and egg value, which is important in the economic evaluation of response, is complex and depends on egg grading systems. This issue is not discussed further here but it is important in practice.

There are indications that the relative magnitude of the responses in rate of lay and egg weight may vary between amino acids. Jensen *et al.* (1990) comment on the absence of response to tryptophan supplementation in their experiments and contrast this with the response to other amino acids. Their observation is well supported by other experiments on tryptophan. No mechanism has been suggested to explain such effects and beyond noting them here the issue is not considered further in this review. Similarly there is no discussion here of the possibility that some amino acids may act as metabolic regulators, with direct effects perhaps on egg production, at levels which transcend the apparent quantitative requirement for protein deposition. Ohtani *et al.* (1989), again working with tryptophan, discuss this issue in relation to their own experiment but the argument that such effects may exist, in the present author's opinion, is not well made.

Egg Production and Amino Acid Nutrition

A typical modern laying hen at maturity weighs about 1.8 kg and has a roughly constant body protein weight of 340 g, of which 60 g is in the feathers and not subject to turnover. After sexual maturity at ∼155 days a laying cycle of about 1 year follows, with ∼300 (i.e. ∼18 kg) eggs being produced. This production contains 2.2 kg egg protein and thus represents about 800% of the feather-free body protein per year or 2.2% per day. Egg protein represents about 30% of daily crude protein intake. Further cycles of production will be initiated after a moult and the use of two cycles is fairly widespread in industrial practice. Of the 7–7.5 g protein (N × 6.25) in an egg, about 3.1 g (42%) is yolk protein, synthesized in the liver, and about 4 g (54%) is egg white (albumen) protein, synthesized mainly in the magnum region of the oviduct. The remaining protein (3–4% of the total) is in the shell and its associated membranes which originate in the isthmus and uterus regions of the oviduct (figures from Svensson, 1964). The approximate distribution of amino acids between yolk and albumen proteins is shown in Table 11.1. When considering amino acid composition of body proteins in birds it is important to distinguish

Table 11.1. Approximate distribution of amino acids between yolk and albumen proteins. Values for yolk and albumen are mean values reported by Lunven *et al.* (1973). Values for whole egg contents assume a 60 g egg containing 19.1, 34.3 and 6.6 g weight and 27, 17 and 5.3 g nitrogen kg^{-1} in yolk, albumen and shell respectively. The values in brackets show the distribution of the total amino acids in egg contents between yolk and albumen in an egg of this composition.

Amino acid	Abbreviation	Yolk (mg g^{-1} N)	Albumen (mg g^{-1} N)	Egg contents (mg g^{-1} N)
Lysine	Lys	477 (53)	378 (47)	425
Methionine	Met	175 (39)	240 (61)	209
Cystine	Cys	163 (45)	178 (55)	171
Threonine	Thr	313 (51)	272 (49)	291
Tryptophan	Trp	121 (48)	116 (52)	118
Arginine	Arg	434 (54)	330 (46)	379
Isoleucine	Ile	348 (48)	331 (52)	339
Leucine	Leu	548 (48)	521 (52)	534
Valine	Val	378 (44)	429 (56)	405
Histidine	His	148 (50)	132 (50)	140
Alanine	Ala	313 (44)	357 (56)	336
Aspartic acid	Asp	613 (46)	628 (54)	621
Glutamic acid	Glu	780 (44)	869 (56)	827
Glycine	Gly	179 (44)	205 (56)	193
Phenylalanine	Phe	261 (39)	368 (61)	318
Proline	Pro	247 (51)	209 (49)	227
Serine	Ser	499 (51)	429 (49)	462
Tyrosine	Tyr	253 (47)	257 (53)	255

between feather and non-feather proteins, which differ markedly in amino acid composition. Further subdivision of body proteins is probably not required for describing nutritional effects. Recent data on the amino acid composition of feather (WPSA, 1992) and body protein (Håkansson *et al.*, 1978) are available.

Classical studies using crystalline amino acid diets (Fisher and Johnson, 1956), showed that arginine, glutamic acid, histidine, isoleucine, leucine, lysine, methionine, phenylalanine, threonine, tryptophan and valine are essential for egg production. This list, but excluding glutamic acid, is widely accepted although the quantitative aspects of phenylalanine-tyrosine and methionine-cystine relationships and the possible need for glycine-serine and proline in very high-producing hens has apparently not been re-examined.

A further remaining uncertainty is the total-N or non-essential amino acid requirement of the high-producing modern hen when furnished with sufficient, but not excess, of essential amino acids. Again the evidence is rather old and relates to levels of egg production lower than those now obtained. Fisher (1976) reviewed a number of experiments, mainly using NH_3^+ as a source of N, and concluded that about 16 g crude protein (N × 6.25) is required each day. In a hen eating 110 g food d^{-1} this equates to 145 g kg^{-1} protein in the feed.

Nutritional Interactions in Egg Production

Well-established interactions amongst amino acids, and between amino acids and vitamins, may all be assumed to apply in the laying hen although few have been the subject of direct investigation. These include methionine–cystine, phenylalanine–tyrosine, methionine–choline–betaine, tryptophan–niacin, etc. Amongst the well-known amino acid interactions the lysine–arginine antagonism appears not to have been investigated but there is some evidence concerning leucine–isoleucine–valine (Bray, 1970). Figure 11.1 shows that although supplements of leucine depressed production in birds given low or marginal levels of isoleucine, the effect was mediated through food intake and the plot of egg output against isoleucine intake is independent of leucine level. Similar results have recently been reported for immature birds (Burnham *et al.*, 1992).

Recently it has been suggested that amino acid utilization in growing birds is diminished by a general effect of excess dietary protein (see Chapters 4 and 10). As a consequence it is suggested that requirements should be expressed as a proportion of crude protein over the full range of both deficient and excess protein levels (Morris *et al.*, 1987). Data on this subject for the laying hen are fragmentary and refer only to methionine and tryptophan.

Experiments on this topic concerned with methionine present considerable problems of interpretation because of the effect of methionine-

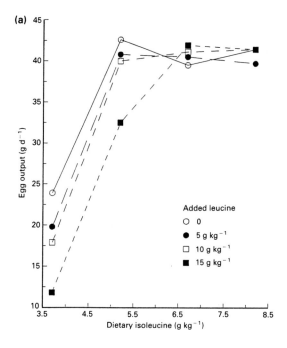

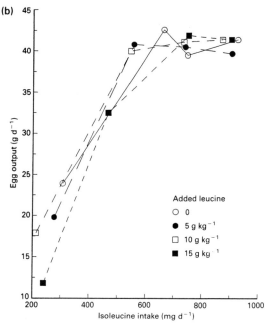

Fig. 11.1. Response of egg output to (a) dietary isoleucine content and (b) isoleucine intake at different levels of isoleucine supplementatiun (from Bray, 1970; Experiment 2). Basal diet, based mainly on maize and soyabean meal contained $10\,g\,kg^{-1}$ leucine by calculation. Note that most of the effects of the leucine-induced imbalance are mediated by changes in food intake.

cystine relationships. Bray (1965) studied the response to methionine at two protein levels, 70 and 100 g kg^{-1}. Methionine utilization was apparently improved at the higher protein level, arguably because of the higher contribution of cystine. When interpreted as a sulfur amino acid response the data are not unambiguous but protein level appeared to be without effect on utilization (Fig. 11.2a). Calderon and Jensen (1990) have re-examined this question and concluded that estimated requirements do increase with protein level but not in direct proportion. However, this work is difficult to interpret critically and does not really resolve the point in question. In particular it is not clear that these authors examined excessive protein levels in their work and there is a lot of uncertainty about methionine being first-limiting. Fisher and Morris (1970) showed that response to methionine was not affected by two protein levels differing by 23 g kg^{-1} over a range of levels. However, this also is not a very critical test because only single levels of methionine supplementation were used.

Similar experiments with tryptophan are reported by Jensen *et al.* (1990). From their Experiment 4 the authors concluded that tryptophan requirements were 1.28, 1.5 and 1.65 g kg^{-1} at 140, 160 and 180 g kg^{-1} crude protein. These values were derived by fitting a segmented ('bent-stick') response model. The plot of egg output to tryptophan intake from this experiment (Fig. 11.2b) suggests that this effect on requirement is a reflection mainly of different levels of production and that there is no effect of the protein levels tested on tryptophan utilization.

From these very limited results it appears that there is no general effect of protein levels on amino acid utilization and requirement in the hen and that the feed formulation rules proposed by Morris *et al.* (1987) should not be followed at this time.

Response of Egg Output to Amino Acid Intake

Figure 11.3 shows data from six experiments using populations of birds and from which the response of egg output (grams egg bird^{-1} day^{-1}) to amino acid intake (mg bird^{-1} day^{-1}) can be calculated. In all these experiments the diets were fed for short periods (10–12 weeks) during 'peak' egg production (20–40 weeks of age). The importance of these constraints is discussed below.

The response is seen to have three components – a rising linear section, a flat plateau and a curved 'diminishing-response' section joining these two. It can also be seen that extrapolation of the responses to zero intake will give a positive intercept on the *x*-axis. Such curves are fairly reproducible in groups of healthy, modern strains of laying hens kept in cages at thermoneutral temperatures. Once the factors which influence the curves are understood and defined then models of appropriate complexity can be derived to predict economically optimum feeding practices.

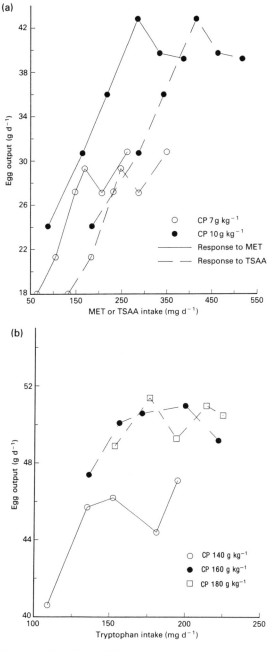

Fig. 11.2. Data illustrating the effect of dietary protein level on the response of egg output to amino acid intake. (a) Responses to methionine and TSAA (total sulfur amino acids) at 70 and 100 g kg^{-1} crude protein (from Bray, 1965, Experiment 1) and (b) responses to tryptophan at three protein levels as shown (from Jensen *et al.*, 1990). The data are interpreted as showing no effect of protein level on amino acid utilization. In the experiment of Bray (1965) it is assumed that TSAA and not methionine is first-limiting.

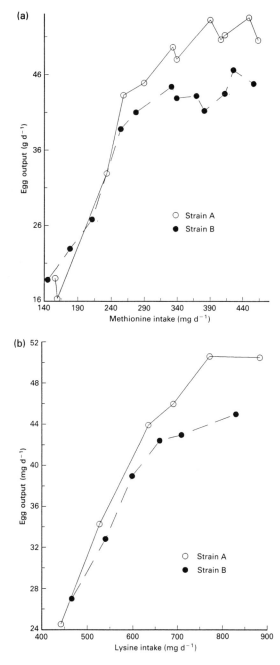

Fig. 11.3. The response of egg output to intake of amino acids in young laying hens. (a) From Fisher (1970): final 2 of 10-week assay from 30 weeks. (b) From Pilbrow and Morris (1974): 36–47 weeks of age, diets fed from 25 weeks.

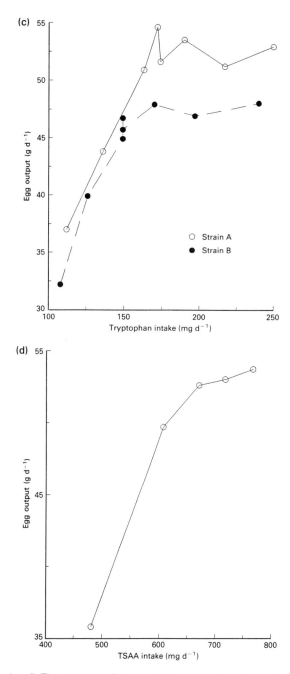

Fig. 11.3. (continued) The response of egg output to intake of amino acids in young laying hens. (c) From Morris and Wethli (1978): final 3 of 9-week assay, from 44 weeks. (d) From Schutte *et al.* (1984): 12-week assay from 28 weeks. (TSAA = total sulfur amino acids.)

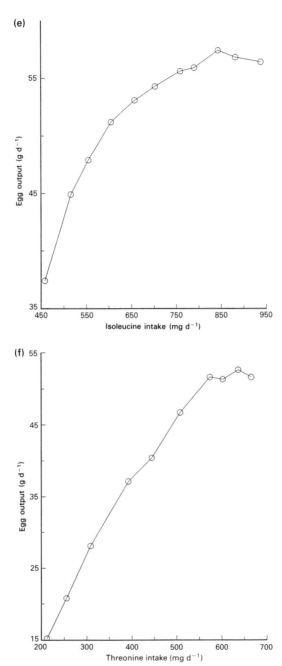

Fig. 11.3. (continued) The response of egg output to intake of amino acids in young laying hens. (e) From Huyghebaert *et al.* (1991): last 4 of 10-week assay from 26 weeks. (f) From Huyghebaert and Butler (1991): as (e) except started at 28 weeks.

The plateau, or upper limit, of the response, will ultimately reflect the genotype of the bird. This may be modified by the environment, of which space/competition, health and temperature are probably the most important components in the present context. For practical purposes the plateau is considered to be flat. If high-quality proteins are used, laying hens will tolerate crude protein levels up to about $300 \, g \, kg^{-1}$ (Frank and Waibel, 1960; Fisher, 1970) or about twice the requirement. The addition of excesses of individual amino acids $(10 \, g \, kg^{-1})$ to a balanced practical feed was also without effect when L-lysine ($\sim 2.3 \times$ requirement), DL-methionine ($\sim 4 \times$), L-threonine ($\sim 3.2 \times$) and L-tryptophan ($\sim 8 \times$) were tested (Koelkebeck *et al.*, 1991).

The intercept value on the x-axis, the input for zero egg output, is, in simple terms, an estimate of the average maintenance requirement. It is estimated rather inefficiently by extrapolation of the response curve. Maintenance is discussed further below; insofar as it is proportional to bodyweight then the hens' genotype will be the main source of variation. The intercept value observed in a single experiment is also strongly influenced by the age of the birds and the duration of the experiment (see below).

The slope of the linear part of the response is influenced by many variable factors and is the most difficult to characterize with the precision required. If the inverse of the slope is compared with the amino acid content of a unit of egg an estimate of the apparent efficiency of amino acid utilization is obtained (see e.g. McDonald and Morris, 1985).

As a product of the primary reproductive system egg protein composition is highly conserved. Thus environmental variation, including diet, mainly affects the number or size of eggs rather than their composition. However, in experimental studies, such as those shown in Fig. 11.3 there are many small but systematic changes in egg composition as the amino acid supply is reduced below the requirement. In most experiments these have been ignored. Figure 11.4 shows the time course of changes in egg production and egg weight when two deficient levels of a methionine-limited protein were fed for 10 weeks. The average weight of eggs laid is seen to decline by 2–3 g during the first 3–4 weeks and then to stabilize. The continuing decline in egg numbers shows that the birds remain in a state of deficiency throughout the experiment. The smaller eggs laid by deficient birds show a reduction in the size of all the major components – shell, albumen and yolk. However, the proportional reduction in yolk size (and to a lesser extent shell) is less than the reduction in total weight, and the reduction in albumen weight is proportionately greater. The resulting increased proportion of yolk in the smaller eggs is an effect of similar magnitude to that observed amongst eggs of different weight laid by fully fed birds at one time (Fisher, 1969). These alterations in the proportion of yolk and albumen are confounded by changes in the gross chemical

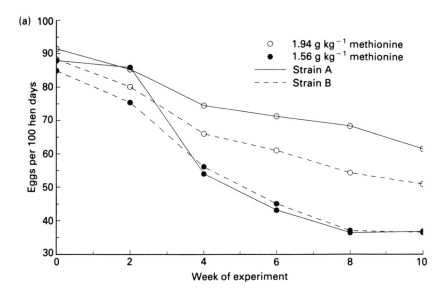

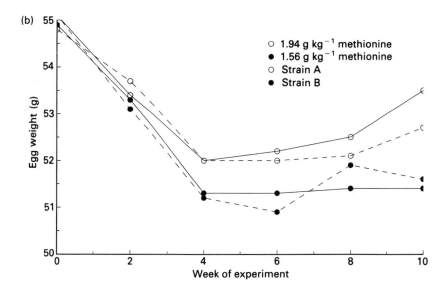

Fig. 11.4. Time course of changes in (a) egg production rate and (b) egg weight when two methionine-deficient feeds are given for 10 weeks from 30 weeks of age. Note: (i) that egg weight ceases to decline although the egg production data show a continuing state of deficiency; and (ii) that the equilibrium egg weight differs between diets although both are deficient. (Data from Fisher, 1970, for two strains.)

composition of the components (Fisher, 1970). With increasing protein deficiency the dry matter content of both components tends to decrease whilst the nitrogen content of the dry matter remains unchanged. No studies on the effects of diet on the amino acid composition of individual egg components have been found. The contribution of these small, and not well characterized, changes to the total amino acid composition of egg have rarely been assessed. A rare example is the work of Kirchgessner and Steinhart (1981). For the purpose of establishing nutritional response and requirements the effects are probably unimportant. For the purpose of determining amino acid utilization under different circumstances the topic merits consideration. In addition to the effects of diet discussed here there may be genetic and age variations in egg composition which will lead to different slopes in the response of egg output to amino acid supply.

Factors concerned with the input scale and which influence the slope of the response curves in Fig. 11.3 are also numerous and have been little investigated in the specific context of the laying hen. The observed slope of the response will obviously reflect the scale on which amino acids are defined – total, digestible or available. There is no reason to think these issues raise different principles in hens in comparison with other mono-gastric animals. Also there is no general body of data which suggests any quantitative uniqueness for this class of stock. When comparing data from different experiments uncertainty about amino acid levels and availability remains a potent source of variation. In studies using amino acid supplementation of a deficient basal diet it is important that the input is defined in availability terms since availability increases systematically with the level of inclusion. Any effect of amino acid level on digestibility or availability will obviously be reflected in the slope of the response.

Mobilization of previously stored tissue amino acids in birds given amino acid-deficient feeds will also affect the slope of the response curve. Since such mobilization undoubtedly occurs (Machlin, 1954; Miller *et al.*, 1960) the slope of the response tends to be underestimated when it is represented as in Fig. 11.3. A possible approach to the estimation of this effect from the results of feeding trials is shown in Fig. 11.5. These data for methionine suggest that about 7 $(=1/0.14)$ mg methionine are required to sustain each gram of bodyweight gain. Given that changes in body composition were ignored in this calculation, this estimate is in approximate agreement with the mean carcass content of methionine – about 3.4 mg g^{-1}. This single result suggests that data on the feather-free body composition could be used to correct for changes in bodyweight although systematic changes in body fat may also need to be considered.

In summary, the response of egg output to amino acid intake is an effective practical representation of experimental results in young laying pullets. Such measures of response may be sufficient for the determination of optimum feeding levels. However, this simple two-dimensional curve

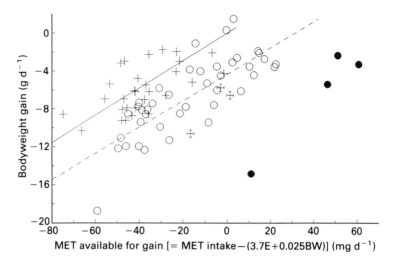

Fig. 11.5. The relationship between bodyweight gain and the amount of methionine calculated to be available for such gain in individual birds fed a methionine-deficient feed (1.56 g kg^{-1}) from 30 to 34 wks (○, ●) and 45 to 49 wks (+, +). The lines are regression lines: for 30–34 wks, $y = -2.23 + 0.0945x$, $r = 0.601$***; for 45–49 wks, $y = -4.18 + 0.1410x$, $r = 0.795$***. Data points marked ● and + were omitted from the calculation of regressions because they deviated widely from the general trend and involved periods of continuous non-laying greater than 10 days (out of 28 total). E = egg output (g d^{-1}) and BW = body weight (kg).

ignores many factors which influence the response and these may be important when the results of different experiments are being compared and when amino acid utilization is being considered.

Formal Description of Response Curves

A useful formal model to describe the responses shown in Fig. 11.3 was proposed by Fisher *et al.* (1973) and Curnow (1973). The so-called 'Reading' model is summarized in Fig. 11.6. This bridges the gap between simple factorial models, such as those proposed by Combs (1960) for methionine, and the widely observed curvilinear response curve. The model has been used to describe and interpret experimental data (see Table 11.2), to determine the effect of genotype (Pilbrow and Morris, 1974) and stage of lay (Wethli and Morris, 1978) on amino acid response, to estimate maintenance requirements and to consider amino acid utilization in feeding trials (McDonald and Morris, 1985). The properties of the model

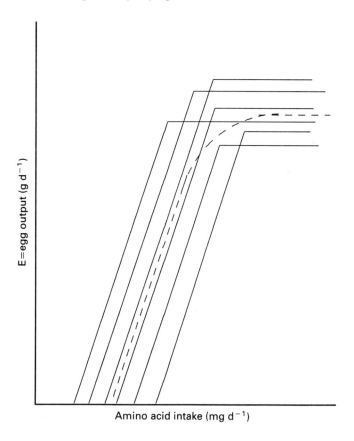

Fig. 11.6. A schematic diagram showing the principle of the 'Reading' model for describing the response of laying hens to amino acid intake (Fisher *et al.*, 1973). In the graph the solid lines show the responses for individual birds, the broken line is the average response of the population. Note that all individuals are assumed to have the same slope of response (= 1/**a** when $E < E_{max}$, or = 0 when $E \geq E_{max}$; **a** mg amino acid g^{-1} egg). The intercept value on the *x*-axis represents the maintenance requirement and is estimated as **b**W where W = bodyweight; **b** mg amino acid kg^{-1} bodyweight. The distributions of E and W are assumed to be normal but may be correlated. Note that the population curve can have the shape of the individual curves only if the correlation of E with W is −1.0 and the standard deviations of E and W are in the ratio b:a (Morris, 1983).

Table 11.2. Literature reports in which the 'Reading' model was used to interpret observations of the response of laying hens to amino acid intake.

Reference	Amino acid	Factors studied
Gous *et al.* (1987)	Lys/Met/Ile	Energy levels
Latshaw (1976)*	Lys	Basal diets
McDonald (1978)*	Lys	Genotypes/diets
Morris (1981)*	Lys	–
Pilbrow and Morris (1974)	Lys	Genotypes
Fisher (1970)	Met	Genotypes/age
Fisher and Morris (1970)	Met	AA balance
Morris and Blackburn (1982)	Met	Shape of response
Morris (1979)*	Ile/Val	–
Huyghebaert *et al.* (1991)	Ile	AA availability
Morris and Wethli (1978)	Trp	Genotypes
Wethli and Morris (1978)	Trp	Age
Du Preez (1980)*	Trp	–
Hyghebaert and Butler (1991)	Thr	AA availability

* Analysis reported only by McDonald and Morris (1985); see this paper for details of these references.

have been evaluated experimentally by Morris and Blackburn (1982) and discussed by Morris (1983).

In the original proposal, and in most published applications, the underlying factorial model was assumed to apply to an individual bird's inputs and outputs averaged over time. Application of similar ideas to growing birds (Emmans and Fisher, 1986) has made it obvious that such a static description of amino acid response can only logically be assumed to apply to one individual at one time. Changes in output levels or in maintenance over time will produce curvature in the response for one animal just as it does for a population of variable animals. In principle this additional constraint should be applied to laying hens as well as the growing animal. In experiments such as those shown in Fig. 11.3 the distinction is irrelevant since the animals are in a quasi steady state with regard to bodyweight and potential output, but the matter is important when longer time periods or longer experiments are considered.

Alternative, or more fully developed, formal approaches to the description of amino acid response in laying hens have not been suggested in the literature. Simple (e.g. Combs, 1960 – methionine) and complex (Hurwitz and Bornstein, 1973) factorial models for calculation of requirements have been suggested. These determine a single point on an assumed linear response and do not include diminishing marginal returns. Alternative empirical curves can be used to describe experimental data but none seems to have useful properties and many have disadvantages (Morris, 1983). Curnow (1986) has suggested that the family of inverse polynomials (Nelder, 1966) would be useful but this has not been developed for laying hens. A useful asymptotic curve (Morgan *et al.*, 1975) has been quite widely used to interpret data on amino acid response in growing birds but not, to the author's knowledge, in laying hens.

Response to Dietary Amino Acid Levels

Although it is convenient to consider amino acid intake on the *x*-axis of response graphs, for the purposes of practical nutrition amino acid concentration in the feed must eventually be specified. This brings into consideration a wide range of factors which influence food intake and also raises the question as to how dietary concentrations of amino acids *per se* influence food intake. In the laying hen it appears that food intake is the main factor which mediates response to amino acid imbalance. The simplest model for requirements is:

$$\text{Dietary requirement } (\text{g kg}^{-1}) = \frac{\text{Intake requirement } (\text{g})}{\text{Food intake } (\text{kg})}$$

Across individual animal genotypes and across groups treated in many different ways there will be a strong correlation between intake requirements for amino acids and food intake under ad lib feeding. This arises because both variables are determined by factors such as egg production, and bodyweight, etc. A consequence is that, for example across genotypes, widely different intake requirements may be met on a single feed because of correlated differences in food intake. It is by ignoring these complex relationships, both within experimental data sets, and in making predictions, that the formal model discussed above is most severely limited.

Any factor which affects food intake independently of intake requirements (or vice versa) will of course have a direct and proportional effect on the response to dietary amino acid concentration and on dietary requirement. A review of all such factors is outside the scope of this chapter but environmental temperature, feed texture, unpalatable ingredients, inter-bird competition and health are the types of influences which will frequently give rise to interactions. These effects are usually more readily understood if nutrient intake and food (energy) intake are considered separately. Interpretation of response to diet composition with ad lib

feeding may be difficult if it is not clear whether a nutrient or energy is
the first-limiting feed resource.

The simple equation given above clearly fails if the dietary level of an
amino acid directly influences food intake. That this is so is apparent from
the proposed response of food intake to dietary amino acid level shown
in Fig. 11.7. At deficient amino acid levels food intake increases with
supplementation to a maximum value and then declines. The maximum
intake will occur at a dietary amino acid level lower than that required to
maximize output. The decline in intake at higher amino acid levels means
that the requirement for maximum feed conversion ratio (unit eggs per
unit food – a key measure of economic response) will be higher than that
for maximum output.

It must be emphasized that experimental measurement of the response
in Fig. 11.7 is very variable and that the data shown have been highly
selected. This variability is compatible, however, with a very simple model
for the underlying mechanism. It can also be noted that the choice of
treatments will strongly influence which parts of the response are observed
and also that random variations in output, especially on the plateau of
response, will have effects on food intake which may mask the underlying
pattern.

Emmans (1987) has proposed that birds have a genetically determined
requirement for nutrients and that they will attempt to eat their 'desired'
food intake in order to meet this requirement for the first-limiting feed
resource. 'Actual' food intake will equal the 'desired' intake unless other
limiting events intervene. In this latter case food intake will be at the level
permitted by the limiting factor. Important limits are seen as the capacity
to lose heat (interaction with environment, feathering, etc.), the bulk of
the feed and the bird's capacity for bulk and, possibly, the capacity to
utilize excess energy for fat deposition. Thus on amino acid-deficient feeds
birds will attempt to consume more feed until diet bulk or the capacity
to dispose of energy becomes limiting. With decreasing levels of deficiency
this balance will be struck at higher levels of food intake, higher limiting
nutrient intakes and higher levels of output. There will be a maximum
intake but this will be at different nutrient levels for each bird and thus
the peak intake will be less well defined in flock experiments. At higher
nutrient levels food intake will decline as requirements are met at lower
intakes. However, for flocks of birds the average decline will be less than
that required to maintain constant intake of the nutrient being investigated
because of variations in requirements and because, in a graded series, the
nutrient under investigation will progressively cease to be the first-limiting
resource and energy balance will become the controlling element. In laying
hens this expectation of complex variation in response will be overlaid by
diurnal variation in food intake and in changing priorities for food intake.
At certain times of the egg-laying cycle calcium supply may be first-limiting

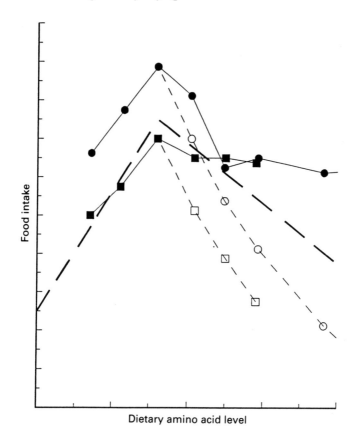

Fig. 11.7. Proposed schematic response of food intake to dietary amino acid level (_ _ _ _) and two illustrative experimental observations. For each of the experiments the broken line from the highest observed intake shows a path giving equal intakes of the limiting amino acid. Data from Morris and Wethli (1978), Experiment 1, Shaver birds, for tryptophan (●—●); and from Pilbrow and Morris (1974), pooled data for medium bodyweight birds, 36–47 weeks of age, for lysine (■—■). Data for dietary amino acid levels were expressed relative to the level giving the highest food intake (=100).

even on treatments designed to study amino acid levels. Conceptually this putative variation between birds and over time in response of food intake to diet composition is quite sufficient to explain both the 'typical' response suggested in Fig. 11.7 and the wide variation in its expression. It is apparent that very simple mechanistic ideas expressed at the level of one animal at one time can lead to extremely complex patterns of response when groups are studied over periods of time. It seems most unlikely that the search for explanatory mechanisms at lower levels of animal organization, for example by looking for physiological mechanisms, has any greater chance of explaining what is observed when chickens are offered different feeds (the paper by Boorman, 1979, still offers the best point to start a study of the literature in this field).

From one individual at one time to populations over time

The formal response model in Fig. 11.6 shows how a linear model for an individual animal can lead, and indeed must lead, to a non-linear response for groups of animals fed at different amino acid levels. It was also suggested that time had an important effect although this may not be apparent if the animals are in a more-or-less steady state.

When measurements are made over longer periods of time, or repeated at intervals on the same birds, there is no longer a steady state and the response to amino acids is found to change systematically. The general pattern is that the response moves to the right as birds age or as longer experiments are considered. The explanation, in simple terms, is probably to be found in the increasing proportion of intermittently- or non-laying birds in the population (Overfield, 1969). This simple explanation is supported by the fact that such changes in response are only occasioned by factors which increase intermittent laying patterns, such as ageing (Jennings *et al.*, 1972) or adverse light patterns (Bray, 1968). Wethli and Morris (1978) also showed, in an important experiment, that the effect of age was reversed by a successful moulting treatment. Such a reversal is not observed if the moult does not restore egg production (Fisher, 1970). The results from Wethli and Morris (1985) are used to illustrate the effects of ageing on the response to tryptophan intake and its reversal by moulting (Fig. 11.8).

It thus seems probable that the response of laying hens to amino acid intake can be satisfactorily described by assuming a simple linear model for one animal at one time. Response curves for populations over time can then be constructed by considering, in particular, the distribution of laying and non-laying periods within groups as well as changes in average egg output and bodyweight. At the present time it appears that all the elements of such a model can be described but they have not been put together. Until this is done and evaluated it is unnecessary to erect more complex theories to explain the results of different experiments. It is also probable that such a model would allow the response shown by different classes of stock, e.g. broiler breeders, turkeys, ducks, to be considered in the same way.

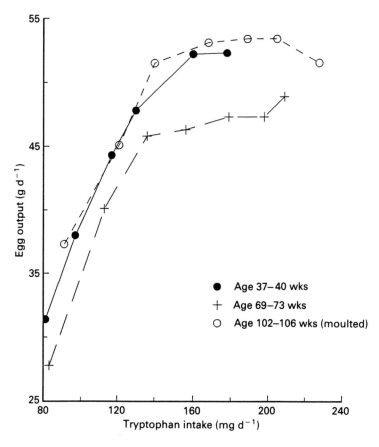

Fig. 11.8. Response of egg output to tryptophan intake at different ages. Note the displacement of the response curve to the right at the end of the laying year (69–73 weeks) and the effect of moulting on the position of the response curve. (From Wethli and Morris, 1978. Reproduced permission of British Poultry Science Ltd.)

Amino Acid Utilization for Maintenance

The maintenance component of the response to amino acids in laying hens is neither well-defined conceptually nor well-established quantitatively. Since, as a first approximation, between 7% (phenylalanine) and 21% (lysine) of the total intake in normally fed birds is accounted for under this heading this lack of precision is of some concern. At first sight it is odd that lysine should be seen as having the highest proportionate maintenance cost since this is classically assumed to be very low (Leveille and Fisher, 1959). However, this is what currently suggested values indicate.

Mechanistic concepts of maintenance have been little discussed and will not be pursued here. Nutritional definitions might include (i) the requirement to maintain the hen in a functioning (normal?) state at zero egg production; or (ii) the difference in requirement between two groups with the same egg output but different body mass. Experimental estimates have been obtained from studies of nitrogen balance in adult roosters and by extrapolation of response curves. The variable evidence was reviewed by Fisher (1983) and the 'average' values suggested at that time are shown in Table 11.3. Also shown are the values suggested by McDonald and Morris (1985), largely on the basis of fitting the 'Reading' model to published data sets.

It seems logical to scale maintenance requirements for amino acids to the feather-free body protein mass (BP). This excludes feather growth or replacement as a component of maintenance, an element which received

Table 11.3. Estimates of requirements of amino acids for maintenance. Presumed input scale is dietary total amino acids.

Amino acid	Requirement estimates (mg kg^{-1} BW d^{-1})		
	A	B	C
Arginine	50	53	-
Histidine	10	16	-
Leucine	?	32	-
Isoleucine	50	67	44.5[1]; 60[2]
Lysine	85	73	73[3]
Methionine	25	31	-
Methionine + cystine	60	80	-
Phenylalanine	?	16	-
Phenylalanine + tyrosine	?	32	-
Threonine	40	32	44.4[4]
Tryptophan	10	11	-
Valine	60	76	-

A - values suggested by Fisher (1983) from a review of the available evidence.
B - from McDonald and Morris (1985). These values formed part of the evidence used to derive the values under A.
C - Recent experimental evidence: [1]Huyghebaert et al. (1991), from response curves; [2]Burnham and Gous (1992), N-balance in adult roosters; [3]Gous et al. (1983), N-balance in adult roosters; [4]Huyghebaert and Butler (1991), from response curves.

emphasis, but not measurement, in early work by Leveille and Fisher (1960). Following Brody (1945), since hens are in a mature state, scaling to $BP^{0.73}$ seems appropriate. In most studies, as in Table 11.3, maintenance requirements have simply been expressed as $mg\,kg^{-1}$ bodyweight.

Earlier experiments using nitrogen balance in mature roosters (Leveille and Fisher, 1960; Ishibashi, 1973) gave rather variable results. The data were interpreted at either zero or 'normal' nitrogen balance giving different results. More robust experiments using adult cockerels have been reported by Gous *et al.* (1983) for lysine and by Burnham and Gous (1992) for isoleucine. The results of these studies, calculated at zero nitrogen balance, were 73 mg lysine and 60 mg isoleucine per kg bodyweight per day (kg BW d^{-1}), and are in reasonable agreement with the values in Table 11.3. Recent estimates based on extrapolated response curves are from Huyghebaert *et al.* (1991) for isoleucine (44.5 $mg\,kg^{-1}$ BW d^{-1}) and Huyghebaert and Butler (1991) for threonine (44.4 $mg\,kg^{-1}$ BW d^{-1}).

In general the distinction between total and available amino acids has not been made consistently in these various experimental sources. An exception is the recent work by Huyghebaert and colleagues. Overall the data available to quantify maintenance requirements of laying hens remain rather variable. An alternative approach suggested by Emmans (1987) for growing birds is to assume a universal requirement for scaled ideal protein requirement (Emmans suggests 0.008 kg protein per kg $BP^{0.73}$, where BP is body protein at maturity). Discussion can then move to consider what is the ideal balance of amino acids for maintenance and how this might be determined.

Amino Acid Utilization for Egg Production

This topic can also be approached from either a mechanistic or empirical point of view, but, as in the case of maintenance, only the latter has been substantially discussed in the literature.

When the model in Fig. 11.6 is fitted to experimental data by iterative approximation (Curnow, 1973) an estimate of the coefficient a (mg amino acid g^{-1} egg) is obtained. In principle this can be compared with the amino acid composition of egg to give an empirical estimate of the efficiency of utilization. The result obtained clearly depends critically on the scale used to describe the feed inputs and, to a lesser extent, on the accuracy of the data for egg composition. Factors which influence the estimate of slope in individual experiments have been alluded to above; of greatest importance is the fact that only experiments covering similar ages and time course of feeding can be compared. To compare a short-term experiment with one covering the whole laying year is clearly meaningless.

McDonald and Morris (1985) interpreted the results of 15 feeding trials in this way. Diet compositions were recalculated from a single table of values for *total* amino acids. All the experiments had a similar age/time structure. The results for 15 individual experiments and the mean values for each of five amino acids are given in Table 11.4. Also shown are the results of two more recent experiments to which the same method of analysis has been applied, one for isoleucine and one for threonine, an amino acid not considered by McDonald and Morris (1985).

Table 11.4. Estimates of the coefficients **a** (mg amino acid g^{-1} egg) and **b** (mg amino acid kg^{-1} BW maintenance). Values determined by McDonald and Morris (1985) for individual experiments and pooled data are given. Also recent evidence published since 1985. (n = no. of data points included in analysis.)

Reference[*]	Amino acid	n	a	b
1	Lysine	36	10.71	53.0
2	Lysine	12	10.49	33.0
3	Lysine	16	7.89	80.5
4	Lysine	24	13.29	37.1
5	Lysine	6	8.24	31.2
6	Lysine	46	10.26	79.5
Pooled 1-6	Lysine	140	9.99	72.6
7	Methionine	17	4.81	32.1
8	Methionine	10	4.48	29.4
9	Methionine	12	4.71	30.2
10	Methionine	18	5.91	9.9
Pooled 7-10	Methionine	57	4.77	30.7
11	Isoleucine	15	9.10	63.7
12	Isoleucine	5	6.42	76.2
Pooled 11–12	Isoleucine	20	7.97	67.2
New A[†]	Isoleucine	11	(tot) 9.48	44.5
			(dig) 8.38	7.1
13	Tryptophan	28	2.73	13.7
14	Tryptophan	8	1.66	16.2
Pooled 13-14	Tryptophan	36	2.62	11.2
New B[‡]	Threonine	10	(tot) 8.70	43.4
			(dig) 7.06	21.8
15	Valine	7	8.90	75.6

[*] Details of the references are given by McDonald and Morris (1985).
[†] New A: Huyghebaert *et al.* (1991).
[‡] New B: Huyghebaert and Butler (1991).
tot = total amino acids; dig = digestible amino acids

By comparing the mean results of the coefficient a in Table 11.4 with assumed egg composition data (egg contains 18 g kg^{-1} N; amino acid data from Lunven *et al.*, 1973), McDonald and Morris obtained estimates of efficiency ranging from 0.74 for methionine to 0.83 for valine. Given the nature of the exercise they argued that the average efficiency of 0.77 could be applied to all amino acids. By means of this assumption they could extrapolate to other amino acids for which no data were available.

Consideration of the more recent data tends to emphasize the limitations of the approach rather than illuminate the correct conclusion. For isoleucine Huyghebaert *et al.* (1991) obtained estimates of 9.48 and 8.38 mg g^{-1} egg for the coefficient a using total and digestible input data respectively. Using the McDonald and Morris (1985) data for egg composition gives efficiencies of 0.63 and 0.72 respectively, considerably outside the range conflated by McDonald and Morris into a mean of 0.77. However, Huyghebaert *et al.* (1991) report a determined isoleucine content of egg of 6.43 mg g^{-1} (cf. 6.0 used by McDonald and Morris, 1985) which when applied to the above estimates of response gives calculated efficiencies of 0.68 and 0.77. The value of 0.68 is still lower than the range observed by McDonald and Morris (1985).

In a similar way the results of Huyghebaert and Butler (1991) for threonine also lead to estimates of efficiency lower than 0.77 for total amino acids. Using the assumptions of McDonald and Morris (1985) leads to a calculated egg content of 5.33 mg g^{-1} and efficiencies of 0.61 and 0.76 for total and digestible threonine respectively in this experiment. Using reported data for egg composition ($5.7 \text{ mg threonine g}^{-1}$ egg) raises these estimates to 0.66 and 0.81.

Thus the more recent results are consistent across two amino acids but differ from the earlier analyses. At present no reasonable explanation can be proposed for these rather different findings although the individual experiments and the assumptions made by McDonald and Morris (1985) have not been re-examined. For isoleucine it should be noted that this is one amino acid for which the duration of hydrolysis is of critical importance in analysis although the effect of this would have to be differential for feeds and eggs to explain the results for efficiency.

In all of these analyses systematic changes in bodyweight (and thus maintenance) and differences in growth across treatments have been ignored. As pointed out these may be important when trying to reconcile results for amino acid utilization. Although the exercise is far from conclusive some approximate calculations can be made on the results of Huyghebaert and Butler (1991) for threonine. The data analysed by the authors are for the final 4 weeks of a 10-week assay. Thus on deficient feeds less bodyweight is maintained (extreme values 1.43 and 1.92 kg) and, in this trial also, there was a variable change in bodyweight (-4.2 to $+3.3 \text{ g d}^{-1}$) during the period reported. The equivalent threonine intakes of these differences in maintenance and growth across treatments were calculated using $40 \, \text{g kg}^{-1} \, \text{d}^{-1}$ for maintenance and $7 \, \text{mg g}^{-1}$ for body-weight gain or loss. The corrections ranged from $-33 \, \text{mg}$ (-5%) at the highest threonine level to $+48 \, \text{mg}$ ($+22.6\%$) at the lowest. Since access to the computer program for the 'Reading' model was not available the effects of these adjustments to intake were estimated by the linear regression of egg output on threonine intake using data for the lowest five threonine levels. Before making adjustment to the intake data the inverse of the regression coefficient (an estimate of the coefficient a) was $7.99 \, \text{mg g}^{-1}$ and after adjustment $9.06 \, \text{mg g}^{-1}$. Applying this magnitude of correction to the a value reported by the authors, $8.7 \, \text{mg g}^{-1}$, gives $((7.99/9.06) \times 8.7) = 7.7 \, \text{mg g}^{-1}$. The corresponding efficiencies are 0.66, as noted above without adjustment, and 0.74 after adjustment. These calculations, which involve many untestable assumptions, therefore bring the efficiency in this study into line with the average suggested by McDonald and Morris (1985) although it is important to note that corrections of this sort were not made by these latter authors. The hypothesis that all amino acids are utilized for egg production with the same efficiencies remains neither proven nor disproven.

Other experiments have been reported in recent years on the response to amino acids in laying hens (see Table 11.5 for a list of these) but, from these, data for the final (quasi-equilibrium) stages of a short-term assay cannot be calculated. For the time being it can be stated that the assumption of constant efficiency is useful, and is probably sufficient for practical nutrition, whilst more critical tests will have to be devised if the assumption is to be convincingly rejected. It should be noted that whilst the argument for using different efficiencies is strong for growing animals (see Baker, 1991, and Chapters 8 and 10) the situation in laying hens is not necessarily the same.

A mechanistic approach to the question of amino acid utilization by the laying hen has not been fully developed in the literature and is largely outside the scope of this chapter. Whilst such an approach could start from a variety of biochemical and physiological mechanisms it appears on *a priori* grounds that it will first be necessary to consider discontinuities and variations in protein synthesis over time and the lack of concordance over time between intake under ad lib feeding and the utilization of amino acids. Any variations in amino acid uptake or supply which exceed or fail to meet requirements for protein synthesis will lead, in the case of an excess, to a diminution of efficiency or, in the case of a deficiency, in the rate of synthesis. Since egg protein synthesis and voluntary food intake vary systematically, both within and between days, it is certain that such events determine to some extent the apparent efficiencies observed in feeding trials.

A very simple example of this may be usefully illustrative. It is suggested above that the utilization of amino acids for egg production is about 0.77 for total amino acids or, say, 0.85 (0.77/0.90) for available amino acids. Estimates are fairly consistent in young birds laying at a high rate. Such birds will be laying in closed cycles, i.e. with inter-clutch intervals of 1 day, and the rate of lay over a period is closely correlated with clutch length. The lowest rate of lay for a given period is 50% (0.5 eggs per day), obtained with regular 1-egg clutches. It seems reasonable to suppose that over the range 0.5 to 1.0 eggs per day protein synthesis will be regular, although fluctuating, and that amino acids may be used with constant efficiency in birds eating ad lib. It is also obvious that when egg production is zero, efficiency is also zero. These two assertions allow the relationship between rate of egg production and amino acid utilization to be modelled as in Figure 11.9. For any fixed time period, rates of lay, especially below 50%, can result from many different patterns of clutch length and inter-clutch intervals; however, this simple model is probably a reasonable approximation to the truth. Figure 11.9 shows data for individual birds fed a methionine-deficient feed at two ages. These suggest that the proposed model goes some way to explain the age effect on apparent amino acid utilization that has been discussed above. The idea also provides a way of approaching amino acid utilization in birds such as broiler breeders and turkeys which, on a flock basis, have a lower and more variable overall rate of lay.

In constructing formal models to calculate amino acid requirements of hens, Hurwitz and Bornstein (1973) associated inefficiency in utilization with diurnal variations in the pattern of egg protein synthesis, an idea first proposed by Moran (1969). Proteins accumulated over long periods of time, such as yolk proteins and ovoglobulin, were assumed to be derived from the food with a given high efficiency. The requirements for proteins synthesized over shorter periods of time, such as ovomucoids and ovomucin, were assumed to create a requirement which could not be met directly from absorbed amino acids and which therefore involved breakdown (turnover) of body protein with lower efficiencies. These concepts were incorporated into different models of utilization which were then tested empirically (Hurwitz and Bornstein, 1977). The ideas in the models were discussed further by Smith (1978a,b).

Fisher (1980) estimated probable protein deposition rates following the assumptions above (Fig. 11.10). During a typical clutch sequence total protein deposition was calculated to vary only between 0.2 and 0.3 $g h^{-1}$ and it was argued that such variation was small when compared with the mechanisms available to buffer the supply of amino acids at sites of synthesis. The calculations made by Fisher (1980) are lent some credence by recent evidence of whole-body protein synthesis rates determined by a primed-continuous infusion of [^{15}N]methionine for 3 h (Hiramoto *et al.*, 1990). These authors found that protein synthesis varied from a low of

Table 11.5. Experiments on amino acid requirements of laying hens reported since 1980.

Reference	Expt no.	n-range[1] (n/g kg⁻¹)	Age/length[2] (wks)	Requirement[3] (g kg⁻¹)	Requirement[3] (mg d⁻¹)	Egg prodn[4]	Notes
Lysine							
Al Bustany and Elwinger (1987a,b)	E1	5; 5.5-9.3	20/60	7.7	692	c. 46 g d⁻¹	Strain effects
	E2	4; 5.7-8.2	20/60	6.4	798	c. 49 g d⁻¹	
Nathanael and Sell (1980)	E1	4; 4.6-7.6	22/20	5.6-6.6	567-702	c. 78%	
	E2	8; 5.7-7.8	42/12	6.6	700	c. 75%	
Uzu and Larbier (1985)	E1	3; 5.8-7.5	20/42	6.5	790	c. 84%	2 CP level
	E2	3; 5.8-7.5	20/42	6.4	720	c. 84%	3 CP levels
Van Weerden and Schutte (1980)		4; 6.4-8.4	26/52	7.4	930	c. 51 g d⁻¹	
Methionine + cystine							
Bertram and Schutte (1992)	E1	4; 6.0-7.5	25/12	6.5	710	c. 97%	Met:Cys ratios
	E2	2; 5.0-6.6	25/12	6.6	720	c. 97%	
Calderon and Jensen (1990)	E1	4; 5.1-8.0	32/4	6.1-6.8	692-789	c. 57 g d⁻¹	3 CP levels
	E2	6; 5.1-7.8	59/5	5.4-7.3	594-803	c. 56 g d⁻¹	3 CP levels
Harms and Miles (1988)		5; 5.0-6.3	34/12	5.3	502-512	c. 48 g d⁻¹	± Fermacto™
Harms and Wilson (1980)		4; 3.55-5.28	24/40	4.7	839	c. 61%	Broiler breeders
Schutte et al. (1983)	E1	3; 5.5-6.5	25/52	6.0	750	c. 51 g d⁻¹	Energy level
	E2	3; 6.0-7.0	25/52	6.5	767	c. 52 g d⁻¹	
Schutte et al. (1984)		5; 5.0-7.0	28/12	6.0	671	c. 53 g d⁻¹	
Vogt and Krieg (1983)		7; 6.0-8.4	22/46	7.2	916	c. 87%	Plus other treatments

Tryptophan

Jensen et al. (1990)						
E1	4; 1.0–1.8	38/12	1.37	123	c. 43 g d^{-1}	
E2	4; 1.0–1.8	30/55	1.18	95	c. 42 g d^{-1}	
E3	5; 1.1–2.3	44/6	1.3–1.64	136–168	c. 56 g d^{-1}	3 CP levels
E4	5; 1.1–2.3	60/6	1.28–1.65	124–162	c. 51 g d^{-1}	3 CP levels
Ishibashi (1985)	20; 0.86–3.2	?/4	1.88	210	c. 85%	
Ohtani et al. (1989)	3; 1.5–2.0	25/58	2.0	239	53.5 g d^{-1}	4 small expts.

[1] Number and range of dietary amino acid levels tested.
[2] Age at start and duration of experiment in weeks.
[3] Requirement reported by authors or estimated by examination, for maximum egg mass.
[4] Approx. level of production at peak of response.

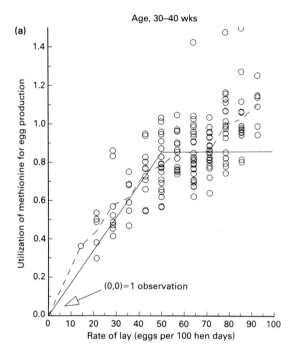

Age, 30–40 wks

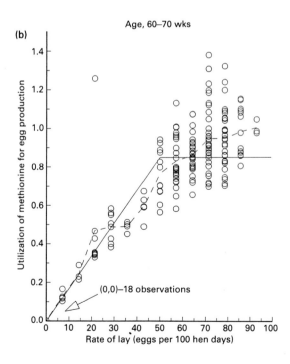

Age, 60–70 wks

1.75 g h^{-1} when a released ovum was in the infundibulum to 2.2 g h^{-1} when the ovum was in the magnum (albumen synthesis). Most of this diurnal variation was found in the oviduct, where synthesis ranged from 0.2 to 0.5 g h^{-1}, a range in agreement with the calculations of Fisher (1980). Protein synthesis in the liver, where yolk precursors are synthesized, was not influenced by the ovulatory cycle and at 0.25 g h^{-1} was about twice the level of yolk protein deposition. The rate of total protein synthesis measured by Hiramoto *et al.* (1990), about 50 g d^{-1}, is some seven times egg protein deposition in a typical bird.

Variations over time also occur in the voluntary intake of feed and hence in amino acid supply to the animal. Between days, total food consumption is about 16.5% higher on a 'laying' day as opposed to a 'non-laying' day (Morris and Taylor, 1967). This effect is generally attributed to the demand for calcium on egg-forming days and is not affected by protein deficiency (Fisher, 1970). On a diurnal basis, and depending on the light pattern, food intake is also highest during periods of egg shell calcification (Mongin and Sauveur, 1979). It may be that during such periods a bird is responding to a relative deficiency of calcium even when being given an amino acid-limited feed. The significance of these patterns of intake for amino acid utilization where a single feed is being given have not been worked out in detail. However, they must contribute to the overall losses when events over a period of time are averaged out. When variation in methionine, lysine or whole protein supply was induced by changing feed composition Shannon (1981) concluded that the period of deficiency against which a hen was buffered was less than 24 h and may be close to 12 h.

In conclusion, the elements of a detailed model of amino acid utilization by the laying hen seem to be available. However, such a model has yet to be constructed. Success in this area would probably not improve the efficiency of practical nutrition but it would be intellectually satisfying.

Fig. 11.9. The relationship between the utilization of methionine and rate of egg production in hens of two ages. Each data point relates to observations over 14 days on individual birds; five successive observations per bird over a 10-week feeding period are shown. All birds received a methionine-deficient feed containing 1.56 g methionine kg^{-1}. Methionine utilization was calculated from the equation: (Met intake − (25BW + 3ΔBW))/3.7, where BW = bodyweight, ΔBW = bodyweight change. The solid line shows the model proposed for this relationship, as referred to in the text. The broken lines join the average data points for each rate of lay. Note that the average response curve for the older flock is displaced to the right (poorer apparent utilization), as shown for tryptophan in Fig. 11.8. This figure shown here suggests that this effect is due to a higher proportion of poor layers (< 50%) in the older flock. (Data from Fisher, 1970.)

C. Fisher

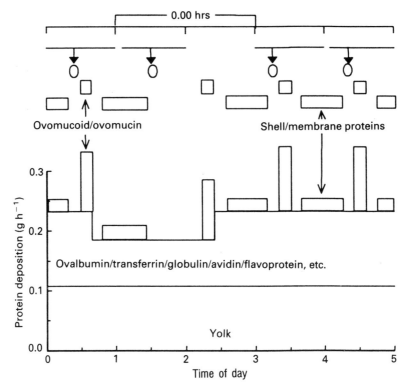

Fig. 11.10. An estimate of the rate of protein synthesis/deposition in laying hens. The scheme is based on the idea originally proposed by Moran (1969) that some egg white proteins (ovalbumin, transferrin, etc.) are synthesized continuously although deposited in the egg over short time periods. Other proteins (ovomucoid, ovomucin and the shell proteins) are assumed to be synthesized and deposited simultaneously. Note that the rates shown relate only to those proteins eventually deposited in the egg and do not refer to total protein synthesis. The upper part of the diagram shows the approximate timing of oviposition (0), egg white and shell secretion on successive days, assuming a sequence: 4-egg clutch, 1-day interval, 3-egg clutch. (From Fisher, 1980; data from Smith, 1978a,b. Reproduced permission of Butterworth-Heinemann Ltd.)

References

Al Bustany, Z. and Elwinger, K. (1987a) Response of laying hens to different lysine intakes. A comparison of some commercial hybrids with strains selected on a low protein diet. *Acta Agriculturae Scandinavica* 37, 27–40.

Al Bustany, Z. and Elwinger, K. (1987b) Shell and interior quality and chemical composition of eggs from hens of different strains and ages fed different dietary lysine levels. *Acta Agriculturae Scandinavica* 37, 175–187.

Baker, D.H. (1991) Partitioning of nutrients for growth and other metabolic functions – efficiency and priority considerations. *Poultry Science* 70, 1797–1805.

Bertram, H.L. and Schutte, J.B. (1992) Evaluation of the sulphur containing amino acids in laying hens. In: *Proceedings of the XIXth World's Poultry Congress*, vol. 3. World's Poultry Science Association, Amsterdam, pp. 606–609.

Boorman, K.N. (1979) Regulation of protein and amino acid intake. In: Boorman, K.N. and Freeman, B.M. (eds) *Food Intake Regulation in Poultry*. British Poultry Science Ltd, Edinburgh, pp. 87–126.

Bray, D.J. (1965) The methionine requirement of young laying pullets. *Poultry Science* 44, 1173–1180.

Bray, D.J. (1968) Photoperiodism and age as factors affecting the protein requirements of laying pullets. *Poultry Science* 47, 1005–1013.

Bray, D.J. (1970) The isoleucine and valine nutrition of young laying pullets as influenced by excessive dietary leucine. *Poultry Science* 49, 1334–1341.

Brody, S. (1945) *Bioenergetics and Growth*. Reinhold, New York.

Burnham, D., Emmans, G.C. and Gous, R.M. (1992) Isoleucine requirements of the chicken – the effect of excess leucine and valine on the response to isoleucine. *British Poultry Science* 33, 71–87.

Burnham, D. and Gous, R.M. (1992) Isoleucine requirements of the chicken – requirement for maintenance. *British Poultry Science* 33, 59–69.

Calderon, V.M. and Jensen, L.S. (1990) The requirement of sulphur amino acids by laying hens as influenced by the protein concentration. *Poultry Science* 69, 934–944.

Combs, G.F. (1960) Protein and energy requirements of laying hens. In: Proceedings Maryland Nutrition Conference, 1960, pp. 28–45.

Curnow, R.N. (1973) A smooth population curve based on an abrupt threshold and plateau model for individuals. *Biometrics* 29, 1–10.

Curnow, R.N. (1986) The statistical approach to nutrient requirements. In: Fisher, C. and Boorman, K.N. (eds) *Nutrient Requirements of Poultry and Nutritional Research*. Butterworths, London, pp. 79–89.

Emmans, G.C. (1987) Growth, body composition and feed intake. *World's Poultry Science Journal* 43, 208–227.

Emmans, G.C. and Fisher, C. (1986) Problems in nutritional theory. In: Fisher, C. and Boorman, K.N. (eds) *Nutrient Requirements of Poultry and Nutritional Research*. Butterworths, London, pp. 9–39.

Fisher, C. (1969) The effects of a protein deficiency on egg composition. *British Poultry Science* 10, 149–154.

Fisher, C. (1970) Studies on the protein and amino acid requirements of the laying hen. Unpublished PhD thesis, University of Reading.

Fisher, C. (1976) Protein in the diets of the pullet and laying bird. In: Cole, D.J.A., Boorman, K.N., Buttery, P.J., Lewis, D., Neale, R.J. and Swan, H. (eds) *Protein Metabolism and Nutrition, EAAP publication No. 16*. Butterworths, Edinburgh, pp. 323–351.

Fisher, C. (1980) Protein deposition in poultry. In: Buttery, P.J. and Lindsay, D.B. (eds) *Protein Deposition in Animals*. Butterworths, London, pp. 251–270.

Fisher, C. (1983) The physiological basis of the amino acid requirements of poultry. In: Aranal, M., Pion, R. and Bonin, D. (eds) *Protein Metabolism and Nutrition. Les Colloques de l'INRA, No. 16.* INRA, Paris, pp. 385–404.

Fisher, C. and Morris, T.R. (1970) The determination of the methionine requirements of laying pullets by a diet dilution technique. *British Poultry Science* 11, 67–82.

Fisher, C., Morris, T.R. and Jennings, R.C. (1973) A model for the description and prediction of the response of laying hens to amino acid intake. *British Poultry Science* 14, 469–484.

Fisher, H. and Johnson, D. (1956) The amino acid requirement of the laying hen. 1. The development of a free amino acid diet for maintenance of egg production. *Journal of Nutrition* 60, 261–273.

Frank, F.R. and Waibel, P.E. (1960) Effect of dietary energy and protein levels and energy source on White Leghorns in cages. *Poultry Science* 39, 1049–1056.

Gillin, E. (1992) Trends in production and trade. In: Poultry International 1992 *International Year Book.* Watt, Mount Morris, Illinois, pp. 46–51.

Gous, R.M., Fisher, C. and Broadbent, L.A. (1983) The maintenance requirement for lysine of adult male fowls. *World's Poultry Science Journal* 39, 239–240 (abstract).

Gous, R.M., Griessel, M. and Morris, T.R. (1987) Effect of dietary energy concentration on the response of laying hens to amino acids. *British Poultry Science* 28, 427–436.

Håkansson, J., Eriksson, S. and Svensson, S.A. (1978) *Influence of Feed Energy Level on Chemical Composition of Tissues and on the Energy and Protein Utilisation by Broiler Chicks.* Swedish University of Agricultural Science, Department of Animal Husbandry Report No. 5, pp. 1–110.

Harms, R.H. and Miles, R.D. (1988) Influence of Fermacto on the performance of laying hens when fed diets with different levels of methionine. *Poultry Science* 67, 842–844.

Harms, R.H. and Wilson, H.R. (1980) Protein and sulfur amino acid requirements of broiler breeder hens. *Poultry Science* 59, 470–472.

Heuser, G.F. (1940) Protein in poultry nutrition – a review. *Poultry Science* 20, 367–368.

Hiramoto, K., Muramatsu, T. and Okumura, J. (1990) Protein synthesis in tissues and in the whole body of laying hens during egg formation. *Poultry Science* 69, 264–269.

Hurwitz, S. and Bornstein, S. (1973) The protein and amino acid requirements of laying hens: suggested models for calculation. *Poultry Science* 52, 1124–1134.

Hurwitz, S. and Bornstein, S. (1977) The protein and amino acid requirements of laying hens: experimental evaluation of models of calculation. 1. Application of two models under various conditions. *Poultry Science* 56, 969–978.

Huyghebaert, G. and Butler, E.A. (1991) Optimum threonine requirement of laying hens. *British Poultry Science* 32, 575–582.

Huyghebaert, G., De Groote, G., Butler, E.A. and Morris, T.R. (1991) Optimum isoleucine requirement of laying hens and the effect of age. *British Poultry Science* 32, 471–481.

Ishibashi, T. (1973) Amino acid requirements for maintenance of the adult rooster. *Japanese Journal of Zootechnical Science* 44, 39–49.

Ishibashi, T. (1985) Tryptophan requirement of laying hens. *Japanese Poultry Science* 22, 256–263.

Jennings, R.C., Fisher, C. and Morris, T.R. (1972) Changes in the protein requirements of pullets during the first laying year. *British Poultry Science* 13, 279–281.

Jensen, L.S., Calderon, V.M. and Mendonca, C.X. (1990) Response to tryptophan of laying hens fed practical diets varying in protein concentration. *Poultry Science* 69, 1956–1965.

Kirchgessner, M. and Steinhart, H. (1981) Amino acid composition of egg contents of laying hens dependent on different protein and energy supplies. *Archiv für Geflügelkunde* 45, 179–185.

Koelkebeck, K.W., Baker, D.H., Han, Y. and Parsons, C.M. (1991) Effect of excess lysine, methionine, threonine, or tryptophan on production performance of laying hens. *Poultry Science* 70, 1651–1653.

Leveille, G.A. and Fisher, H. (1959) Amino acid requirements for maintenance in the adult rooster. II. The requirements for glutamic acid, histidine, lysine and arginine. *Journal of Nutrition* 69, 289–294.

Leveille, G.A. and Fisher, H. (1960) Amino acid requirements for maintenance in the adult rooster. IV. The requirements for methionine, cystine, phenylalanine, tyrosine and tryptophan; the adequacy of the determined requirements. *Journal of Nutrition* 72, 8–15.

Lunven, P., Le Clement de St Marcq, C., Carnovale, E. and Fratoni, A. (1973) Amino acid composition of hen's egg. *British Journal of Nutrition* 30, 189–194.

Machlin, L.J. (1954) Methionine metabolism in the laying hen 1. Effect of change in the dietary protein or tryptophan level on deposition of S^{35} in the egg. *Poultry Science* 33, 201–205.

McDonald, M.W. and Morris, T.R. (1985) Quantitative review of optimum amino acid intakes for young laying pullets. *British Poultry Science* 26, 253–264.

Miller, E.C., O'Barr, J.S. and Denton, C.A. (1960) Studies on a short-term procedure for determining amino acid requirements of laying hens. *Poultry Science* 39, 1438–1442.

Mongin, P. and Sauveur, B. (1979) The specific calcium appetite of the domestic fowl. In: Boorman, K.N. and Freeman, B.M. (eds) *Food Intake Regulation in Poultry*. British Poultry Science Ltd, Edinburgh, pp. 171–189.

Moran, E.T. (1969) Levels of dietary protein needed to support egg weight and laying hen production. *Feedstuffs (Minneapolis)* 41 (31 May), 26–28.

Morgan, P.H., Mercer, L.P. and Flodin, N.W. (1975) General model for nutritional responses of higher organisms. *Proceedings of the National Academy of Sciences of the USA* 72, 4327–4331.

Morris, B.A. and Taylor, T.G. (1967) The daily food consumption of laying hens in relation to egg formation. *British Poultry Science* 8, 251–257.

Morris, T.R. (1983) The interpretation of response data from animal feeding trials. In: Haresign, W. (ed.) *Recent Developments in Poultry Nutrition*. Butterworths, London, pp. 13–24.

Morris, T.R. and Blackburn, H.A. (1982) The shape of the response curve relating protein intake to egg output for flocks of laying hens. *British Poultry Science* 23, 405–424.

Morris, T.R. and Gous, R.M. (1988) Partitioning the response to protein between egg number and egg weight. *British Poultry Science* 29, 93–99.

Morris, T.R. and Wethli, E. (1978) The tryptophan requirements of young laying pullets. *British Poultry Science* 19, 455–466.

Morris, T.R., Al-Azzawi, K., Gous, R.M. and Simpson, G.L. (1987) Effects of protein concentration on responses to dietary lysine by chicks. *British Poultry Science* 28, 185–195.

Nathanael, A.S. and Sell, J.R. (1980) Quantitative measurement of the lysine requirement of the laying hen. *Poultry Science* 59, 594–597.

Nelder, J.A. (1966) Inverse polynomials, a useful group of multi-factor response functions. *Biometrics* 22, 303–315.

Ohtani, H., Saitoh, S., Ohkawara, H., Akiba, Y., Takahashi, K., Horiguchi, M. and Goto, K. (1989) Production performance of laying hens fed L-tryptophan. *Poultry Science* 68, 323–326.

Overfield, N.D. (1969) Calcium requirements for good egg shell quality. *NAAS Quarterly Review* 86, 84–91.

Pilbrow, P.J. and Morris, T.R. (1974) Comparison of lysine requirements amongst eight stocks of laying fowl. *British Poultry Science* 15, 51–73.

Schutte, J.B., van Weerden, E.J. and Bertram, H.L. (1983) Sulphur amino acid requirements of laying hens and the effect of excess dietary methionine on laying performance. *British Poultry Science* 24, 319–326.

Schutte, J.B., van Weerden, E.J. and Bertram, H.L. (1984) Protein and sulphur amino acid nutrition of the hen during the early stage of lay. *Archiv für Geflügelkunde* 48, 165–170.

Shannon, D.W.F. (1981) The effect of short-term changes in methionine supply on egg production. *Poultry Science* 60, 1729–1730.

Smith, W.K. (1978a) The amino acid requirements of laying hens: models for calculation. 1. Physiological background. *World's Poultry Science Journal* 34, 81–96.

Smith, W.K. (1978b) The amino acid requirements of laying hens: models for calculation. 2. Practical application. *World's Poultry Science Journal* 34, 129–136.

Svensson, S.A. (1964). Composition and energy content of eggs, growing chicks and hens, with some notes on preparation and method of analysis. *Lantbrukshogskolaus Annalar* 30, 405.

Uzu, G. and Larbier, M. (1985) Lysine requirement in laying hens. *Archiv für Geflügelkunde* 49, 148–150.

Van Weerden, E.J. and Schutte, J.B. (1980) Lysine requirement of the laying hen. *Archiv für Geflügelkunde* 44, 36–40.

Vogt, H. and Krieg, R. (1983) Einfluss von Rohprotein- und Methionin-plus Cystingehalt im Futter auf des Leistungsvermogen von Legehennen. *Archiv für Geflügelkunde* 47, 248–253.

Wethli, E. and Morris, T.R. (1978) Effect of age on the tryptophan requirement of laying hens. *British Poultry Science* 19, 559–565.

WPSA (1992) *European Amino Acid Table*. Working Group No. 2 (Nutrition) of the World's Poultry Science Association, Beekbergen, The Netherlands, pp. 1–123.

Modelling Amino Acid Metabolism in Ruminants

12

R.L. BALDWIN, C.C. CALVERT, M.D. HANIGAN
AND J. BECKETT
*Department of Animal Science, University of California,
Davis, California 95616-8521, USA.*

Introduction

The development and documentation of models are most often intellectually demanding, laborious and tedious processes. Consider a model in which 30 transactions among ten state variables are addressed. In such a model, defensible equation forms for each of the transactions (30) must be formulated based upon concepts and data relating to each transaction. Data sufficient to estimate, on average, two parameter values for each equation and to set initial values for each of the state variables must be collected; e.g. 70 numerical values must be established. At a minimum, 100 publications would have to be consulted in formulation of the first version of the model. Then a test data set to challenge the model must be formulated based upon another, preferably, 100 plus independent experimental studies. In this context, it is very daunting to attempt summarizations and evaluations of models of protein and amino acid metabolism in ruminants. Only a few, more or less, representative models can be considered and only then in part and possibly and unfairly out of context. These considerations were clearly in our minds when we selected models, or rather parts thereof, for consideration in this chapter.

We elected to start with models of ruminant digestive processes varying in complexity but all leading to estimates of protein availability to the ruminant animal. We started with the current US National Research Council model for lactating dairy cows (NRC, 1989), which is simple, static and highly empirical, then considered the Cornell model, which is static but incorporates a number of mechanistic elements (Search: Agriculture, 1990) and ended with consideration of the UC Davis model (Baldwin *et al.*,

1987), which is dynamic and incorporates mechanistic elements similar to those in the Cornell model.

A major concern in the formulation of models of amino acid and protein metabolism over time is that accurate estimates of protein synthesis and degradation and of reutilization of amino acids arising from degradation in protein resynthesis be available. The specific methods and analytical models used to interpret the data obtained impact on estimates of protein synthesis. Therefore, a section of the chapter is dedicated to consideration of potential sources of error in estimates of protein synthesis and the complexity of data and analytical models required to overcome the problems. This section is followed by a discussion of the basic elements and equations in dynamic models of amino acid metabolism in recent models of whole animal, liver and mammary gland metabolism.

Digestive Elements

NRC model

Rumen digestion of nitrogenous feed components and protein available to the animal in the NRC model (NRC, 1989) is predicted using static equations. Inputs required to predict protein availability at the small intestine include total digestible nutrients (TDN) or net energy for lactation (NE_L), forage intake, concentrate intake, protein concentration of the diet, and rumen degradability of the protein source. Bacterial nitrogen (BCP) is the sole variable function in the model and, in the beef model, is predicted from TDN, forage intake (FI), and concentrate intake (CI) where:

$$BCP = 6.25 \ TDN(8.63 + 14.6 \ FI - 5.18 \ FI^2 + 0.59 \ CI) \qquad (12.1)$$

The dairy model predicts *BCP* from NE_L where:

$$BCP = 6.25(11.45 \ NE_L - 30.93) \qquad (12.2)$$

Bacterial nitrogen production is subsequently used to calculate rumen available protein (RAP) using the assumption that 90% of RAP is used for bacterial protein synthesis. It is assumed that the remaining 10% of RAP is lost from the rumen as ammonia (rumen excreted protein, REP). These relationships are based on the assumptions that RAP is always adequate to support rates of microbial growth as predicted from TDN intake and that protein availability in the rumen does not affect rates of microbial growth. These simplifications fail to capture interactions between protein and carbohydrate availability. The assumption that REP is a constant proportion of RAP is probably inadequate as well, since this flux is largely dependent upon rumen ammonia concentrations.

Once RAP is calculated, the proportion of potentially digestible intake protein that escapes rumen degradation is calculated by difference from dietary protein:

$$IP = \frac{UIP + RAP}{1.15} \qquad (12.3)$$

where IP is dietary intake crude protein which is either degraded in the rumen, with partial or total conversion to bacterial and protozoal crude protein, or passed from the rumen as undegraded intake protein or UIP (NRC, 1985). The 1.15 divisor is used to accommodate an estimate of urea derived from dietary protein which is recycled back into the rumen. Since urea recycling is largely determined by circulating concentrations of urea, the assumption that 15% of intake protein is recycled is clearly an oversimplification.

After BCP yield has been estimated, total absorbed protein is calculated using the assumptions that 80% of BCP is true protein (BTP); that 80% of BTP is digested and absorbed (DBP); and that 80% of the potentially digestible protein that escapes degradation in the rumen is digested and absorbed (digested undegraded protein, DUP). The remaining 20% is excreted in faeces (IUP).

In the NRC systems, microbial growth is a function of either TDN or NE_L and dietary protein passage as a constant percentage of insoluble protein. Fluid and particle dynamics of the rumen do not play a role in the kinetics of digestion and passage of protein. The sole variable function is estimation of BCP. Other functions such as urea recycling, ammonia loss from the rumen, and interactions between carbohydrate and protein availability and microbial growth are simplified to simple algebraic functions of TDN and feed intake.

Cornell model

Bacterial yield in the Cornell model (Search:Agriculture, 1990) is defined as a static function of feed inputs. Inputs required to predict bacterial yield include: (i) a complete fibre analysis (Van Soest); (ii) total nitrogen and NPN; (iii) acid-detergent insoluble nitrogen; (iv) neutral-detergent insoluble nitrogen; (v) protein solubility; (vi) solvent soluble fat; and (vii) ash. These inputs are used to predict the growth rates of three separate pools of microbes as defined by the following equations:

$$Bact_j = NSCBact_j + SCBact_j \qquad (12.4)$$

where $SCBact_j$ = the yield of structural carbohydrate (SC)-fermenting bacteria from the jth feedstuff (g d^{-1}) and $NSCBact_j$ = the yield of non-structural carbohydrate (NSC)-fermenting bacteria from the jth feedstuff (g d^{-1}). The NSC bacterial pool is further divided into starch-fermenting and sugar-fermenting pools. The efficiency of bacterial yield is calculated according to the following equation:

$$\frac{1}{Y_{aj}} = \frac{Km_a}{Kd_{aj}} + \frac{1}{YG_a} \tag{12.5}$$

where Km_a = the maintenance requirement of the ath pool of bacteria (a = SC-fermenting, sugar-fermenting, or starch- and pectin-fermenting; g carbohydrate $\times$ g $^{-1}$ bacteria $\times$ h^{-1}), Kd_{aj} = the rate of fermentation or growth of the ath pool of bacteria on the jth feed (h^{-1}), and YG_a is the theoretical maximal yield of the ath pool of bacteria (g bacteria g^{-1} carbohydrate). This type of equation captures the principle that slowly fermented feeds will have a greater proportion of fermentable substrate utilized to maintain bacterial populations than feeds that are fermentable at higher rates. Similar equations are used for sugar- and starch-fermenting bacteria. Nitrogen availability affects the bacterial yield of $NSCBact$ according to the following equations:

$$Y_{aj} = Y_{aj} \times (1 + IMP_j \times 0.01) \tag{12.6}$$

$$IMP_j = \exp(0.404 \times \ln(Ratio_j \times 100) + 1.942) \tag{12.7}$$

$$Ratio_j = \frac{RDPep_j}{RDSugar_j + RDStarch_j + RDPep_j} \tag{12.8}$$

$$RDPep_j = RDPB_{bj} + RDPB_{bj} + RDPB_{bj} \tag{12.9}$$

$$RDPB_{bj} = DietPB_{bj} \times \frac{KD_{bj}}{Kd_{bj} + Kp_j} \tag{12.10}$$

where $RDPB_{bj}$ = rumen degradable protein (b = slow, intermediate, or fast rates of degradation) of protein in the jth feed, Kd_j = factorial degradation of the bth protein compartment of the jth, feed, and Kp_j = the rate of passage from the rumen of protein in the jth feed. The quantity of protein escaping rumen degradation or bypassing the rumen is a function of rates of degradation in and passage from the rumen, which are specified by input constants. NSCBact utilize peptides generated by protein degradation in support of their growth while growth is dependent upon carbohydrate availability as described above. Equations describing peptide uptake are of the same form as those used for protein degradation where uptake is a constant function of rumen available peptides. Peptides generated in excess of requirements for NSCBact growth are subject to deamination on passage from the rumen. Therefore, when NSCBact growth rates are slow, rates of peptide use for microbial growth are low and rates of deamination are high. Nitrogen retained by microbes is calculated using the assumptions that microbes are 10% nitrogen and a maximum of 66% of the nitrogen requirement of NSCBact can be provided by peptides. Peptides taken up in excess of this requirement are deaminated and can be used to meet ammonia requirements of growing microbes. Effects of ionophores are accommodated by reducing the rate of peptide uptake by the NSCBact pool in a stepwise manner. SCBact utilize all of their nitrogen as ammonia.

Rumen ammonia is calculated using the following static, factorial, empirical equation:

$$RAN = (Y + RDPA + NPN) - (PEPUPNR + NSCAMMNR + SCAMMNR)$$
$$(12.11)$$

where Y = recycled nitrogen ($Y = 121.7 - 12.01\ X + 0.3235X^2$; X = IP and is expressed as percent of DM), $RDPA$ = rumen degraded protein, NPN = nonprotein nitrogen derived from the diet, $PEPUPNR$ = peptide nitrogen retained by microbes, $NSCAMMNR$ = ammonia nitrogen retained by NSCBact, and $SCAMMNR$ = ammonia nitrogen retained by SCBact.

Bacterial nitrogen is algebraically divided into true protein, cell wall-associated nitrogen, and nucleic acid nitrogen. These various nitrogenous fractions and fractions derived directly from feed leave the rumen and are digested and absorbed as separate entities allowing specification of unique digestion coefficients for each pool of nitrogen.

This approach overcomes a number of deficiencies inherent in the NRC model. These include linkage of microbial yields, rates of particle and fluid passage, and chemical composition of diets; division of microbial populations by substrate; allowance for variance in ruminal and postruminal digestion of nitrogenous components due to changes in rates of particle passage; and estimates of rumen ammonia concentration.

UC Davis model

The dynamic model of Baldwin *et al.* (1987) is more complex than the NRC and Cornell models, and, therefore, requires more inputs. These include soluble sugars, organic acids, pectins, soluble and insoluble starch, plant lipids, feed fats, soluble and insoluble protein, non-protein nitrogen, hemicellulose, cellulose, lignin, and soluble and insoluble ash. The various feed components enter large particle, small particle or soluble pools of the rumen depending on their chemical and physical characteristics. Rates of rumination influence rates of conversion of large into small particles. Components of small particles are hydrolysed to chemical constituents by associated microbes or passed from the rumen. The inclusion of three pools for feed inputs and association and growth of microbes in association with small particles is required to explicitly accommodate digestion lag times as observed *in vivo*.

Cellulose and hemicellulose digestion are dependent upon microbial attachment and are influenced by rumen pH and peptide concentrations. As concentrations of cellulolytic microbes increase, rates of attachment to small particle-structural carbohydrate increase and rates of hydrolysis of feed particles increase. This results in concomitant increases in rates of entry of peptide and amino acid nitrogen into the soluble pool. Utilization of peptide N is handled somewhat differently as compared to the Cornell model. Peptides and amino acids act to enhance theoretical maximal yields of microbes and are used for SC-bacterial growth with a maximal yield of SC-bacteria from peptide and amino acid nitrogen of 50%. The following equations describe the microbial growth components of the model.

$$\frac{dMi}{dt} = U_{MiG} - U_{MiP} \qquad (12.12)$$

dMi/dt represents the change in the microbial pool (kg) with respect to time. The size of the Mi pool is determined at any time by integrating the equation. Mi partitioning between large particle, small particle, and soluble pools is based on the proportion of DM present in each pool. U_{MiG} represents the input of microbes to the pool due to microbial growth where:

$$U_{MiG} = YATP \times ATP_G \times NH_4Adj \times FatAdj \qquad (12.13)$$

YATP represents the theoretical maximal yield of bacteria per unit of ATP derived from available nutrients:

$$YATP = 0.012 + \frac{RYATP}{1 + \dfrac{kRAa}{cRAa}} \qquad (12.14)$$

where *RYATP* represents the maximal additional yield of Mi that could be realized with infinitely high concentrations of amino acids and peptides (*cRAa*) and *kRAa* represents the apparent affinity constant for rumen amino acids and peptides.

ATP_G represents ATP available for growth after maintenance needs have been met where:

$$ATP_M = \frac{Mi}{FATP_M} \qquad (12.15)$$

$FATP_M$ represents the moles of ATP used per kilogram of Mi per day for maintenance where $FATP_M$ is equal to 20 if pH $\geq$ 6.2, 40 if pH $\leq$ 5.4, and $20 + (20(0.8 - (RpH\text{-}5.4) / 0.8))$ if 5.4 <pH> 6.2.

NH_4Adj and *FatAdj* represent the adjustment factors associated with rumen ammonia concentrations and the additional fat added to the diet. The NH_4Adj function is Michaelis–Menten in nature with a V_{max} of 1 and ruminal ammonia as a substrate. The *FatAdj* function is a linear equation with an intercept of 1 and a slope of 0.3. Therefore, maximal yield of microbes is reduced when ammonia concentrations are low and enhanced when fat is added to the diet.

U_{MiP} represents the outflow of microbes from the pool due to passage from the rumen. Microbial outflow is associated with passage of the soluble and small particle pools and is calculated from liquid and particulate passage rates, respectively.

Rumen amino acid concentrations are estimated using the following equation:

$$\frac{dRAa}{dt} = U_{Ps,Aa} + U_{Pi,Aa} + U_{SPs,Aa} - U_{Aa,VFA} - U_{Aa,Mi} - U_{Aa,P}$$
$$(12.16)$$

where $U_{Ps,Aa}$, $U_{Pi,Aa}$, and $U_{SPs,Aa}$ represent fluxes into the rumen amino acid (RAa) pool from soluble feed protein degradation, degradation of insoluble feed protein in the small particle pool, and salivary protein degradation, respectively. $U_{Aa,VFA}$, $U_{Aa,Mi}$, and $U_{Aa,P}$ represent effluxes from the pool due to degradation of amino acids to volatile fatty acids and ammonia, use for microbial growth, and passage from the rumen in the soluble phase. Insoluble protein degradation is a linear function of protein concentration and microbial concentration in the small particle pool. Added dietary fat inhibits protein degradation. Degradation of amino acids (Aa) to VFA is a function of amino acid concentrations. This equation utilizes saturation kinetic arguments. The maximal rate of degradation is affected by microbial concentration in the soluble pool.

Rumen ammonia concentrations are calculated directly from rates of protein hydrolysis, feed and salivary NPN entry, microbial utilization, and loss from the rumen. Urea entry is calculated from circulating concentrations of urea.

These equations capture the concepts that low pH inhibits microbial growth by increasing the maintenance requirement of the microbes, and high peptide concentrations enhance rates of growth of SC-fermenting microbes.

The primary advantage of the UC Davis model is the dynamic tracking of the rumen environment, most particularly the microbial pool size. Dynamic solutions of rumen functions provide an explicit means of simulating reductions in digestibility associated with increased DMI, effects of particle size and solubility on protein and carbohydrate availability in the rumen, and effects of diet form and composition on microbial growth and yield. Additionally, they allow utilization of measurements made in non-steady-state conditions. This provides another dimension to previously utilized methods and, thereby, more power in terms of model parametization.

Analytical Models of Protein Turnover in Animals and Tissues

Protein synthesis measurements

Characterization of tissue protein turnover and metabolism must encompass both protein synthesis and degradation. Classical techniques for studying protein metabolism use radioisotopes or other markers and various mathematical models to describe the rates of protein synthesis and degradation. However, due to the difficulty in measuring rates of protein degradation, most work has focused on protein synthesis. In the live animal, protein degradation rates are usually inferred from measurements of protein synthesis and accretion. Protein synthesis rate can be defined and expressed as the amount of protein synthesized per unit time or as fractional synthesis rate (FSR), the fraction of total protein synthesized per unit time. Both are measured using precursor-product analysis of the precursor and protein pools of amino acids (Zak *et al.*, 1979) and depend upon accurate identification of the precursor and product pool. It is well recognized that neither extracellular (EC) nor intracellular (IC) pools of amino acids are representative of the direct precursor pool of amino acids used for protein synthesis. Rather, amino acids which have been esterified to aminoacyl-tRNA are used to synthesize protein. Unfortunately, this pool is very small relative to total free amino acids in the cytoplasm and has a half-life of 2 s or less (Airhart *et al.*, 1974). Thus, measurement of the specific radioactivity of an amino acid in this pool is both technically difficult and time consuming.

The vast majority of published rates of protein synthesis are based on the assumption that specific radioactivities (SRA) of amino acids in the aminoacyl-tRNA (SA_t), the EC (SA_e) and the IC (SA_i) pools are similar. Only rarely have investigators measured SA_t or attempted to quantitate the specific radioactivity differences between amino acid pools used for calculation of protein synthetic rates. However, when such measurements have been made, inconsistent results have been obtained. In some *in vitro* studies, SA_t has been shown to be intermediate to SA_i and SA_e (Airhart *et al.*, 1974, 1981; McKee *et al.*, 1978; Hammer and Rannels, 1981), while in others SA_t was lower than SA_i and SA_e (Airhart *et al.*, 1981; Hall and Yee, 1989; Opsahl and Ehrhart, 1987; Schneible and Young, 1984; Hildebran *et al.*, 1981). The interpretation of such data is difficult. Some data support the view that tRNAs are charged by intracellular (IC) amino acids, some that extracellular (EC) amino acids charge tRNAs, and other data that tRNA is acylated by amino acids from protein degradation which do not mix with the IC pool. Hod and Hershko (1976) used the time course of label incorporation into protein to describe a model where the precursor of protein synthesis, aminoacyl-tRNA, is charged from both the IC and EC pools of amino acid. Similar models have been proposed by Khairallah and Mortimore (1976) and Khairallah *et al.* (1977). However, there has been no attempt to quantitatively evaluate these models and theories. Additionally, these models do not account for observations when SA_t is less than either SA_i or SA_e.

Measurements of protein synthetic rates *in vivo* are complicated by the fact that specific radioactivities of radiolabelled amino acids generally change during the course of experiments. In the pulse dose method, a tracer dose of radiolabelled amino acid injected intravenously causes a rapid rise in SA_e and SA_i to a peak followed by a convex decay with time while protein specific radioactivity is a function of the kinetics associated with the change in radiolabel of the amino acid in the precursor pool, the number of proteins in the pool, and the rate of turnover of each of those individual proteins in the total pool of protein. This method has numerous disadvantages, which include the large number of time points and animals required to describe the complex and rapidly changing curves and the complexity of the analytical models needed to interpret the data (Haider and Tarver, 1969; Garlick, 1980).

To reduce the complexity of mathematical models necessary to calculate protein synthesis, techniques were developed which simplified specific radioactivity curves. One such method was the constant infusion of radiolabelled amino acid, as developed by Stephen and Waterlow (1965; Waterlow and Stephen, 1966), wherein the infusion of a tracer amino acid causes the amino acid specific radioactivity in plasma and tissues to reach a plateau. However, the plateau reached in tissues is lower than that in plasma due to the intracellular contribution of amino acids from protein degradation. The use of the constant infusion method requires several assumptions including (i) that either SA_e or SA_i is quantitatively equal to SA_t; and (ii) that there is no recycling of the labelled amino acid from the protein pool back to the precursor pool. This method also suffers from the disadvantage that many measurements are required to accurately define the shape of the curve during the rise to plateau. However, if the infusion is long enough, the rise to plateau can be estimated by a single exponential (Garlick, 1978). This method addresses the problem of reutilization by assuming that, at plateau, a steady state exists and SA_i represents the contribution of proteolysis and extracellular amino acids.

The protein fractional synthetic rate (FSR) in a tissue is calculated as described by Waterlow *et al.* (1978) using the model:

$$\frac{S_B}{S_f} = \left(\frac{\lambda_f}{\lambda_f - k_s}\right)\left(\frac{1 - e^{-k_s t}}{1 - e^{-\lambda_f t}}\right) - \left(\frac{k_s}{\lambda_f - k_s}\right) \qquad (12.17)$$

where S_B is the SRA of the protein-bound amino acid, S_f is the SRA of the free amino acid pool in the tissue, k_s is the FSR per day, t is the period of infusion (days) and λ_f is the rate constant describing the rise to plateau of the SRA of free amino acid in the tissue.

Because $\lambda_f = Rk_s$, where R is the concentration ratio, protein-bound amino acid:free amino acid in the tissue and in some tissues, $\lambda_f = \lambda_p$, where λ_p is the rate constant in the formula describing the rise to the plateau of amino acid specific radioactivity in plasma (Sp):

$$Sp = Sp_{max}(1 - e^{-\lambda - p t}) \qquad (12.18)$$

where Sp_{max} is the plasma amino acid plateau specific radioactivity and t is the period of infusion (d).

While this is the case in perfused liver and lung (Khairallah and Mortimore, 1976; Kelley *et al.*, 1984; Watkins and Rannels, 1980), the relationship between SA_e, SA_i and SA_t *in vivo* have not been reported.

The flooding or loading dose method seeks to circumvent the need to measure aminoacyl-tRNA by decreasing the differences in specific radioactivities among amino acid pools. First used by Henshaw *et al.* (1971), this method combines a large dose of labelled and unlabelled amino acid (McNurlan *et al.*, 1979; Garlick, 1980). The flooding dose is used to expand the intracellular pool of tracer amino acid such that SA_i becomes equivalent to SA_e (Garlick *et al.*, 1980), supposedly forcing SA_t to be equivalent to SA_e and SA_i. Using the flooding dose method protein fractional synthesis rate (k_s) is calculated as:

$$k_s = \frac{S_B}{S_f t} \qquad (12.19)$$

where S_B is the specific radioactivity of the amino acid in protein, S_f is the specific radioactivity of the amino acid in the precursor of amino acids used for protein synthesis, and t is expressed as days (McNurlan *et al.*, 1979). This assumes S_f does not change over time. If, however, S_f changes over time k_s is calculated using S_f, which represents the average S_f over time during which incorporation occurred (Bernier and Calvert, 1987).

However, SA_t is not necessarily similar to SA_e or SA_i, even when a flooding dose is given. Airhart *et al.* (1981) examined the effect of extracellular amino acid concentration on SA_i and SA_t in chick skeletal muscle cells *in vitro*. They found that a flooding concentration (5.0 mM) of radiolabelled leucine forced SA_e and SA_i to equality, while cells in a physiologically normal extracellular concentration (0.2 mM) of leucine had an SA_i that was 41.4% of SA_e. SA_t was less than half SA_e at both 0.2 and 5.0 mM extracellular leucine. They concluded that expansion of cellular amino acid pools was not successful in flooding the aminoacyl-tRNA pool with radioactive amino acid, and proposed that this failure was due to direct charging of tRNA with amino acids arising from protein degradation without mixing with a common intracellular pool. Similarly, Schneible and Young (1984) found that they could not force leucyl-tRNA specific radioactivities to a level equal to either SA_e or SA_i in cultured chick muscle cells. They concurred with Airhart *et al.* (1981) that accurate quantitation of protein synthesis required measurement of SA_t. Barnes (1990) reached the same conclusion using the flooding dose method with growing chicks. He reported that failure to consider SA_t resulted in underestimation of FSR by 3.9, 1.7, 1.3 and 2.25 fold in the gastrointestinal tract, liver, pectoralis muscle and thigh muscle, respectively.

To examine the possibility that tRNA is directly charged by amino acids resulting from protein degradation prior to the mixing of these amino acids with the general IC pool, Barnes *et al.* (1992) prelabelled chicken HD11 macrophage proteins with [^{3}H]leucine. After removal of [^{3}H]leucine, the cells were washed and incubated in media with either 0.23 mM or 2.3 mM cold leucine. SA_e was constant for the first 30 minutes and increased 58% and 32% for the 0.23 mM and 2.3 mM levels of leucine, respectively, during the second 30 minutes. SA_i did not change with time for the tracer level of leucine; however, for the flooding dose, SA_i decreased 40% during the course of the experiment. SA_t was not affected by increasing extracellular leucine concentration and remained well above SA_i or SA_e. However, when HD11 cells which had not been prelabelled were incubated with either a tracer or flooding dose of [^{3}H]leucine, SA_e and SA_i were higher compared to SA_t. These data strongly implicate tRNA charging with amino acids released from protein degradation.

In light of the previous discussion, the limitations of protein synthesis measurements obtained from SA_e, or SA_i must be resolved if truly quantitative rates of protein synthesis are to be obtained. It is documented from direct measurements that tracer amino acid specific radioactivities in extracellular, intracellular and tRNA pools can be very dissimilar. Previously discussed data strongly suggest that neither the IC or EC pools are the sole source of amino acids used to charge the tRNA pool. Thus protein synthesis rates calculated from either SA_e or SA_i will not provide an accurate estimate of synthesis rate.

Cellular metabolism

Evidence strongly suggests that amino acids are channelled from either extracellular sources or from protein degradation directly to protein synthetic machinery without mixing with intracellular amino acids and being exposed to enzymes involved in amino acid oxidation. Thus, the extent of amino acid channelling may be a critical determinant of the fate (oxidation versus protein synthesis) of amino acids arriving from the digestive tract and those arriving from the turnover of existing proteins.

Amino acid channelling

The organization of multienzyme complexes can result in compartmentalization or channelling of specific metabolites from one enzyme to another without equilibration with other pools of that metabolite. Channelling of metabolites is important for many metabolic pathways, including fatty acid synthesis, glycolysis and the urea cycle (Srere, 1987; Watford, 1989; Srere and Ovadi, 1990). Channelling has the potential of increasing the efficiency of a metabolic pathway by limiting loss of substrate. Additionally, the transport of substrates across cell membranes often depends on the coupling of an enzyme of that metabolite's utilization to the transport protein. Consequently, the channelling of substrates through complex metabolic pathways often begins at the cell membrane. One prerequisite for channelling is the existence of a structural organization for the components of the pathway that can lead to the catalysis of sequential reactions without the dissociation of intermediates. A second prerequisite for channelling is the demonstration of distinct pools of substrate and intermediates that do not freely mix with the respective general intracellular pools. Evidence supporting these two prerequisites implicates the channelling of amino acids arising from proteolysis towards protein synthesis.

The structural organization that implicates the channelling of aminoacyl-tRNA to protein synthesis is well documented. Aminoacyl-tRNA synthetases are enzymes that activate amino acids and esterify them to tRNAs. Many of the aminoacyl-tRNA synthetases exist in a multienzyme complex. Presently, ten of the 21 synthetase activities have been isolated from the multienzyme complex (isoleucine, leucine, lysine, methionine, arginine, proline, phenylalanine, glutamine, glutamic acid, aspartate) and it is suspected that additional enzymes are lost in the purification procedure (Schimmel, 1987; Yang and Jacobo-Molina, 1990). The enzyme complex has been identified in many cell lines as well as myoblasts (Shi *et al.*, 1991). The synthetase complex is physically associated with elongation factor eEF1 (Sarisky and Yang, 1991), one of the primary regulators of rate of protein translation. Several of the enzymes in this multienzyme complex have amino terminal hydrophobic regions that associate with lipids, presumably in one of the cell membranes (Huang and Deutscher, 1991). With gentle homogenization of chick embryos, aminoacyl-tRNA synthetases purify in the microsomal fraction (Norton *et al.*, 1965). Although it has not been shown which membrane the complex is associated with, the linkage between tRNA charging enzymes and amino acid transport across membranes is implicated. Quay *et al.* (1975) have shown that the transport of leucine is linked to and regulated by leucyl-tRNA synthetase activity.

Distinct substrate pools have also been identified. Sivaram and Deutscher (1990) provided evidence for two pools of arginyl-tRNA, one which is free in the cytosol and is involved in the post-translational modification of proteins, and a second which is a component of the aminoacyl-tRNA synthetase complex. Further, neither free arginyl-tRNA nor free phenylalanyl-tRNA are used for ribosomal protein synthesis, but the tRNAs formed in the multienzyme complex are efficiently used (Negrutskii and Deutscher, 1991). This suggests that aminoacyl-tRNAs formed in the multi-enzyme complex are transferred directly from the synthetase to the elongation factor and the ribosome without mixing with the total fluid of the cell, representing the channelling of aminoacyl-tRNA for protein synthesis.

As previously described, distinct substrate pools of free amino acids are apparently used for the charging of tRNA. This pool is separate from the total intracellular pool of free amino acids. The extent and time course of dilution of the radio-tracer's specific activity by unlabelled free amino acids in the cell or from proteolysis, as reflected in aminoacyl-tRNA, indicates the source of amino acids used for protein synthesis. Further work is needed to determine the extent of this amino acid partitioning and the method of its regulation.

Putative charging of tRNA from amino acids channelled directly from protein degradation is understandable both energetically and nutritionally. By directly reutilizing a significant proportion of the amino acids released from protein degradation, the cell can prevent the efflux of essential amino acids from the cell and minimize their loss to catabolic pathways. Efficient reutilization of amino acids could result in decreased energy requirements associated with their transport into the cell. Additionally, efficient channelling of amino acids arising from the diet or other tissues across the cell membrane directly to the aminoacyl-tRNA synthetases would preclude the possibility of loss of the amino acid to oxidation.

Amino acid recycling

Since it appears likely that amino acids are preferentially channelled from protein degradation to the immediate precursor pool for protein synthesis, the assumption that there is no recycling of the tracer amino acid, which is required if one uses the precursor-product relationship to quantitate protein synthesis from a radiolabelled amino acid, is not true. The fact that, as a part of cellular metabolism, amino acids are recycled through the precursor pool and back to protein impacts on the determined rate of protein synthesis. If one considers rate of protein synthesis in a whole body, it must be recognized that the body pool of proteins consists of a multitude of protein pools with half-lives ranging from minutes to weeks or months. For proteins with relatively short half-lives, the issue of recycling is very important and can impact on estimates of protein synthesis rates. The more rapidly a protein turns over the more rapidly the specific radioactivity of the amino acid in the protein approaches equality with the specific radioactivity of the precursor pool. Thus, specific radioactivity of the tracer amino acid released from the fast turnover protein approaches equality with that of the precursor pool. As the specific radioactivity of the amino acid in the protein approaches the specific radioactivity of the precursor pool, it becomes impossible to measure incorporation of label into that protein (as an increase in protein radioactivity). Thus, over time, measured rates of protein synthesis decrease. The longer the time period allowed for incorporation of labelled amino acids into the protein pool, the lower the estimate of protein synthesis rate. Short time incorporation experiments bias the results towards fast turnover times with rapidly turning over proteins exerting a significant influence on the result. On the other hand, longer incorporation times bias the results toward proteins with slower turnover rates since specific radioactivity of the amino acids in the fast turnover proteins comes into equality with the precursor pool amino acid specific radioactivity such that no measurable incorporation is now taking place. Using the data of Bernier and Calvert (1987), it can be demonstrated that the protein FSR calculated at 30 minutes post injection of a flooding dose of [1-^{14}C]leucine is 33.6% that calculated at 2 minutes post injection.

Conclusions

Obviously, if one is to obtain data which truly reflect rate of protein synthesis in any given system, accurate and precise measures of precursor specific radioactivities are essential. The relationships among the extracellular, intracellular and tRNA pools must be described in greater detail. The complexity of amino acid pools and relationships among them precludes the interpretation of tracer data using simple empirical and static models. In order to adequately describe and analyse amino acid tracer data, dynamic (time-variant) and mechanistic (causal, theoretical) models must be developed. Such models must represent the dynamic relationships among EC, IC, tRNA and protein-bound amino acids, as well as the formation of catabolic intermediates.

In order to accommodate the several confounding factors discussed above, a generally applicable analytic model of protein synthesis must incorporate each of the entities and transactions depicted in Fig. 12.1. The number of measurements required to solve such a complex model would include: extracellular amino acid specific radioactivity; intracellular amino acid specific radioactivity; aminoacyl-tRNA amino acid specific radioactivity; specific radioactivity of the amino acid in the protein pool; fluxes of amino acid from the intracellular pool to the aminoacyl-tRNA pool to protein and from protein to the intracellular aminoacyl-tRNA pool; and size of protein pool, aminoacyl-tRNA pool and intracellular amino acid pool. Such measurements may be prohibitively expensive. However, until results from such an experiment are available estimates of protein synthesis and degradation based upon simpler data sets and analytical models must be interpreted with great care. Further, these measurements must be made at varying concentrations of extracellular amino acids.

Models of Amino Acid Metabolism in Intact Animals

Gill *et al.* (1989) developed a dynamic, mechanistic model depicting the processes associated with growth, particularly protein turnover and ATP-dependent ion transport. Ten tissues are represented in the model. They

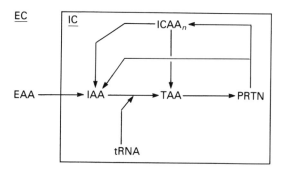

Fig. 12.1. Diagrammatic representation of the amino acid pools in the cell. EC: extracellular space; EAA: extracellular amino acids; IC: intracellular space; IAA: intracellular amino acids; $ICAA_n$: nth pool of intracellular amino acids; TAA: amino acids acylated to tRNA; PRTN: cellular protein.

are: adipose, CNS, GIT, heart, kidney, liver, muscle, pancreatic and salivary glands, reticuloendothelial system and skin. For each tissue there is an amino acid pool and a protein pool; each tissue interacts with a blood amino acid pool.

The primary driving variable in this model is amino acid transport across the gut wall. Digestible organic matter intake is the secondary driving variable. The major (simplifying) assumption in the model is that ATP is not limiting in any tissue; therefore, protein metabolism is independent of energy status.

Amino acid uptake by each tissue and IC amino acid efflux are represented by mass-action equations:

$$P_{Xa,BaXa} = k_{BaXa}V_X C_{Ba} \qquad (12.20)$$

where Ba represents the blood amino acid, Xa represents the tissue amino acid pool, $P_{Xa,BaXa}$ is the rate of transfer of amino acids from blood to tissue, k_{BaXa} is a rate constant for amino acid uptake in tissue, V_X is the volume of tissue X, and C_{Ba}, is the concentration of amino acids in blood. Tissue intracellular amino acid efflux is similarly represented by mass action kinetics for all tissues except the gastrointestinal tract and reticuloendothelial system. Those two tissues have blood amino acid concentration inhibition of amino acid efflux, represented as:

$$U_{Xa,XaBa} = \frac{K_{XaBa}Q_{Xa}}{1 + \dfrac{C_{Ba}}{J_{Ba,XaBa}}} \qquad (12.21)$$

where $U_{Xa,XaBa}$ represents the rate of amino acid efflux, K_{XaBA} is a rate constant for amino acid efflux, Q_{Xa} is the quantity of tissue, C_{Ba} is the concentration of the blood amino acid, and $J_{BA,XaBa}$ is the inhibition constant for the efflux of amino acids from the tissue with respect to blood amino acid.

Protein synthesis in the several tissues is represented using equations of the form.

$$\text{rate} = K_{XaXp}Q_{Xp}/(1.0 + K_{Xa,XaXp}/C_{Xa}) \qquad (12.22)$$

where $K_{XaXp}Q_{Xp}$ is the maximum rate of conversion of amino acid into protein as a function of the quantity of protein (Q_{Xp}) in that tissue; $K_{Xa,XaXp}$ is the affinity constant of the protein synthesis system for intracellular amino acid, and C_{Xa} is the concentration of amino acid. Equations for protein degradation are of the form:

$$\text{rate} = K_{XpXa}Q_{Xp}/(1.0 + C_{Xa}/J_{Xa,XpXa}) \qquad (12.23)$$

where the maximum rate is defined by a constant (K_{XpXa}) and quantity of protein (Q_{Xp}) and $J_{Xa,XpXa}$ is defined as an inhibition constant for an effect of intracellular concentrations of amino acids on degradation rates.

Two important assumptions of the model are: (i) inter-organ transport of specific amino acids in relation to the nitrogen economy of the ruminant is not essential to the model description, and as such, is not represented; and (ii) the portion of amino acid flux from the blood into the cell that incurs an energy cost is assumed to equal the requirement for net accretion. Therefore, amino acid efflux is negligible.

A specific description of liver amino acid metabolism follows.

Liver amino acid metabolism

Three equations represent inputs to the liver amino acid pool. These include:

1. Liver protein degradation – a mass action equation rather than the form defined by equation 12.23 describes the addition of amino acids to the liver pool from protein degradation:

$$P_{La,LpLa} = Y_{La,LpLa} \times U_{Lp,LpLa} \qquad (12.24)$$

where $P_{La,LpLa}$ is the rate of production of liver amino acid by liver protein degradation, $Y_{La,LpLa}$ is the yield of amino acids from liver protein breakdown and $U_{Lp,LpLa}$ is the utilization of liver protein for breakdown.

2. Portal vein – mass action flux of amino acid absorbed (driving variable) multiplied by the yield of amino acid entering the liver amino acid pool of those absorbed:

$$P_{La,GaBa} = Y_{La,GaBa} \times U_{Ga,GaBa} \qquad (12.25)$$

where $P_{La,GaBa}$ represents rate of incorporation of amino acid into the liver amino acid pool from portal vein blood, $Y_{La,GaBa}$ is the yield of amino acid absorbed by liver from portal vein, and $U_{Ga,GaBa}$ is the utilization of absorbed amino acid into portal blood.

3. Hepatic artery – uptake of plasma amino acids entering the liver via the hepatic artery. Separation of amino acids entering the liver from the two vascular sources was not undertaken due to inadequate data. Therefore, the contribution of the hepatic artery was set arbitrarily low.

Two equations represent the amino acids leaving the liver amino acid pool. These include:

1. Liver protein synthesis – Michaelis–Menten type kinetics with the concentration of the liver amino acids being the substrate concentration (equation 12.22).

2. Efflux to the blood – mass action kinetics represent the loss of amino acids to the blood amino acid pool (equation 12.21).

Cant et al. (unpublished) represented amino acid dynamics in the lactating cow model of Baldwin et al. (1987). Methionine, lysine and phenylalanine plus tyrosine were considered to be possible limiting amino acids for protein accretion and milk protein synthesis. The dynamics of metabolism of these three amino acids were separated from the total amino acid pool represented in the base model.

Limiting amino acid uptakes for protein synthesis were represented by Michaelis–Menten type kinetics:

$$\text{Uptake} = V_{max} * K_{inh} / (1 + (K_{aa,j \to k} / [\text{limiting amino acid}])^{exp}) \quad (12.26)$$

where V_{max} is a variable estimate of the biosynthetic capacity of the mammary gland dependent upon stage of lactation, K_{inh} is the inhibition of amino acid uptake by retained milk, $K_{aa,j \to k}$ are the affinities for the limiting amino acids $j \to k$, and exp is an empirically derived (steepness) parameter which allows the modeller to fit sigmoidal data.

The limiting amino acid for protein synthesis is determined by calculating potential uptakes of each of the three potential amino acids and considering the smallest of the three estimates to indicate the limiting amino acid. Protein synthesis is then calculated by dividing the availability of the limiting amino acid by the fraction of that amino acid in proteins, whether considering protein accretion in lean body or viscera, or milk protein synthesis.

Amino acid absorption in this model is calculated as the sum of microbial protein passage and adjusted dietary amino acid intake and the (limiting) amino acids in these proteins. The addition of the limiting amino acids to the pool from protein degradation was the product of the fraction of the protein made up of that amino acid and rate of protein degradation. Amino acids taken up in each tissue in excess of protein synthesis were considered to be catabolized. Degradation of limiting amino acids in liver are represented by Michaelis–Menten equations.

Tissue Models of Amino Acid Metabolism

Liver

Modelling analyses of amino acid metabolism in liver have been very limited in number and scope. We expect that primary reasons for this paucity of models are an extreme lack of amino acid input/output data on liver and an almost total lack of information on the regulation of amino acid metabolism in ruminant liver. In other words, information required to develop and defend equation forms and parametize the equations developed is severely lacking. Often, one can defend extrapolation of information on regulatory mechanisms and resultant equation forms based on non-ruminant (rodent) studies to the ruminant on the grounds that the same qualitative mechanisms exist across species even though major quantitative differences may, or probably, exist. However, in the case of gluconeogenesis in rodents as compared to ruminants, such an extrapolation is clearly inappropriate. Rates of gluconeogenesis in ruminants are high in the fed state and decrease during fasting, while in rodents the opposite is true. In rodents, glucocorticoids regulate the activities of a number of gluconeogenic enzymes, but this is not true in ruminants (Ely and Baldwin, 1976). Also, glucagon is a major (acute) effector of rates of gluconeogenesis in rodents while its effects in ruminants are marginal or absent (Looney *et al.*, 1987).

In introducing the nitrogen metabolism section of their liver model, Freetly *et al.* (1993) stated that 'currently data are not available to rigorously parametize the model for protein synthesis and metabolism. However, omission of such provisions would clearly lead to erroneous solutions. . . . Thus, these components were included even though a number of assumptions were required.'

Freetly *et al.* (1993) considered the turnover of liver and export proteins (albumin, fibrinogen and globulins) separately. The data and comments of Lobley *et al.* (1978, 1980) led to adoption of a fractional synthesis rate (FSR) of $0.15 \, d^{-1}$ for liver proteins in the reference lactating cow, which led to the estimate that 318 g of liver protein are turned over per day. Based on the data of Payne *et al.* (1973, 1974), Rowlands *et al.* (1975) and Swanson (1977), total export protein synthesis was set at $101 \, g \, d^{-1}$. Thus, total protein synthesis in liver in the reference cow was set at $419 \, g \, d^{-1}$ or 3.81 mol of amino acid incorporated into protein per day. If the fractional synthesis rate data of Southorn *et al.* (1992) for sheep are scaled to bodyweight raised to the 0.75 power and applied to the 550 kg reference cow, estimates of protein synthesis range from 306 to $760 \, g \, d^{-1}$. The mid-range value of $452 \, g \, d^{-1}$ is in reasonable agreement with the $419 \, g \, d^{-1}$ estimate used by Freetly *et al.* (1993) but indicates that this is a 'soft' estimate subject to considerable variance and error. In view of the fact that available data are very limited and kinetic data relating amino acid concentrations in blood to rates of protein synthesis and amino acid catabolism in liver are not available, Freetly *et al.* (1993) simply set rates of liver protein synthesis ($U_{Aa,Pt}$) and degradation ($U_{Pt,Aa}$) constant at 2.89 mol of amino acids per day and the rates of export protein synthesis ($U_{Aa,Ept}$) constant at $0.918 \, mol \, d^{-1}$ (total synthesis, as above, set at 3.81 mol amino acid incorporation into protein/day). The variable aspect of amino acid metabolism thus became amino acid catabolism ($U_{Aa,CAT}$):

$$U_{Aa,CAT} = U_{Aa,UP} + U_{Pt,Aa} - U_{Aa,Pt} - U_{Aa,ePt} \qquad (12.27)$$

where $U_{Aa,UP}$ is amino acid uptake by liver and is a constant specified as an input which can be varied in sensitivity analyses, as can be the rates of protein synthesis and degradation.

There is considerable uncertainty about liver uptake of amino acids from arterial and portal blood in view of the contention of Webb *et al.* (1992) that concentrations of di- and tripeptides in portal blood are very high, presumably because of incomplete hydrolysis of these in the lumen of the intestine and, after absorption, in intestinal cells. Reynolds *et al.* (1986, 1988) reported net uptakes of α-amino nitrogen by liver of 4.0–4.2 mol d^{-1}. These were considered minimal estimates by Freetly *et al.* (1993) based on the di- and tripeptides arguments of Webb *et al.* (1992). Based on the estimates of microbial protein yields, passage of dietary proteins from the rumen (rumen by-pass proteins) and respective digestion coefficients for these in the lower digestive tract, and utilization of amino acids for milk protein synthesis in the reference cow (in nitrogen balance), it was calculated that liver uptake of amino acids by liver should be 5.328 mol d^{-1}. This estimate implies that 23% of amino acids available to the liver arise from liver and blood proteins degraded in liver and di- and tripeptides from portal blood while 77% of amino acids available to the liver are blood amino acids. Since 3.81 mol of these amino acids are (re)incorporated into protein, 1.52 (5.328–3.81) mol of amino acids are catabolized per day in the reference cow's liver (AaCAT).

Lacking data to the contrary, Freetly *et al.* (1993) were forced to assume that the relative concentrations of amino acids in blood entering the liver were relatively constant and that liver uptakes of the several amino acids were proportional to their concentrations in blood. Sensitivity to these assumptions was relatively small in terms of overall liver metabolism. These assumptions allowed utilization of the stoichiometric estimates of the products of amino acid catabolism developed by Krebs (1964) in calculating entry rates to the pyruvate, dicarboxylic acid, α-ketoglutarate, propionyl CoA, mitochondrial acetyl-CoA, and ketone body pools.

The Freetly *et al.* (1993) model must be considered very tentative, in general, because of the lack of available data and concepts noted above. The amino acid and protein metabolism elements, in particular, must be considered particularly weak in the sense that key elements, such as protein synthesis and degradation, amino acid uptake and catabolism, and the stoichiometries of amino acid metabolism, are not dynamic variables as they should be and are not based on 'hard' data. One can hope that the relatively recent development of methods for portal/arterial venous difference studies of liver metabolism coupled with mechanistic studies in isolated hepatocytes will lead to advances in our understanding of liver metabolism in the near future.

Mammary glands

Nitrogen metabolism in the mammary gland model is partially deaggregated. Discrete pools are defined for nitrogen carriers such as alanine, aspartate and asparagine (Asp), and glutamate (Glu) with fluxes to and from these pools dictated by extracellular concentrations of the respective amino acids, intracellular concentrations of NH_3, and intracellular concentrations of their respective α-keto-acids. The equation describing Glu is:

$$\frac{dGlu}{dt} = U_{eGlu,Glu} + U_{Akg,Glu} + U_{AA,Glu} - U_{Glu,eGlu} - U_{Glu,Akg}$$
$$- U_{Glu,AA} - U_{Glu,Prt} \tag{12.28}$$

where $U_{eGlu,Glu}$ and $U_{Glu,eGlu}$ describe the exchange reactions between extracellular and intracellular Glu; $U_{Akg,Glu}$ and $U_{Glu,Akg}$ describe interconversions of α-keto-glutarate (Akg) and intracellular Glu; $U_{Aa,Glu}$ and $U_{Glu,Aa}$ describe interconversions between Glu and glutamine and conversion of other amino acids, such as proline, to Glu; and $U_{Glu,Prt}$ describes the incorporation of Glu into milk protein. With the exception of $U_{Glu,Prt}$, all of the above equations are of the Michaelis–Menten form where the primary substrate is the amino acid or α-keto-acid. Ammonia and NADH serve as secondary and tertiary substrates for $U_{Akg,Glu}$, and Glu serves as a secondary substrate for conversion of oxaloacetate and pyruvate to Asp and alanine, respectively, thereby linking transamination to ammonia concentrations. Integration of this equation with respect to time yields the concentration of Glu which is subsequently used to calculate fluxes from the pool.

Possible rate-limiting essential amino acids (methionine, lysine, tyrosine plus phenylalanine, and threonine) are tested for limiting status after uptakes are predicted from circulating concentrations. Uptake equations for these primary essential amino acids were derived from *in vivo* data (Hanigan *et al.*, 1992). Uptake of each amino acid was linearly correlated to arterial concentrations, but the intercept for each equation was negative. Negative uptakes of these amino acids have not been observed, as would be predicted given the gland's inability to synthesize them. However, incorporation of the linear equation form would have resulted in a potential source of error. Rather, data were used to define a sigmoidal equation form:

$$A - V = \frac{V_{max}}{1 + [k_{EAA}/EAA]^{\text{Exp}}} \qquad (12.29)$$

where $A - V$ is arteriovenous difference at uptake; k_{EAA} is apparent affinity for essential amino acids; and EAA is the blood concentration of essential amino acids.

Since the data were inadequate to define all three parameters required for this equation, V_{max} was fixed from *in vitro* observations, and the remaining parameters were derived from the data. This approach prevents the possible violation of physiology should a particular dietary treatment result in an unusually low essential amino acid concentration. It also provides for a smooth transition from a state of uptake to non-uptake as opposed to the abrupt shift that would occur if a linear equation were used and a switch was incorporated to prevent output of essential amino acids when essential amino acid concentrations dropped below critical concentrations. The rate-limiting amino acid determines the rate of protein synthesis. This rate is used to calculate incorporation of the remaining amino acids (essential and non-essential) into protein. Excess essential amino acids other than methionine, lysine, tyrosine plus phenylalanine and threonine are degraded. This approach is justified by the high correlations among net uptakes of these critical amino acids and allows one to avoid defining intracellular pools for each of the essential amino acids.

The remaining amino acids are aggregated into a common pool (Aa) with stoichiometries of degradation calculated from input/output data collected *in vivo*. These data were also used to parametize uptake equations for each of the amino acids.

Carbon skeletons derived from deaminated amino acids represent approximately 5.5% of carbon flow in the TCA cycle and, therefore, have a significant impact on energy balance and patterns of carbon labelling when the various metabolites are traced. The latter effect is particularly acute when examining rates of acetate label incorporated into the glycerol moiety of milk triacylglycerol as the majority of the carbon skeletons enter the tricarboxylic acid cycle after citrate resulting in net efflux of oxaloacetate from the cycle to phosphoenolpyruvate. Label incorporated into phosphoenolpyruvate is readily converted to α-glycerol phosphate. Failure to consider amino acid degradation also results in overestimates of oxidation of other energy sources such as acetate and glucose to meet energy requirements of the gland.

References

Airhart, J., Vidrich, A. and Khairallah, E.A. (1974) Compartmentation of free amino acids for protein synthesis. *Biochemistry Journal* 140, 539–545.

Airhart, J., Arnold, J.A., Bulman, C.A. and Low, R.B. (1981) Protein synthesis in pulmonary alveolar macrophages. *Biological Biophysics Acta* 653, 108–117.

Baldwin, R.L., Thornley, J.H.M. and Beever, D.E. (1987) Metabolism of the lactating cow II. Digestive elements of a mechanistic model. *Journal of Dairy Research* 54, 107–131.

Barnes, D.M. (1990) Determination of aminoacyl-tRNA specific activity for the measurement of protein synthesis in vivo and in vitro. PhD dissertation, University of California, Davis.

Barnes, D.M., Calvert, C.C. and Klasing, K.C. (1992) Source of amino acids for tRNA acylation, implications for measurement of protein synthesis. *Biochemistry Journal* 283, 583–589.

Bernier, J.F. and Calvert, C.C. (1987) Effect of a major gene for growth on protein synthesis in mice. *Journal of Animal Science* 65, 982–995.

Ely, L.O. and Baldwin, R.L. (1976) Effects of adrenalectomy upon ruminant liver and mammary function during lactation. *Journal of Dairy Science* 59, 491–503.

Freetly, H.C., Knapp, J.R., Calvert, C.C. and Baldwin, R.L. (1993) Development of a mechanistic model of liver metabolism in the lactating cow. *Agricultural Systems* 41, 157–195.

Garlick, P.J. (1978) An analysis of errors in estimation of the rate of protein synthesis by constant infusion of labeled amino acids. *Biochemistry Journal* 176, 402–405.

Garlick, P.J. (1980) Protein turnover in the whole animal and specific tissues. In: Florkin, M., Neuberger, A. and Van Deena, L.L.H. (eds) *Comprehensive Biochemistry* Vol. 19B. Elsevier, London, pp. 77–152.

Garlick, P.J., McNurlan, M.A. and Preedy, V.R. (1980) A rapid and convenient technique for measuring the rate of protein synthesis in tissues by injection of [3H]phenylalanine. *Biochemistry Journal* 192, 719–723.

Gill, M., France, J., Summers, M., McBride, B.W. and Milligan, L.P. (1989) Mathematical integration of protein metabolism in growing lambs. *Journal of Nutrition* 119, 1269–1286.

Haider, M. and Tarver, H. (1969) Effect of diet on protein synthesis and nucleic acid levels in rat liver. *Journal of Nutrition* 99, 433–445.

Hall, G.E., and Yee, J.A. (1989) Parathyroid hormone alteration of free and tRNA-bound proline specific activities in cultured mouse osteoblast-like cells. *Biochemical and Biophysical Research Communications* 161, 994–1000.

Hammer, J.A. and Rannels, D.E. (1981) Protein turnover in pulmonary macrophages. *Biochemistry Journal* 198, 53–65.

Hanigan, M.D., Calvert, C.C., DePeters, E.J., Reis, B.L. and Baldwin, R.L. (1992) Kinetics of amino acid extraction by lactating mammary glands in control and sometribove-treated holstein cows. *Journal of Dairy Science* 75, 161–173.

Henshaw, E.C., Hirche, C.A., Morton, B.E. and Hiatt, H.H. (1971) Control of protein synthesis in mammalian tissues through changes in ribosome activity. *Journal of Biological Chemistry* 246, 435–446.

Hildebran, J.N., Airhart, J., Stirewalt, W.S. and Low, R.B. (1981) Prolyl-tRNA-based rates of protein and collagen synthesis in human lung fibroblasts. *Biochemistry Journal* 198, 249–258.

Hod, Y. and Hershko, A. (1976) Relationship of the pool of intracellular valine to protein synthesis and degradation in cultured cells. *Journal of Biological Chemistry* 251, 4458–4467.

Huang, S. and Deutscher, M.P. (1991) The NH_2-terminal extension of rat liver arginyl-tRNA synthetase is responsible for its hydrophobic properties. *Biochemical and Biophysical Research Communications* 180, 702–708.

Kelley, J., Stirewalt, W.S. and Chrin, L. (1984) Protein synthesis in rat lung. *Biochemistry Journal* 222, 77–83.

Khairallah, E.A. and Mortimore, G.E. (1976) Assessment of protein turnover in perfused rat liver. *Journal of Biological Chemistry* 251, 1375–1384.

Khairallah, E.A., Airhart, J., Bruno, M.K., Puchalsky, K. and Khairallah, L. (1977) Implications of amino acid compartmentation for the determination of rates of protein catabolism in livers of meal fed rats. *Acta Biologica et Medica Germanica* 36, 1735–1745.

Krebs, H.A. (1964) The metabolic fate of amino acids. In: Munro, H.N. and Allison, J.B. (eds) *Mammalian Protein Metabolism*. Academic Press, New York, pp. 125–176.

Lobley, G.E., Reeds, P.J. and Pennie, K. (1978) Protein synthesis in cattle. *Nutrition Society Proceedings* 37, 96A.

Lobley, G.E., Milne, V., Lovie, J.M., Reeds, P.J. and Pennie, K. (1980) Whole body tissue protein synthesis in cattle. *British Journal of Nutrition* 43, 491–502.

Looney, M.C., Baldwin, R.L. and Calvert, C.C. (1987) Gluconeogenesis in isolated lamb hepatocytes. *Journal of Animal Science* 64, 283–294.

McKee, E.E., Cheung, J.Y., Rannels, D.E. and Morgan, H.E. (1978) Measurement of the rate of protein synthesis and compartmentation of heart phenylalanine. *Journal of Biological Chemistry* 253, 1030–1040.

McNurlan, M.A., Tompkins, A.M. and Garlick, P.J. (1979) The effect of starvation on the rate of protein synthesis in rat liver and small intestine. *Biochemical Journal* 178, 373–379.

Negrutskii, B.S. and Deutscher, M.P. (1991) Channeling of aminoacyl-tRNA for protein synthesis in vivo. *Proceedings of National Academy of Sciences of the USA* 88, 4991–4995.

Norton, S.J., Key, M.D. and Scholes, S.W. (1965) *Archives of Biochemistry and Biophysics* 109, 7.

NRC (National Research Council) (1985) *Ruminant Nitrogen Usage*. National Academy Press, Washington, DC.

NRC (National Research Council) (1989) *Nutrient Requirements of Dairy Cattle*, 6th ed. National Academy Press. Washington, DC.

Opsahl, W.P. and Ehrhart, L.A. (1987) Compartmentalization of proline pools and apparent rates of collagen and non-collagen protein synthesis in arterial smooth muscle cells in culture. *Biochemical Journal* 243, 137–144.

Payne, J.M., Rowlands, G.J., Manston, R. and Dew, S.M. (1973) A statistical appraisal of the results and metabolic profiles test on 75 dairy herds. *British Veterinary Journal* 129, 370–385.

Payne, J.M., Rowlands, G.J., Manston, R., Dew, S.M and Parker, W.H. (1974) A statistical appraisal of the results of 191 herds in the B.V.A./A.D.S.A. joint exercise in animal health and production. *British Veterinary Journal* 130, 34–43.

Quay, S.C., Kline, E.L. and Oxender, D.L. (1975) Role of leucyl-tRNA synthetase in regulation of branched-chain amino-acid transport. *Proceedings of National Academy of Sciences of the USA* 72 (10), 3921–3924.

Reynolds, C.K., Huntington, G.B., Tyrrell, H.F., Reynolds, P.J. and Elsasser, T.H. (1986) Net portal-drained visceral and hepatic flux of nutrients in lactating cows. *Federation Proceedings* 45, 240 (abstract).

Reynolds, C.K., Huntington, G.B., Tyrrell, H.F. and Reynolds, P.J, (1988) Net portal-drained visceral and hepatic metabolism of glucose, L-lactate, and nitrogenous compounds in lactating Holstein cows. *Journal of Dairy Science* 71, 1803–1812.

Rowlands, G.J., Manston, R., Pocock, R.M. and Dew, S.M. (1975) Relationships between stage of lactation and pregnancy and blood composition in a herd of dairy cows and the influences of seasonal changes in management on these relationships. *Journal of Dairy Research* 42, 349–362.

Sarisky, V. and Yang, D.C.H. (1991) Co-purification of the aminoacyl-tRNA synthetase complex with the elongation factor eEF1. *Biochemical and Biophysical Research Communications* 177, 757–763.

Schimmel, P. (1987) Aminoacyl tRNA synthetases: general scheme of structure-function relationships in the polypeptides and recognition of transfer RNAs. *Annual Review Biochemistry* 56, 125–158.

Schneible, P.A. and Young, R.B. (1984) Leucine pools in normal and dystrophic chicken skeletal muscle cells in culture. *Journal of Biological Chemistry* 259 (3), 1436–1440.

Search: Agriculture (1990) Number 34. Cornell University Agriculture Experiment Station, Ithaca, NY, pp. 1–128.

Shi, M.H., Tsui, F.W.L. and Rubin, L.A. (1991) Cellular localization of the target structures recognized by the anti-Jo-1 antibody: immunofluorescence studies on cultured human myoblasts. *Journal of Rheumatology* 18, 252–258.

Sivaram, P. and Deutscher, M.P. (1990) Existence of two forms of rat arginyl-tRNA synthetase suggests channeling of aminoacyl-tRNA for protein synthesis. Proceedings of National Academy of Sciences of the USA 87, 3665–3669.

Southorn, B.G., Kelly, J.M. and McBride, B.W. (1992) Phenylalanine flooding dose procedure is effective in measuring intestinal and liver protein synthesis in sheep. *Journal of Nutrition* 122, 2398–2407.

Srere, P.A. (1987) Complexes of sequential metabolic enzymes. *Annual Review of Biochemistry* 56, 89–124.

Srere, P.A. and Ovadi, J. (1990) Enzyme-enzyme interactions and their metabolic Role. *FEBS Letters* 268, 360–364.

Stephen, J.M.L. and Waterlow, J.C. (1965) Protein turnover in the rat measured with ^{14}C-lysine. *Journal of Physiology* (London) 178, 40–41.

Swanson, E.W. (1977) Factors for computing requirements for protein for maintenance of cattle. *Journal of Dairy Science* 60, 1583–93.

Waterlow, J.C. and Stephen, J.M.L. (1966) Adaptation of the rat to a low-protein diet: the effect of a reduced protein intake on the pattern of incorporation of ^{14}C-lysine. *British Journal of Nutrition* 20, 461–471.

Waterlow, J.C., Garlick, P.J. and Millward, D.J. (1978) *Protein Turnover in*

Mammalian Tissue and in the Whole Body. Elsevier, North Holland, Amsterdam.

Watford, M. (1989) Channeling in the urea cycle: a metabolon spanning two compartments. *Trends in Biochemical Sciences* 14(8), 313–314.

Watkins, C.A. and Rannels, D.E. (1980) Measurement of protein synthesis in the rat lung perfused in situ. *Biochemistry Journal* 188, 269–278.

Webb, K.E., Matthews, J.C. and DiRienzo, D.B. (1992) Peptide absorption: a review of current concepts and future perspectives. *Journal of Animal Science* 70, 3248–3257.

Yang, D.C.H. and Jacobo-Molina, A. (1991) Organization of mammalian protein biosynthetic machinery revealed from the amino acyl-tRNA synthetase complex. In Srere, P.A., Jones, M.E. and Mathews, C.K. (eds) *Structural and Organizational Aspects of Metabolic Regulation*. Alan R. Liss, New York, pp. 199–214.

Zak, R., Martin, A.F. and Blough, R. (1979) Assessment of protein turnover by use of radioisotopic tracers. *Physiological Reviews* 59, 407–447.

Amino Acid Nutrition in Sheep

13

X.B. CHEN AND E.R. ØRSKOV
*The Rowett Research Institute, Greenburn Road,
Bucksburn, Aberdeen, AB2 9SB, UK.*

Introduction

This chapter discusses our views on protein/amino acid nutrition of ruminants in the light of results mainly obtained in our laboratory. Although the title appears specific for sheep, the nutritional principles apply equally to other ruminants. The discussion covers two aspects: (i) the flow of protein to the small intestine (*intestinal protein flow*); and (ii) the amount of amino acids that must be absorbed to meet the needs of maintenance and production (*amino acid requirement*). In adult ruminants, one cannot consider amino acid requirement in isolation from protein flow. Dietary protein provision is not equal to the protein flow in the intestine. Understanding of what determines the protein flow is therefore important in diet formulation. Indeed, it is in the area of assessment of protein flow that major progress has been made in the last two decades of ruminant protein nutrition research. Our understanding of animal requirement is relatively limited since it is only recently that it has been possible to measure and manipulate the protein flow to the small intestine of ruminants.

Amino acid nutrition of the newborn

The newborn ruminant is very similar to a monogastric animal. While the reticulorumen and omasum obviously are present, they only begin to function when solid food is first eaten, in the lamb after 2–3 weeks of age. Even then, ewe's milk remains the lamb's main food for several more weeks and provides a valuable supplement to the microbial protein from the rumen for months thereafter. The milk escapes the reticulorumen by way of the oesophageal groove, so passing directly to the abomasum and the intestine.

Protein Flow to the Small Intestine

Assessment of intestinal protein flow

In recent years it has been recognized in the new protein evaluation systems for ruminants – e.g. ARC (1984) and AFRC (1992) in the UK, AAT-PBV system (Madsen, 1985) in Scandinavia, NRC (1985) in the USA, the PDI system (Vérité and Peyraud, 1989) in France – that intestinal amino acid absorption cannot be assessed from the crude protein or digestible crude protein contents of the feed, as dietary protein bears little relationship with the type and amount of protein reaching the small intestine (see discussion by Ørskov, 1992). Intestinal protein flow is assessed from the contributions of undegraded dietary protein and microbial protein.

Dietary protein

A large part of the dietary protein and non-protein nitrogen is degraded to ammonia in the rumen and depending on fermentation conditions a variable amount of the ammonia together with ammonia from endogenous sources is incorporated into microbial protein. Degradation of dietary protein depends on how readily it is fermented and on its retention time in the rumen. Roughages have a very long retention time and usually very little of the roughage protein leaves the rumen undegraded (Ørskov, 1992). Protein from animal sources (e.g. fish meal) is less rapidly degraded and may pass undegraded to the intestine in substantial amounts.

Microbial protein

Microbial protein is by far the largest source of protein passing to the small intestine. It is part of microbial biomass from bacteria, protozoa and fungi. Since only a small part of the energy of the diet is available for microbial growth under the anaerobic conditions prevailing in the rumen, there is a general relationship between the supply of fermented energy (largely carbohydrates) and the yield of microbial biomass. However, as will be discussed later, this is also influenced by rumen retention time of live organisms and by breakdown of microbial cells in the rumen.

Determination of intestinal protein flow

Dietary protein

The flow of dietary protein to the small intestine can be estimated from its ruminal degradability. Of the many methods developed for measurement of protein degradability, the nylon bag method is most widely used.

Feed samples contained in nylon bags are incubated in the rumen for different time intervals (2–36 hours) and the loss of protein is measured (Mehrez and Ørskov, 1977). The degradation kinetics can be described by the equation proposed by Ørskov and McDonald (1979), which defines degradability (p, %) thus:

$$p = a + b \left(1 - e^{-ct}\right) \qquad (13.1)$$

Here t is the duration of incubation (h). The biological meaning of the fitted parameters are: a, degradation at time zero representing loss from the bag of the water-soluble fraction of the feed protein; b, potential degradability of the water-insoluble fraction; and c, the degradation rate constant. Further refinement of these parameters (McDonald, 1981) takes account of the lag time between the start of incubation and that of degradation of the water-insoluble but potentially fermentable materials.

The potential degradability of a given feed protein is represented by the asymptote $a + b$. The actual extent of degradation when the feed is consumed by the animal will be less than this if the protein leaves the rumen before its potential degradation has been achieved. The term 'effective degradability' was therefore introduced. Effective degradability can be calculated as:

$$p = a + \frac{bc}{c + k} \qquad (13.2)$$

where k is the rumen solid outflow rate. This equation is not suitable for large particles. This is because large particles tend to be selectively retained in the rumen and their outflow rate cannot be measured accurately. Many protein evaluation systems still adopt static values for degradability (i.e. potential degradability) of given feeds. However, since outflow rate can vary between 1 and 10% h^{-1} with different feeding regimes and can thus have a great influence on the effective degradability, the use of static values is not really appropriate.

There is still a major problem in determining the degradability of protein contained in roughages. This is largely due to the fact that microbes adhere to fibrous particles and introduce a large analytical error in measurement of dietary nitrogen loss. In fact, due to a long rumen retention time, very little dietary protein contained in roughages normally escapes rumen degradation except the lignin-bound fraction, which is totally indigestible.

The supply of dietary amino acids to the small intestine can only be approximately calculated from the outflow of undegraded protein from the rumen and the amino acid composition of the original feed protein, even though it is recognized that the amino acid composition of the undegraded fraction of dietary protein may differ to some extent from that of the original feed protein. When protein is degraded, not all amino acids are deaminated at the same rate. Amino acids that are more resistant to ruminal degradation, such as phenylalanine (Kowalczyk and Jaczewska, 1991) will be enriched in the undegraded protein flowing to the small intestine.

Microbial protein

The commonly used methods for measurement of microbial protein flow to the intestine are based on determination of a microbial marker. This is either naturally present as a microbial cell constituent, e.g. RNA, 2,6-diaminopimelic acid (DAPA), 2-aminoethylphosphonic acid (AEPA), D-alanine, or is introduced into microbial cells, e.g. ^{35}S, ^{15}N, ^{32}P. The marker:N ratios in rumen microbial isolate (R_1) and duodenal digesta (R_2) and the daily flow of digesta into the small intestine (F) are measured. The intestinal flow of microbial N is then calculated as:

$$\frac{R_1 \times F}{R_2} \tag{13.3}$$

The technical details of these methods have been given by various authors: Ling and Buttery (1978: DAPA, AEPA), Zinn and Owens (1986: RNA), Mathers and Miller (1980: ^{35}S). These methods have three major shortcomings. Firstly, they need post-ruminal cannulation of animals. Secondly, the experimental procedure is complicated and thus very liable to error. Thirdly, microbial marker concentrations vary between different species of microorganisms although to what extent the marker concentration in the mixed microbial population varies is not so clear. Comparisons between methods are not reassuring. These procedures have been reviewed by Smith (1975), Buttery and Cole (1977), Stern and Hoover (1979) and Robinson et al. (1992). In recent years, little research effort has been devoted to the further development of these methods.

A lot of interest in recent years has been devoted to the development of a new method based on urinary excretion of purine derivatives (PD), namely, allantoin, uric acid, xanthine and hypoxanthine. The idea is not new (Topps and Elliott, 1965; Rys et al., 1975) but its application has awaited a better understanding of the metabolism of purine derivatives in ruminants. Its attractive feature is that it can be applied to intact animals as it only requires urine collection. The non-invasive nature and simplicity give it a potential for use under farm conditions to monitor the microbial protein supply status. The principle is that PD excretion is directly related to the absorption of exogenous purines, which are largely of microbial origin, for the normally small intake of dietary nucleic acids is extensively degraded in the rumen. On the assumption that the purine:N ratio in mixed rumen microbes is constant, microbial N absorption can be calculated from purine absorption estimated from PD excretion. Chen *et al.* (1990b) observed the following relationship between PD excretion (Y, mmol d^{-1}) and purine absorption (X, mmol d^{-1}) in sheep:

$$Y = 0.84\,X + (0.150\,W^{0.75}\,e^{-0.25X}) \tag{13.4}$$

This relationship takes account of the small contribution of endogenous PD (expression within the parentheses), and the loss $(1 - 0.84 = 0.16)$ of PD by routes other than excretion in urine (Chen *et al.*, 1990c). In this equation the net endogenous contribution declines from $0.150\,\text{mmol}\,\text{kg}^{-1}\,W^{0.75}\,\text{d}^{-1}$ as a greater amount of exogenous purines becomes available for utilization by the animal since sheep can utilize some of the absorbed exogenous purines to offset the loss of endogenous purines. Balcells *et al.* (1991) also observed a similar relationship for the absorption and excretion of purines despite slightly different values for the equation. Further work is needed to confirm the parameters used in the model, particularly the proportion of plasma PD to be excreted in urine (the 0.84 in equation 13.4). More extensive studies are also needed to examine the variability of the purine:N ratio, and to verify the results based on this method. At this stage, the method appears sound for comparative purposes.

The following points have important relevance to this technique.

1. Sheep differ from cattle in purine metabolism. Endogenous PD excretion per kg metabolic weight ($W^{0.75}$) is three times higher for cattle than for sheep (Chen *et al.*, 1990a).

2. Sheep can utilize the exogenous purines for nucleic acid synthesis whereas cattle are able to do so only to a very limited extent. Therefore the equations relating PD excretion to purine absorption are different for sheep and cattle (Chen *et al.*, 1990b; Verbic *et al.*, 1990; Balcells *et al.*, 1992).

3. Sheep urine contains all four purine derivatives, allantoin, uric acid, xanthine and hypoxanthine, and the proportion of each component varies with purine inputs (Chen *et al.*, 1990b). Cattle urine contains allantoin and uric acid but negligible amounts of xanthine and hypoxanthine (Chen *et al.*, 1990d; Verbic *et al.*, 1990). The ratio of allantoin to the total appears constant within the same animal but not between animals. On the other hand, red deer urine contains lower proportions of allantoin and uric acid (Maloiy *et al.*, 1970). Therefore, on the basis of a general absorption-excretion relationship to be used across different animals or different feeding conditions, the sum of all components would give a more accurate estimation of purine absorption than any single component.

Factors affecting intestinal protein flow

Dietary protein

Dietary proteins vary significantly in their degradation characteristics. Even with the same protein, degradability can be affected by the processing conditions, i.e. degree and duration of heating during pelleting. The largest difference can be found between vegetable and animal proteins. The insoluble fraction in fish meal is known to have a low degradability, but the proportion of soluble fraction, which is rapidly degraded, varies widely according to processing conditions. Blood meal and other albumin-containing materials are also low in degradability. In fact, this property has been successfully used in a method to protect vegetable proteins (e.g. soyabean meal) from ruminal degradation, by coating the vegetable protein with fresh blood and drying (Ørskov *et al.*, 1980). It is not intended here to review the methods used for protecting protein. Formaldehyde treatment and some form of controlled heat treatment have so far been the most successful commercially. These aspects of protein protection have been dealt with in detail by Ørskov (1992).

It must be borne in mind, as discussed above, that dietary factors which increase rumen outflow rate will decrease the effective degradability of most proteins by reducing the time they are exposed to microbial attack.

Microbial protein

Microbial protein production has been reviewed rather extensively (e.g. ARC, 1984). Factors affecting microbial protein supply include: availability of fermentable carbohydrates and nitrogen, synchronization of energy and nitrogen supply, supply of sulfur, rumen outflow rate, protozoal activity, etc. Our understanding of how these factors influence microbial protein flow from the forestomach is more qualitative than quantitative, due to the technical difficulties with the methods used for measurement of intestinal flow of microbial protein. The factors can be grouped into three categories:

1. *Factors affecting the synthesis of microbial protein.* As stated above, microbial protein synthesis is a linear function of fermentable energy supply (e.g. digestible organic matter fermented in the rumen, DOMR). We can call this linear factor 'efficiency of microbial protein synthesis' expressed as microbial N kg^{-1} DOMR. Microbial protein synthesis also depends on availability of degradable and endogenous nitrogen and sulfur, and on the synchronization of microbial nutrient supply. Extensive discussion of factors of this category was given by Hespell and Bryant (1979).

2. *Factors affecting the breakdown of microbial protein within the rumen.*

The breakdown includes lysis of microbial cells and engulfment of bacteria by protozoa. Protozoal engulfment appears to be the major factor contributing to the breakdown of bacterial protein (Wallace and McPherson, 1987).

3. *Factors affecting the physical passage of microbial protein from the rumen.* The driving force is the digesta outflow rate, and therefore factors that alter the outflow rate, e.g. level of feed intake, roughage proportion (Owens and Goetsch, 1986), feeding of salts to increase drinking water intake (Harrison *et al.*, 1975; Hadjipanayiotou *et al.*, 1982), salivary secretion (Froetschel *et al.*, 1989), could all indirectly influence the amount of microbial protein entering the abomasum. Some microbes are freely suspended in rumen fluid, others are attached to feed particles, and so both fluid and particulate outflow rates need to be considered.

As shown in Fig. 13.1, synthesis, breakdown and outflow form the three elements of the kinetics of microbial protein in the rumen. With a given diet, the amount of microbial protein synthesized (S, $g\,N\,d^{-1}$) is given by $S = eI$, where I is the fermentable OM intake ($g\,DOMR\,d^{-1}$) and e the efficiency of microbial protein synthesis ($g\,N\,kg^{-1}\,DOMR$). The outflow (Y, $g\,N\,d^{-1}$) can be calculated as:

$$Y = \frac{Sk}{(k + \delta)} \qquad (13.5)$$

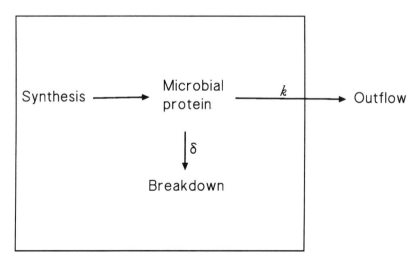

Fig. 13.1. A simplified diagram to illustrate the kinetics of microbial protein in the rumen. k is the digesta outflow rate and δ the degradation rate constant of microbial protein.

where δ (fraction h^{-1}) is the degradation rate and k (fraction h^{-1}) the digesta outflow rate. For simplicity, liquid and solid phases are not differentiated, and therefore k is only a conceptual term. Expressed as amount of microbial protein reaching the small intestine per unit DOMR intake, the 'efficiency of microbial protein supply (E, g N kg^{-1} DOMR)' becomes a function of e, δ and k:

$$E = \frac{ek}{(k + \delta)} \qquad (13.6)$$

Assuming that e is constant for a given diet, the influences of k and δ can be simulated (see Fig. 13.2). Figure 13.3 shows the experimental data obtained by Chen *et al.* (1992b) illustrating the influence of DM intake (of the same diet) per kg bodyweight (BW) on E. In this work, the DMI/ BW ratio is effectively an indicator of the rumen outflow rate k and interestingly its influence on E showed a similar pattern to that theoretically

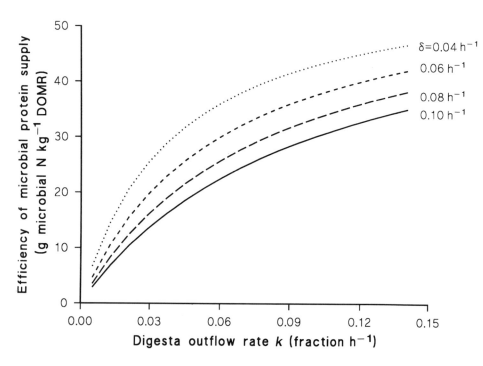

Fig. 13.2. The influences of digesta outflow rate (k) and microbial degradation rate (δ) on the efficiency of microbial protein supply (E), as simulated based on the equation $E = ek/(k + δ)$, assuming the e (the efficiency of microbial protein synthesis) is constant at 60 g N kg^{-1} DOMR.

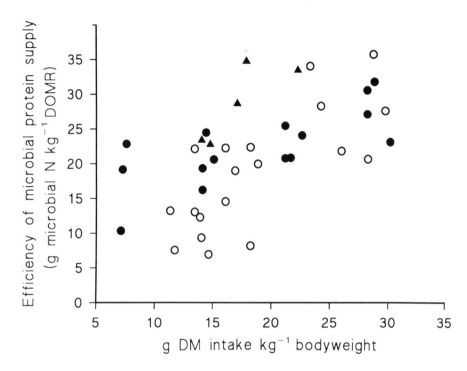

Fig. 13.3. The efficiency of microbial protein supply (g N kg^{-1} DOMR) shown against dry matter intake (DMI) corrected for bodyweight (g DMI kg^{-1} BW) by sheep. ●, different DMI given to sheep of similar BW; ○, same DMI given to sheep of different BW (22-70 kg); and ▲, different DMI in sheep of different BW. (From Chen *et al.*, 1992b. Copyright *Journal of Animal Science*, reproduced with permission.)

simulated (Fig. 13.2). It is important to note that while the efficiency of microbial protein synthesis (*e*) for a given diet could be constant, the efficiency of microbial protein supply (*E*) is not constant but can be influenced significantly by rumen outflow rate and microbial cell degradation rate.

The efficiency of microbial protein synthesis is a characteristic of the diet and is influenced by factors of the first category (supply of microbial nutrients) although an increased digesta passage rate may also improve this efficiency due to the fact that more of the available energy is used for microbial growth rather than for maintenance (Hespell and Bryant, 1979). It is possible that provided fermentable energy is not limiting, the nature of carbohydrate does not affect this efficiency as much as the nature of rumen degradable N (RDN) does. For example, there did not seem to be any difference between ammonia-treated straw, barley and sugar beet as the source of fermentable carbohydrate (Chen *et al.*, 1992a). Amino acids, peptides or protein appear to support a greater efficiency than urea as an

RDN source (Maeng *et al.*, 1976; Maeng and Baldwin, 1976; Stock *et al.*, 1986; Argyle and Baldwin, 1989; Chen *et al.*, 1992c).

Amino Acid Requirement

Estimates of amino acid requirement for production can be obtained based on information of the amino acid composition and the quantity of protein produced. This calculation, however, makes no allowance for amino acids required by metabolic processes that do not lead to net protein synthesis. A method for determination of optimal amino acid proportion of the total requirement will be discussed.

Quantity required

The amino acid requirement of an animal can be assessed by a factorial method, as in the new protein evaluation systems (e.g. ARC, 1984). The method takes into account the needs for maintenance and for deposition of productive proteins and the efficiency with which absorbed amino acids are used to meet these needs. For a sheep, the components of requirement are tissue maintenance, growth, milk, wool and fetuses.

Maintenance

It is of interest to ask if microbial protein alone is sufficient for tissue maintenance requirement if sheep are fed at energy maintenance level (about $450 \, kJ \, kg^{-1} W^{0.75}$ daily). If the estimate of ARC (1984) is used, microbial nitrogen supply is $1340 \, mg \, N \, MJ^{-1}$ of ME intake. The net absorption of microbial amino acid nitrogen would be

$$1340 \times 0.80 \times 0.85 \times 0.80 \times 0.450 = 328 \, mg \, kg^{-1} W^{0.75} \text{ per day}$$

assuming that microbial N contains 80% amino acid N, that the amino acid N is 85% digestible in the small intestine and used at an efficiency of 80% (Storm *et al.*, 1983). Experiments on animals nourished by intragastric nutrition indicated that endogenous nitrogen loss was generally between 300 and $400 \, mg \, kg^{-0.75}$, being higher for lambs than for mature sheep (see review by Ørskov, 1992). Therefore, microbial protein alone would be about sufficient to meet the maintenance requirement of mature animals but not of young animals. Consequently, the young may lose lean tissue. The situation, of course, gets worse if the animals are fed below energy maintenance. In practice, however, microbial protein will always be supplemented by some undegraded dietary protein.

Pregnancy

Robinson (1985) calculated that ewes carrying two or more lambs would produce insufficient microbial protein to meet fetal demand from about 4 weeks before parturition even if the ewes consumed sufficient food to meet the energy need for pregnancy and maintenance. The deficiency is often exaggerated because some intermediaries of protein turnover are being used as glucose precursors to meet the additional energy requirement of the fetus. Such ewes would generally mobilize body tissue to support the fetal demand. There is, however, a limit to the extent to which this can take place without a serious influence on fetal viability, udder development and subsequent milk yield. Thus ewes carrying two or more lambs require some additional undegraded dietary protein to meet the demand for fetus and udder development or they need an energy input in excess of requirement to produce more microbial protein. The latter is often difficult to achieve in late pregnancy as the fetuses restrict the capacity of the rumen and thus feed intake.

Lactation

Protein need in early lactation is very high; for example, a ewe yielding 3 l of milk daily containing 48 g protein l^{-1} produces a total of 144 g protein. If we assume that a further 10 g is required for wool growth and 47 g for tissue maintenance in a 60-kg ewe, the total requirement amounts to about 200 g net protein daily. Even if the ewe is consuming twice as much as is needed for energy maintenance, a level of intake still insufficient to meet the total energy requirement for maintenance and lactation, intestinal flow of microbial protein is only 88 g. Thus there is a very large deficit. Lactating ewes have the ability for a short time to mobilize body tissue protein, amounting to about 25 g d^{-1}, to support lactation (Robinson, 1985). The rest would have to come from dietary protein. It is obviously possible for ewes to consume more than twice their energy maintenance, but if their protein need was to be met wholly from microbial protein the animals would have to consume more than 4.5 times their energy maintenance. Robinson (1985) calculated that allowing for body-weight, the milk protein synthesis required of a ewe producing 3 l of milk is similar to that of a 600-kg Holstein cow giving 46 kg of milk per day. Unfortunately, ewes seldom get the attention and management given to dairy cows. The deficiency of protein supply in early lactation often leads to a very rapid decline in milk production and a poor protein status. If the ewe is to be mated early so as to lamb three times in two years, this can seriously influence both ovarian cycling and ovulation rate. If on the other hand the ewe is lambing only once per year, there may be opportunities for restoration of status before re-mating.

Growth

Protein requirement for growth is beginning to receive new attention. It has been realized that much earlier work measured response to intestinal protein flow and not host animal requirement. Balch (1967) proposed that an animal's capacity for protein deposition (and thus requirement for protein) was determined by energy intake (see Fig. 13.4), and the concept was supported by a large slaughter trial carried out by Andrews and Ørskov (1970). This concept has led to the use of dietary protein:energy ratio as an expression of protein requirement. It is now clear that the influence of dietary energy supply on growth was confounded by the synthesis of microbial protein that is dietary energy-dependent. Using sheep sustained by intragastric infusion of all nutrients yet without microbial fermentation in the rumen, Chowdhury *et al.* (1991) provided the animals with graded levels of casein, with or without provision of energy (from VFA). Their results indicated that, so long as the animals had a sufficient store of fat, dietary energy level had little influence on their capacity for protein deposition (Fig. 13.5). The degree of fatness that can be regarded as 'sufficient' remains to be defined, and in fact the variability in fat store may explain

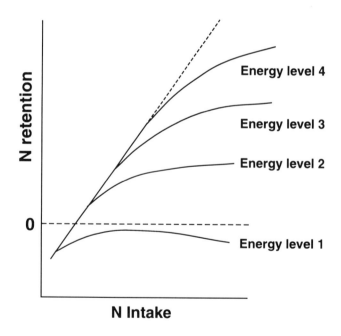

Fig. 13.4. Generalized illustration of the effect of N intake on N retention at three levels of energy intake. Based on the concept suggested by Balch (1967).

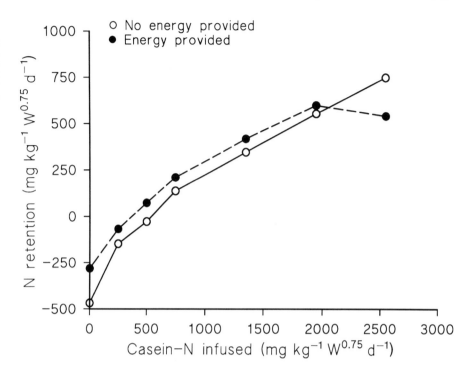

Fig. 13.5. Effect of casein N infusion on N retention in sheep at two levels of energy input: (○) zero energy provided; (●) 250 kJ kg^{-1} W$^{0.75}$ provided daily as volatile fatty acids. (Based on results of Chowdhury *et al.*, 1991.)

the variable effects of dietary energy supply on nitrogen retention reported by different laboratories; for instance, Lindberg and Jacobsson (1990) observed a clear effect of energy level on nitrogen retention at a high level of casein input (1500 mg N kg^{-1}W$^{0.75}$). However, the important implication from the observation of Chowdhury *et al.* is that body fat can be utilized efficiently to fuel protein deposition. Endogenous energy supply has not been taken into account in Balch's concept about energy-protein relationships. Expression of dietary protein:energy ratio to indicate protein requirement of the host animal is therefore inappropriate unless one also allows for endogenous energy, i.e. body fat.

The influence of body fat reserves on protein deposition presents us with a new dimension for assessing an animal's protein requirement, namely, how does nutritional status (or previous nutrition) of an animal influence its current protein need? Ørskov *et al.* (1979) showed that lambs of similar breeds and body weights but different ages had quite different responses to the same protein input. The older animals showed a greater ability to deposit protein and retain nitrogen. In a more controlled experiment using sheep nourished by intragastric infusions of all nutrients (Hovell *et al.*, 1987), the animals were subjected to protein depletion by zero-and low-protein nutrition for up to 40 days and then returned to a high level of protein intake. At the same nitrogen input, nitrogen retention was greater after protein depletion than before depletion (see Table 13.1). This process, which represents a form of compensatory growth (reviewed by Allden, 1970), was sustained for 20 days. Results of these two experiments indicated that animals previously on a low plane of nutrition would have a higher requirement for dietary protein supply. This situation is particularly pertinent in tropical countries where animals are subjected to seasonal fluctuation of nutrition. The influence of protein status of animals therefore should not be neglected in accurate assessment of the animals' need for dietary protein. Unfortunately, there are very few data to describe this influence; moreover, there is not even an established method for quantitative definition of animal's protein status.

Optimal composition of absorbed amino acids

The concept of an 'ideal protein' has been used to refer to the protein that provides absorbed amino acids in the proportion that gives maximum efficiency of utilization, i.e. 1.0. It is important to remember that the optimal composition is specific to a particular type of production. Therefore the optimal composition will differ for maintenance, tissue growth, wool growth, milk production, etc. Moreover it may also vary between species of ruminants. Table 13.2 compares the amino acid compositions of milk, tissue, wool and fetuses in sheep. Based on this comparison, one could expect that the requirement for sulfur-containing amino acids is greater for wool growth than for tissue growth. Clearly, if microbial protein is the only source of amino acids, there could be a deficiency in sulfur-containing amino acids for sheep with a high propensity for wool growth. The amino acid composition required for tissue maintenance is not known, but since the turnover of protein is mainly in tissues, it is possibly similar to that

Table 13.1. Response of N retention to infusion of casein-N before and after a 40-day protein depletion in sheep. (From Hovell *et al.*, 1987.)

	Energy input[*] $(kJ \, kg^{-1} \, W^{0.75})$	Casein-N input $(mg \, N \, kg^{-1} \, W^{0.75})$	N retention $(mg \, N \, kg^{-1} \, W^{0.75})$
Before depletion	650	2430	724 ± 66
After depletion	657	2430	1103 ± 159

[*] Energy supply was in the form of volatile fatty acids infused into the rumen.

Table 13.2. Amino acid composition (g amino acid 16 g^{-1} N) of sheep milk, lean tissue, wool and fetuses compared with that of microbial protein.

Amino acid	Ewe's milk[*]	Lean of lamb	Sheep wool	Sheep fetuses[†]	Microbial protein
Isoleucine	5.7	4.6	3.4	2.5	5.8
Leucine	10.1	7.2	9.4	6.3	8.0
Lysine	8.7	9.8	3.0	5.1	9.2
Methionine	2.7	2.6	0.7	1.4	2.5
Cysteine	0.7	1.3	10.9	1.4	1.4
Phenylalanine	4.7	3.8	4.0	3.5	5.3
Tyrosine	5.3	3.5	6.1	2.4	4.9
Threonine	4.6	4.6	6.5	3.7	5.7
Tryptophan	1.5	1.3	1.7	-	1.5
Valine	7.8	4.8	5.1	4.1	5.8
Arginine	3.2	6.1	10.1	5.3	5.3
Histidine	2.9	3.2	0.8	2.2	2.1
Alanine	5.1	5.8	4.1	5.6	6.8
Aspartic acid	6.3	9.1	7.1	6.5	11.9
Glutamine acid	19.4	16.8	15.3	10.5	12.4
Glycine	0.8	5.0	6.0	6.8	5.4
Proline		4.6	7.1	6.2	3.6
Serine	0.5	4.3	9.8	3.7	4.7

[*] Data from Deutsche Forschungsanstalt für Lebensmittelchemie (1981).
[†] Data from Robinson *et al.* (1985). Others from Ørskov (1992).

needed for tissue growth. This point was indirectly supported by the work of Storm *et al.* (1983), who observed the same utilization efficiency for microbial protein no matter whether the animal was gaining or losing nitrogen.

Identification of the optimal amino acid composition required for a particular productive process can be achieved indirectly by identifying the limiting amino acids in a reference protein whose amino acid composition is known, for example, microbial protein. In the work of Storm *et al.* (1983) on growing animals, the efficiency of utilization of absorbed microbial amino acids was 0.80, indicating that some amino acids are deficient relative to the optimal composition and that the first-limiting amino acid was present at only 80% of requirement. The classical approach to identify the limiting amino acids would be to supplement the amino acid in question at increasing levels and to observe the improvement in the productive parameter (i.e. N retention). Once the desired level of supplementation for the first amino acid is fixed, the procedure is then repeated for a second amino acid, and so on. However, this approach does not work in practice. It cannot identify an amino acid that is not the first limiting.

Even if the amino acid in question is the first limiting, the response of the productive parameter to supplementation will cease once this amino acid has become the second limiting. Therefore, the desired level of supplementation does not reveal the true degree of limitation of this amino acid.

A novel approach was proposed and tested by Storm and Ørskov (1984). The procedure involves two steps: supplementation of all amino acids (step 1), and omission of an amino acid in the mixture (step 2). Step 2 is repeated for each amino acid. This approach can detect not only the type, but also the relative degree of limitation of the limiting amino acid. In order to calculate the level of supplementation required, it is essential to acquire an accurate measurement of the intestinal flow of amino acids. The authors used this approach to identify the limiting amino acids in microbial protein for tissue growth using animals nourished by intragastric infusions of liquid nutrients. The procedure and results are illustrated in Fig. 13.6. A basal infusion of microbial protein was provided in the control period as the sole amino acid source. A mixture of amino acids with the same composition as that of absorbed microbial protein was then supplemented at a level equivalent to 0.25 of the absorbed microbial amino acids, so as to correct for the deficit of the first-limiting amino acid in the microbial protein (step 1). The 0.25 was calculated as $(1.0 - 0.80) \div 0.80$, based on microbial amino acids being utilized at an efficiency of 0.80. The nitrogen retention was increased from the control level (referred to as 'C' in Fig. 13.6) to a level (referred to as 'H') as if the microbial amino acids were to be utilized at an efficiency of 1.0. When methionine was omitted from the amino acid mixture (step 2), nitrogen retention dropped back from H to C. When lysine, histidine and arginine were omitted individually in turn, the drop in nitrogen retention from H was 0.73, 0.32 and 0.26 of the full reduction (H–C) respectively. No significant reduction was noted with the omission of other essential and non-essential amino acids. Clearly, methionine, lysine, histidine and arginine were the first four limiting amino acids and the concentrations of these amino acids were 0.80, 0.85, 0.94 and 0.95, respectively, of the required optimum level. If the concentrations of these amino acids in microbial protein are A_m, A_l, A_h and A_a, then optimum concentrations can be calculated as $1.20 A_m$, $1.15 A_l$, $1.06 A_h$, and $1.05 A_a$, i.e. 3.0, 9.3, 1.8 and 5.2 g/ 100 g of total amino acids for the four amino acids respectively. This is now the optimum composition of these essential amino acids for growth. The same approach may be taken to identify that for lactation, wool growth, etc. For instance, Fraser et al. (1991) used this approach to study the optimum composition for lactation in dairy cows and found a higher histidine requirement for lactation than for tissue growth. In current systems of protein evaluation, ruminant's amino acid requirement is still expressed on the basis of 'total amino acids'. It is possible for future systems

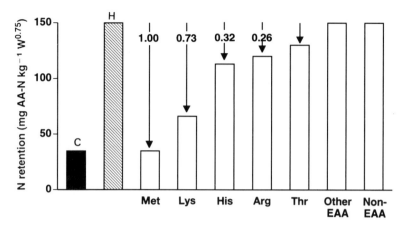

Fig. 13.6. An illustration of the procedure used by Storm and Ørskov (1984) for the identification of the limiting amino acids in microbial protein in growing lambs. ■ basal infusion of microbial protein (C); ▧ microbial protein plus mixture of amino acids (H); □ are amino acid removed from the above amino acid mixture. The relative extent of reduction (↓) in N retention reflects the order of limitation of these amino acids (the values are fractions of the full reduction from H to C).

to specify requirements for individual essential amino acids once more experimental data have been accumulated.

Methods for assessing requirement

Nitrogen balance studies have been, and will remain for some time to come, the basis for assessment of an animal's protein or amino acid requirement. An extensive review of several aspects of nitrogen balance technique was made by Manatt and Garcia (1992). The technique involves determination of the animal's productive response to nutrient inputs. The complexity of digestive physiology of ruminants presents a technical difficulty in nitrogen balance studies: there is a lack of accurate measurement, and lack of control of the amount of protein (both dietary and microbial) actually absorbed by the animal in feeding experiments. This may result in an inaccurate estimation of tissue nitrogen retention calculated as the difference between nitrogen input and output. Also, for the same reason, progress in defining the optimal amino acid compositions for different productive purposes is slow, although the methodology of Storm and Ørskov (1984) has been available for several years.

At this point, it is relevant to consider in more details the intragastric nutrition technique (Ørskov *et al.*, 1979) which has been mentioned in earlier sections. With this technique, animals are nourished entirely by liquid nutrients infused directly into the rumen (volatile fatty acids, buffer and minerals) and abomasum (casein and vitamins). There is no microbial fermentation in the rumen and nutrient inputs can be accurately controlled. The use of this technique significantly enhances the precision of nitrogen balance studies and provides a powerful tool to assess the host animal's requirement without interference from microbial inputs.

An inherent weakness with the nitrogen balance technique is ignorance of the intermediary processes between input and output. What is happening in the body is not known. The use of tracer techniques for determination of protein turnover, amino acid oxidation and endogenous recycling of nitrogen, in combination with nitrogen balance, will provide useful information on the dynamics of amino acid metabolism in the body. Recent reviews of findings based on tracer techniques have been made by Battaglia (1992) and Lobley (1992).

Conclusion

The evolution of the concept that protein supply to the ruminant should be assessed from undegraded dietary protein and microbial protein, rather than from dietary protein intake, represents major progress in the field of ruminant protein nutrition. Reasonably accurate estimates of the dietary contribution to intestinal protein flow can be acquired from degradability measurement using the nylon bag technique with adjustment for rumen digesta outflow rate. The development of simple, non-invasive techniques, e.g. that based on purine derivative excretion measurement, for estimation of microbial protein flow will provide a key for future quantitative studies of microbial protein production.

Our understanding of host animal amino acid requirement is inadequate at present largely due to the fact that in the past it has not been possible to measure and manipulate protein flow to the small intestine accurately enough. Recent findings on the use of endogenous fat as a fuel for protein deposition urges us to re-evaluate some established concepts concerning the relationship between protein requirement and dietary energy supply. More information is needed on amino acid requirement (quantity and composition) for different productive purposes, under different physiological and nutrition conditions and in different ruminant species. The prospect is optimistic because more accurate control of the protein supply to the intestine, and the use of new research tools, such as intragastric nutrition, methods for determination of optimal amino acid composition and tracer kinetics techniques, have paved the way for greater progress.

References

AFRC (Agriculture and Food Research Council) (1992) Nutritive requirements of ruminant animals: protein. AFRC Technical Committee on Responses to Nutrients, Report No. 9. *Nutrition Abstracts and Reviews (series B)* 62, 788–835.

Allden, W.G. (1970) The effects of nutritional deprivation on the subsequent productivity of sheep and cattle. *Nutrition Abstracts and Reviews* 40, 1167–1184.

Andrews, R.P. and Ørskov, E.R. (1970) The nutrition of the early weaned lamb. I. The influence of protein concentration and feeding level on rate of gain in body weight. *Journal of Agricultural Science, Cambridge* 75, 11–18.

ARC (Agricultural Research Council) (1984) *The Nutrient Requirements of Ruminant Livestock*, Supplement No. 1. Commonwealth Agricultural Bureaux, Slough.

Argyle, J.L. and Baldwin, R.L. (1989) Effects of amino acids and peptides on rumen microbial growth yields. *Journal of Dairy Science* 72, 2017–2027.

Balcells, J., Guada, J.A., Castrhillo, C. and Gasa, J. (1991) Urinary excretion of allantoin and allantoin precursors by sheep after different rates of purine infusion into the duodenum. *Journal of Agricultural Science, Cambridge* 116, 309–317.

Balcells, J., Parker, D.S. and Seal, C.J. (1992) Purine metabolite concentration in portal and peripheral blood of steers, sheep and rats. *Comparative Biochemistry and Physiology* 101B, 633–636.

Balch, C.C. (1967) Problems in predicting the value of non-protein nitrogen as a substitute for protein in rations for farm ruminants. *World Review of Animal Production* 3, 84–91.

Battaglia, F.C. (1992) New concepts in fetal and placental amino acid metabolism. *Journal of Animal Science* 70, 3258–3263.

Buttery, P.J. and Cole, D.J.A. (1977) Chemical analysis: sources of error. *Proceedings of the Nutrition Society* 36, 211–217.

Chen, X.B., Ørskov, E.R. and Hovell, F.D.DeB. (1990a) Excretion of purine derivatives by ruminants: endogenous excretion, differences between cattle and sheep. *British Journal of Nutrition* 63, 121–129.

Chen, X.B., Hovell, F.D.DeB., Ørskov, E.R. and Brown, D.S. (1990b) Excretion of purine derivatives by ruminants: effect of exogenous nucleic acid supply on purine derivative excretion by sheep. *British Journal of Nutrition* 63, 131–142.

Chen, X.B., Hovell, F.D.DeB. and Ørskov, E.R. (1990c) Excretion of purine derivatives by ruminants: recycling of allantoin into the rumen via saliva and its fate in the gut. *British Journal of Nutrition* 63, 197–205.

Chen, X.B., Mathieson, J., Hovell, F.D.DeB. and Reeds, P.J. (1990d) Measurement of purine derivatives in urine of ruminants using automated methods. *Journal of the Science of Food and Agriculture* 53, 23–33.

Chen, X.B., Abdulrazak, S.A., Shand, W.J. and Ørskov, E.R. (1992a) The effect

of supplementing straw with barley or unmolassed sugar beet pulp on microbial protein supply in sheep estimated from urinary purine derivative excretion. *Animal Production* 55, 413–417.

Chen, X.B., Chen, Y.K., Franklin, M.F., Ørskov, E.R. and Shand, W.J. (1992b)
The effect of feed intake and body weight on purine derivative excretion and
microbial protein supply in sheep. *Journal of Animal Science* 70, 1534–1542.

Chen, X.B., Gu, C.X., Zhang, W.X. and Ørskov, E.R. (1992c) Rumen microbial
protein supply to sheep given diets containing either urea or casein as the main
N source. *Animal Production* 54, 505–506.

Chowdhury, S.A., Ørskov, E.R., Hovell, F.D.DeB. and Mollison, G. (1991) Pro-
tein utilization during energy undernutrition in the ruminants. In: Eggum,
B.O., Boisen, S., Børsting, C., Danfær, A. and Hvelplund, T. (eds) *Protein
Metabolism and Nutrition: Proceedings of the 6th International Symposium
on Protein Metabolism and Nutrition, vol. 2*. National Institute of Animal
Science, Research Centre, Foulum, Denmark, pp. 154–157.

Deutsche Forschungsanstalt für Lebensmittelchemie (1981) *Food Composition and
Nutrition Tables 1981/82*. Wissenschaftliche Verlagsgesellschaft mgH,
Stuttgart, pp. 16–17.

Fraser, D.L., Ørskov, E.R., Whitelaw, F.G. and Franklin, M.F. (1991) Limiting
amino acids in dairy cows given casein as the sole source of protein. *Livestock
Production Science* 28, 235–252.

Froetschel, M.A., Amos, H.E., Evans, J.J., Croom, W.J.Jr and Hagler, W.M.Jr
(1989) Effects of a salivary stimulant, slaframine, on ruminal fermentation,
bacteria protein synthesis and digestion in frequently fed steers. *Journal of
Animal Science* 67, 827–834.

Hadjipanayiotou, M., Harrison, D.G. and Armstrong, D.G. (1982) The effects
upon digestion in sheep of the dietary inclusion of additional salivary salts.
Journal of the Science of Food and Agriculture 33, 1057–1062.

Harrison, D.G., Beever, D.E., Thomson, D.J. and Osbourn, D.F. (1975) Manipulation of rumen fermentation in sheep by increasing the rate of flow of water
from the rumen. *Journal of Agricultural Science, Cambridge* 85, 93–101.

Hespell, R.B. and Bryant, M.P. (1979) Efficiency of rumen microbial growth:
influence of some theoretical and experimental factors on Y_{ATP}. *Journal of
Animal Science* 49, 1640–1659.

Hovell, F.D.DeB., Ørskov, E.R., Kyle, D.J. and MacLeod, N.A. (1987) Undernutrition in sheep. Nitrogen repletion by N-depleted sheep. *British Journal
of Nutrition* 57, 77–88.

Lindberg, J.E. and Jacobsson, K.-G. (1990) Nitrogen and purine metabolism at
varying energy and protein supplies in sheep sustained on intragastric infusion. *British Journal of Nutrition* 64, 359–370.

Ling, J.R. and Buttery, P.J. (1978) The simultaneous use of RNA, DAPA and
2-aminoethyl-phosphonic acid as markers of microbial nitrogen entering the
duodenum of sheep. *British Journal of Nutrition* 39, 165–179.

Lobley, G.E. (1992) Control of the metabolic fate of amino acids in ruminants:
A review. *Journal of Animal Science* 70, 3264–3275.

Kowalczyk, J. and Jaczewska, A. (1991) The rate of amino acid disappearance from
rumen liquid content. In: Eggum, B.O., Boisen, S., Børsting, C., Danfær,
A. and Hvelplund, T. (eds) *Protein Metabolism and Nutrition: Proceedings*

of the 6th International Symposium on Protein Metabolism and Nutrition, vol. 2. National Institute of Animal Science, Research Centre, Foulum, Denmark, pp. 104–106.

Madsen, J. (1985) The basis for the proposed Nordic protein evaluation system for ruminants. The AAT-PBV system. *Acta Agriculturæ Scandinavica* Suppl. 25, 9–20.

Maeng, W.J. and Baldwin, R.L. (1976) Factors influencing rumen microbial growth rates and yields: effect of amino acid additions to a purified diet with nitrogen from urea. *Journal of Animal Science* 59, 648–655.

Maeng, W.J., Van Nevel, C.J., Baldwin, R.L., Morris, J.G. (1976) Rumen microbial growth rates and yields: effect of amino acids and protein. *Journal of Animal Science* 59, 477–480.

Maloiy, G.M.O., Kay, R.N.B., Goodall, E.D. and Topps, J.H. (1970) Digestion and nitrogen metabolism in sheep and red deer given large or small quantities of water and protein. *British Journal of Nutrition* 24, 843–855.

Manatt, M.W. and Garcia, P.A. (1992) Nitrogen balance: concepts and techniques. In: Nissen, S. (ed) *Modern Methods in Protein Nutrition and Metabolism*. Academic Press, London, pp. 9–66.

Mathers, J.C. and Miller, E.L. (1980) A simple procedure using ^{35}S incorporation for the measurement of microbial and undegraded food protein in ruminant digesta. *British Journal of Nutrition* 43, 503–514.

McDonald, I. (1981) A revised model for estimation of protein degradability in the rumen. *Journal of Agricultural Science, Cambridge* 96, 251–252.

Mehrez, A.Z. and Ørskov, E.R. (1977) A study of the artificial fibre bag technique for determining the digestibility of feeds in the rumen. *Journal of Agricultural Science, Cambridge* 88, 645–650.

NRC (National Research Council) (1985) *Nutrient Requirement of Sheep*. Sixth Revised Edition. National Academy Press, Washington, DC.

Ørskov, E.R. (1992) *Protein Nutrition in Ruminants*, 2nd edn. Academic Press, London.

Ørskov, E.R. and McDonald, I. (1979) The estimation of protein degradability in the rumen from incubation measurements weighted according to rate of passage. *Journal of Agricultural Science, Cambridge* 92, 499–503.

Ørskov, E.R., Grubb, D.A., Wenham, G. and Corrigall, W. (1979) The sustenance of growing and fattening ruminants by intragastric infusion of volatile fatty acid and protein. *British Journal of Nutrition* 41, 553–558.

Ørskov, E.R., Mills, C.F. and Robinson, J.J. (1980) The use of whole blood for the protection of organic materials from degradation in the rumen. *Proceedings of the Nutrition Society* 39, 60A.

Owens, F.N. and Goetsch A.L. (1986) Digesta passage and microbial protein synthesis. In: Milligan, L.P., Grovum, W.L. and Dobson, A. (eds) *Control of Digestion and Metabolism in Ruminants*. Prentice-Hall, Englewood Cliffs, NJ, pp. 196–226.

Robinson, J.J. (1985) Nutritional requirements of the pregnant and lactating ewe. In: Land, R.B. and Robinson, D.W. (eds) *Genetics of Reproduction in Sheep*. Butterworths, London, pp. 361–370.

Robinson, J.J., McDonald, I., Brown, D.S. and Fraser, C. (1985) Studies on reproduction in prolific ewes. 8. The concentrations and rates of accretion of amino acids in the foetuses. *Journal of Agricultural Science, Cambridge* 105, 21–26.

Robinson, P.H., Okine, E.K. and Kennelly, J.J. (1992) Measurement of protein digestion in ruminants. In: Nissen, S. (ed) *Modern Methods in Protein Nutrition and Metabolism.* Academic Press, London, pp. 121–144.

Rys, R., Antoniewicz, A. and Maciejewicz, J. (1975) Allantoin in urine as an index of microbial protein in the rumen. In: *Tracer Studies on Non-protein Nitrogen for Ruminants.* International Atomic Energy Agency, Vienna, pp. 95–98.

Smith, R.H. (1975) Nitrogen metabolism in the rumen and the composition and nutritive value of nitrogen compounds entering the duodenum. In: McDonald, I.W. and Warner, A.C.I. (eds) *Digestion and Metabolism in the Ruminant.* The University of New England Publishing Unit, Armidale, NSW, pp. 399–415.

Stern, M.O. and Hoover, W.J. (1979) Methods for determining and factors affecting microbial protein synthesis. A review. *Journal of Animal Science* 49, 1590–1603.

Stock, R., Klopfenstein, T., Brink, D., Britton, R. and Harmon, D. (1986) Whey as a source of rumen-degradable protein. I. Effects on microbial protein production. *Journal of Animal Science* 63, 1561–1573.

Storm, E. and Ørskov, E.R. (1984) The nutritive value of rumen micro-organisms in ruminants. 4. The limiting amino acids of microbial protein in growing sheep determined by a new approach. *British Journal of Nutrition* 52, 613–620.

Storm, E., Ørskov, E.R. and Smart, R. (1983) The nutritive value of rumen microorganisms in ruminants. 2. The apparent digestibility and net utilization of microbial N for growing lambs. *British Journal of Nutrition* 50, 471–478.

Topps, J.H. and Elliott, R.C. (1965) Relationships between concentrations of ruminal nucleic acid and excretion of purine derivative by sheep. *Nature* 205, 498–499.

Verbic, J., Chen, X.B., MacLeod, N.A. and Ørskov, E.R. (1990) Excretion of purine derivatives by ruminants: effect of microbial nucleic acid infusion on purine derivative excretion by steers. *Journal of Agricultural Science, Cambridge* 114, 243–248.

Vérité, R. and Peyraud, J.-L. (1989) Protein: the PDI systems. In: Jarrige, R. (ed) *Ruminant Nutrition. Recommended Allowances and Feed Tables.* INRA, John Libbey Eurotext, London and Paris, pp. 33–48.

Wallace, R.J. and McPherson, C.A. (1987) Factors affecting the rate of breakdown of bacterial protein in rumen fluid. *British Journal of Nutrition* 58, 313–323.

Zinn, R.A. and Owens, F.N. (1986) A rapid procedure for purine measurement and its use for estimating net ruminal protein synthesis. *Canadian Journal of Animal Science* 66, 157–166.

Amino Acid Requirements of the Veal Calf and Beef Steer

14

A.P. WILLIAMS

Cnwc y Deri, Heol Smyrna, Llangain, Carmarthen, Dyfed, SA33 5AD, UK.

Introduction

There have been many reviews on the amino acid requirements of ruminants, of which the most recent are those of Buttery and Foulds (1985), Asplund (1986), Owens and Pettigrew (1989) and Rohr and Lebzien (1991). Since much of the early work was with sheep, and the dairy cow has recently attracted considerable interest, the growing ruminant bovine has tended to be overshadowed in such reviews. Moreover, the veal calf has been almost completely neglected by reviewers since it does not fit easily into non-ruminant (Fuller, 1991) or ruminant (Rohr and Lebzien, 1991) categories. It clearly belongs to the first group since it was shown by Blaxter and Wood (1952) to be as dependent on its dietary supply of amino acids as monogastric animals.

This neglect is somewhat surprising since the topic has been studied extensively for the last two decades because of the interest in rearing calves, for long periods, on liquid diets containing considerable amounts of proteins less expensive than milk protein. In the last five years this has become increasingly important as the introduction of quotas has led to a reduction in milk surpluses. Although there has been a decline in EC veal production of about 5–7% during this period (Gill, 1989), it is still quite substantial with 800 000 tons produced in 1989 (Best, 1991). Therefore the current review will cover studies designed to estimate the amino acid requirements of the young non-ruminant and ruminant bovine with a critical assessment of the methods used for their determination.

The Veal Calf

Since the veal calf is essentially a monogastric animal it should be possible to determine its amino acid requirements by methods used for these animals. These involve interpretation of changes in some measure of response, e.g. weight gain, nitrogen retention (NR), plasma amino acids (PAA) or plasma urea (PU), that occur when the amount of the limiting amino acid is varied. These observations can then be divided into groups corresponding to treatments that are (i) deficient; (ii) adequate. The dietary intake of the limiting amino acid at the intersection of the straight lines fitted through these two groups of points is then taken as the requirement. In monogastric animals purified or semi-purified diets are used to produce the necessary deficiency in one of the essential amino acids (EAA). Such diets would be expensive for an animal as large as the veal calf, which is often reared on liquid diets to 240 kg liveweight in 26 weeks, i.e. at a growth rate of 1.3 kg d^{-1}. There are also practical difficulties in the use of such diets since, although the digestive system of the veal calf is well adapted to the digestion of cow's milk, the use of non-milk proteins can give rise to digestive disorders resulting in diarrhoea and loss of appetite (Kolar and Wagner, 1991). Such proteins can also affect the curd formation that occurs in the abomasum when milk is fed and may result in different absorption rates of EAA in the small intestine (Jenkins and Emmons, 1982). These effects have also been observed for milk which is either pasteurized or pasteurized and acidified (Scanff *et al.*, 1991).

Early studies on the EAA requirements of the veal calf simply involved supplementing milk replacer diets with either lysine, methionine or both and measuring the effects on weight gain, feed conversion efficiency or NR. The results were understandably variable and would only have provided qualitative data on requirements and are therefore not considered in detail here.

The first quantitative studies were reported in the 1970s with the development of diets that were markedly limiting in one EAA and in which almost all (Williams and Smith, 1975; Foldager *et al.*, 1977) or all (Tzeng and Davis, 1980) of the amino acids were provided in crystalline form. However, the earliest quantitative study was that of Patureau-Mirand *et al.* (1973a), who studied the response in plasma methionine and PU concentrations and in growth rate to supplementation of milk protein diets (26.4% crude protein (CP)) with graded amounts of DL-methionine. Supplementation had no significant effect on growth but from plasma methionine responses it was estimated that the methionine requirement was $580 \text{ mg kg}^{-1} \text{ W}^{0.75} \text{ d}^{-1}$ for calves growing at about 1.0 kg d^{-1}. A similar study in which graded amounts of lysine were used to supplement a milk protein diet was reported by Patureau-Mirand *et al.* (1973b). In this the estimated lysine requirement, based on plasma lysine responses, was

1300 mg kg^{-1} W$^{0.75}$ d^{-1} for 50–130 kg calves growing at the same rate. In a later study, Patureau-Mirand *et al.* (1976) reported that for 150 kg calves growing at 1.5 kg d^{-1} the lysine requirement was 1150 mg kg^{-1} W$^{0.75}$ d^{-1}. The requirements for all the EAA were estimated by Patureau-Mirand *et al.* (1974) from the accumulation of amino acids in the blood of calves given dietary proteins of different amino acid composition and the results are given in Table 14.1.

Williams and Smith (1975) studied the response in plasma methionine and PU to supplementation of a methionine-deficient semi-synthetic cow's milk diet with increasing amounts of L-methionine. For 50–60 kg calves growing at 0.25 kg d^{-1} the estimated methionine and cystine requirements were 230 and 16 mg kg^{-1} W$^{0.75}$ d^{-1} respectively, with good agreement between plasma methionine and PU results. In this study only two calves were used and no direct measurements of animal performance were made. It was therefore decided to extend the study to estimate the lysine requirements from plasma lysine, PU, growth, NR and apparent digestibility of nitrogen (ADN) responses. In this study (Williams and Hewitt, 1979) ten calves (50–58 kg liveweight, growing at 0.25 kg d^{-1}) were given diets consisting of diluted whole milk (to retain some of the clotting characteristics) and wheat gluten and supplemented with appropriate nutrients including amino acids, but not lysine. Wheat gluten was chosen because it has a very low lysine content (Table 14.2), is readily digested and has some clotting ability (Van Kempen and Huisman, 1991). These synthetic diets were supplemented with L-lysine in amounts ranging from 0 to 7.5 g d^{-1}. The lysine requirement was estimated in two ways. First, two straight lines were fitted to the values to relate response to dietary intake. One line was limited to diets that appeared to be deficient, the other to diets considered to have adequate lysine. The lysine intake at the intersection of these lines (Fig. 14.1) was taken to be the requirement. The lysine requirements based on plasma lysine, PU, growth, NR and ADN were 7.5, 8.5, 8.1, 7.2 and 7.6 g d^{-1} respectively. The lysine requirement based on growth rate was considered to be unreliable because of the difficulty in measuring accurately a growth rate as low as 0.25 kg d^{-1}. In the second approach the lysine requirements were estimated by fitting quadratic relationships to the values and identifying the ranges of lysine intake that were associated with small and large changes in response. The requirements, taken to be the intake between these ranges, based on plasma lysine, PU, NR and ADN were 6–7, 8–9, 8–9 and 7–8 g lysine d^{-1} respectively. The growth data were not used. The mean lysine requirement, 7.8 g d^{-1} or 429 mg kg^{-1} W$^{0.75}$ d^{-1}, was the same by both approaches although there were large differences between the methods for the requirement based upon plasma lysine and NR responses.

Having established the requirements for lysine, which has no function other than the synthesis of tissue protein, Williams and Hewitt (1979)

Table 14.1. The essential amino acid requirements (mg kg^{-1} W$^{0.75}$ d^{-1}) of the veal calf.

Amino acid	Patureau-Mirand et al. (1974)	Foldager et al. (1977)	Williams and Hewitt (1979)	Tzeng and Davis (1980)	Van Weerden and Huisman (1985)
Methionine	580	190	115	} 650	310
Cystine	70	70	88		110
Lysine	1300	780	429	692	1100
Threonine	800	410	269	432	550
Valine	900	470	264	421	<550
Isoleucine	850	330	187	302	680
Leucine	1300	730	467	747	790
Tyrosine	} 900	310	165	271	} <800
Phenylalanine		360	242	390	
Histidine	400	160	165	271	<280
Arginine	550	750	467	756	<300
Tryptophan	-	120	55	86	<100
Liveweight (kg)	50	43	54	50	55-75
Growth rate (kg d^{-1})	1.0	0.25	0.25	0.7	0.88-1.0

Table 14.2. Essential amino acid (EAA) requirements (g d^{-1}) of veal calves (50–58 kg liveweight) growing at 0.25 kg d^{-1} and EAA contents (g d^{-1}) of cow's milk and milk replacer diets.

Amino acid	EAA requirements	Cow's milk	Diets (g kg^{-1} CP)		
			500 cow's milk protein; 500 isolated soyabean protein	250 cow's milk protein; 750 wheat gluten	500 cow's milk protein; 500 whey protein
Methionine	2.1	3.6	2.8	2.6	2.5
Cystine	1.6	1.1	1.4	2.4	1.7
Lysine	7.8	11.7	10.0	4.5	8.5
Threonine	4.9	6.4	6.2	4.7	6.2
Valine	4.8	9.2	9.4	6.9	7.4
Isoleucine	3.4	8.2	7.8	6.4	6.6
Leucine	8.4	13.6	13.0	11.5	11.0
Tyrosine	3.0	7.1	7.6	5.6	4.8
Phenylalanine	5.4	6.8	7.8	8.2	4.8
Histidine	3.0	3.8	3.8	3.4	2.6
Arginine	8.5	5.0	8.2	5.6	3.3
Tryptophan	1.0	2.0	2.0	1.5	1.8

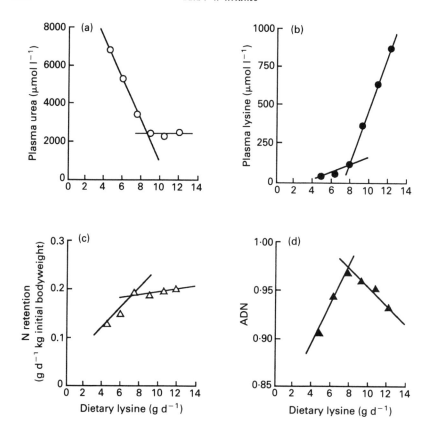

Fig. 14.1. Effect of dietary lysine supplementation on (a) plasma urea; (b) plasma lysine; (c) nitrogen retention and (d) apparent digestibility of N in the veal calf. (From Williams and Hewitt, 1979. Reproduced permission by Cambridge University Press.)

calculated the requirements for other EAA from the ratio of EAA:lysine in the body. The EAA composition of the bodies of newborn (Williams, 1984) and 7-week-old (Williams, 1978) veal calves shown in Table 14.3 was used in this calculation and the EAA requirements are given in Table 14.1. It was recognized that this factorial method cannot give precise values for requirements, for reasons that will be discussed more fully later. However, it was concluded from comparison of the requirements with the quantities of EAA in cow's milk and three milk replacer diets (Table 14.2) that cow's milk would provide adequate and in some cases excessive amounts of all EAA except for methionine and cystine. If whey protein replaced 50% of the cow's milk only histidine would be slightly deficient. When whey protein was replaced by an equivalent amount of isolated soya-bean protein the diet would be markedly deficient in sulfur amino acids

Table 14.3. The essential amino acid composition (g kg^{-1} crude protein) of the muscle and whole body of veal calves.

	Muscle: Van Weerden and Huisman (1985)	Whole body		
		Williams (1984)	Williams (1978)	Foldager *et al.* (1977)
Age	-	39	69	280 (fetus)*
Methionine	} 35	17	17	17
Cystine		13	13	-
Lysine	84	63	64	70
Threonine	43	37	40	37
Valine	51	39	39	42
Isoleucine	52	29	28	29
Leucine	73	64	69	65
Tyrosine	} 75	26	25	28
Phenylalanine		35	36	32
Histidine	32	23	25	15
Arginine	65	68	70	68
Tryptophan	13	8	8	11

* Fetus-280 days from conception.

(SAA) and supplementation would be required. The wheat gluten diet used in the determination of lysine requirements would not only be deficient in lysine but would also provide barely adequate amounts of threonine and histidine.

Foldager *et al.* (1977) used a mixture of a commercial milk replacer, containing dried skim milk and whey, and crystalline amino acids to produce a methionine-deficient diet. These diets, supplemented with increasing levels of L-methionine, were given to 43 kg calves at a rate sufficient to maintain an increase in bodyweight of approximately 0.25 kg d^{-1}. The methionine requirements, estimated from regression analysis of plasma methionine, growth rate and NR, were 190 mg kg^{-1} W$^{0.75}$ d^{-1}, although plasma methionine concentrations did not show the typical break-point response and dietary methionine was not significantly related to growth rate or NR. Concentrations of PU were also measured but could not be used to estimate methionine requirements. Cystine requirements were calculated to be 70 mg kg^{-1} W$^{0.75}$ d^{-1} based on the assumption, from data on poultry and rats, that cystine can supply 55% of the total SAA requirements. Since then Buttery *et al*. (1984) have shown for beef cattle the proportion of methionine that can be used for cyst(e)ine synthesis is 0.24, so these figures would need to be corrected. Foldager *et al*. (1977) used the value obtained for methionine requirements to estimate those of the other EAA by the factorial method (Table 14.1). It can be seen that, with the exception of histidine, these values are much higher than reported by Williams and Hewitt (1979) for similar calves. This is due to the use of methionine, instead of lysine, as the reference amino acid since methionine is not only used for tissue protein synthesis but as a methyl donor and in the synthesis of S-containing compounds (Ericson, 1961). The fact that there is little variation in the amino acid composition of animal carcasses within and between species and therefore little difference in the requirements for growing animals (Smith, 1980) supports the view that Foldager *et al*. (1977) have overestimated requirements for their calves.

A valid criticism of the studies of Williams and Smith (1975), Foldager *et al*. (1977) and Williams and Hewitt (1979) is that the calves were growing at a much slower rate than in commercial veal production. To overcome this Van Weerden and Huisman (1977, 1980, 1985) used 55–70 kg calves growing at an average rate of 0.88–1.0 kg d^{-1}. Initially the limiting EAA were determined in nitrogen balance studies in which the calves were given skimmed milk protein (SMP) diets with the CP reduced from 25 to 16%. The limiting amino acids were methionine, cystine, lysine, threonine, isoleucine and leucine. From quantitative studies with SMP diets supplemented with a mixture of 12 synthetic EAA, with one being omitted in succession, it was shown that the methionine plus cystine requirement was 9.2 g d^{-1} or 0.74% of the diet and the lysine requirement was 23 g d^{-1} or 1.81% of the diet. These levels of SAA and of lysine could be provided by SMP diets containing 21 and 22% CP respectively, without supplementation (Van Weerden and Huisman, 1985). Upper and lower limits for the other EAA were estimated either from the nitrogen balance studies or by the factorial method for threonine, isoleucine and leucine and the results given in Table 14.1.

Tzeng and Davis (1980) determined the methionine requirements of 50 kg veal calves growing at 0.7 kg d^{-1} and given diets in which all the nitrogen was provided by synthetic amino acids. These diets simulated the composition of cow's milk but the amounts of methionine or lysine were varied. Growth, NR and plasma methionine or lysine responses were used to estimate requirements. Good agreement between values estimated from growth and NR were reported but these were higher than those from PAA responses. The methionine or methionine plus cystine requirement, since cystine was not included in the diets, was 7.5 g d^{-1} or 650 mg kg^{-1} $W^{0.75}$ d^{-1}. The lysine requirement was about 12 g d^{-1} or 692 mg kg^{-1} $W^{0.75}$ d^{-1}. Tzeng and Davis (1980) considered that the accuracy of these estimates may have been limited by the number (four) of dietary levels of methionine and lysine used. This should normally be about six or seven (Williams and Hewitt, 1979) which would have avoided the large increment in dietary lysine, from 1.82 to 3.02%, used by Tzeng and Davis (1980). In this review, for comparative purposes, the lysine requirement of Tzeng and Davis (1980) has been used to calculate the requirement for the other EAA by the factorial method using the carcass composition data of Williams (1978) and these results are given in Table 14.1. Although every care has been taken in making such calculations, when using other researchers' data the reader is advised to check the original paper since mistakes can occur (e.g. Lewis and Mitchell, 1976) and data may be misinterpreted.

Comparison of the five published results of the EAA requirements of veal calves (Table 14.1) is difficult because of the wide range of growth rates. Tzeng and Davis (1980) observed that their estimates of total SAA requirements were considerably higher than those of Williams and Hewitt (1979), due possibly to their use of DL- rather than L-methionine, the differences in the growth rates of the calves and in the ADN of the diets. Making certain assumptions, their corrected requirements were much closer to those of Williams and Hewitt (1979). One of these assumptions was that D-methionine cannot be used by the calf thereby reducing the availability of methionine by 50%. Since Patureau-Mirand *et al.* (1973a) also used DL-methionine this is an important point to consider although evidence on the value of D-methionine is contradictory and may vary between species (Borg and Wahlstrom, 1989). For example, Walker and Kirk (1975) reported that supplements of DL-methionine were equivalent to L-methionine in improving nitrogen balance in preruminant lambs. However, Komarek and Jandzinski (1978) have suggested that, in the dairy cow, D-methionine is used as a methyl donor, whereas the L-isomer can also be used for protein synthesis. An obvious solution to the problem is to use only L-methionine in any studies on requirements.

Any correction applied to compensate for the apparent digestibilities of the EAA is also important. Van Weerden and Huisman (1985) and Moughan *et al.* (1989) reported that the apparent digestibilities ranged from 94.1% (arginine) to 96.8% (histidine) when calves were given SMP diets. Added synthetic amino acids would clearly be completely digestible, but when SMP is replaced by other, less digestible proteins then this should be taken into account in calculating requirements.

Although direct comparison of the EAA requirements is difficult there are certain results in Table 14.1 that are clearly incorrect. An example is the low histidine requirement reported by Foldager *et al.* (1977), which is due to the very low histidine content (Table 14.3) of the fetal calf carcass used in their calculations. Van Weerden and Huisman (1985) used the EAA concentrations of calf muscle rather than those of the whole body to estimate the requirements of some EAA by the factorial method. From

Table 14.3 it can be seen that there are considerable differences in the EAA composition of muscle and the whole body. Substitution of the isoleucine content of the whole body for that of muscle would result in a 29% reduction in the leucine requirements, emphasizing the significance of using whole body values (Williams *et al.*, 1954).

Conclusions

For most diets given to veal calves only the SAA, lysine and possibly threonine are likely to be limiting. Estimates of the requirements of all other EAA by the factorial method (Table 14.1) are sufficiently precise for most purposes since they are generally present in excess in most commercial diets. These have included such diverse protein sources as fish, faba bean, rapeseed, alfalfa, wheat gluten, potato, yeast, pea and various soyabean products (Otterby and Linn, 1981, Van Kempen and Huisman, 1991; Kolar and Wagner, 1991). Degussa (1988) have published recommended amino acid allowances (% of diet) for calves up to 16 weeks of age given milk replacer diets of SMP or whey protein plus either casein or soyabean protein. These were: methionine 0.48–0.55; methionine plus cystine, 0.72–0.80; lysine, 1.45–1.80 and threonine, 0.78–1.00. These are probably based on the results of Van Weerden and Huisman (1985), whose data are used in Germany and Holland. In France the recommended allowances are based on the data of Patureau-Mirand *et al.* (1974), and in the UK and USA commercial companies use the data of Williams and Hewitt (1979) (Dr J.W. Sissons, personal communication). Therefore, in spite of the neglect of the veal calf by reviewers, the present state of knowledge of its amino acid requirements must be considered to be reasonably good and further studies would be difficult to justify on the grounds of expense.

The Beef Steer

Although it is unlikely that there are any fundamental differences between the EAA requirements of the veal calf and the beef steer (Buttery and Foulds, 1985) there are considerably greater difficulties in measuring the requirements of the ruminant animal (Lewis, 1992). In the veal calf EAA requirements have been estimated, with the use of synthetic and semi-synthetic diets, by varying the dietary supply of the limiting amino acid and measuring growth rate of PAA responses. That is not possible in the beef steer because dietary protein is extensively degraded in the rumen and the amino acids entering the duodenum consist largely of those provided by microbial protein synthesized in the rumen with smaller, variable amounts from undegraded dietary protein and endogenous protein. Therefore the amino acid supply to the animal's small intestines has to be manipulated either by infusion of the limiting amino acid into the abomasum or duodenum or by addition to the diet of the limiting amino acid, protected in some way from degradation in the rumen. The former method requires animals fitted with abomasal or duodenal cannulas, which are also necessary to be able to determine, from the flow and composition of the abomasal or duodenal digesta, the quantities of amino acids entering the small intestines. Since the amino acid composition of duodenal digesta is generally considered to be fairly constant (Buttery and Foulds, 1985) this represents another problem in the determination of ruminant EAA requirements. As for the veal calf, it is necessary that the deficiency of the limiting amino acid is such that it is possible to have at least three supplementary points below, as well as above the requirement. One method of overcoming this problem has been described by Storm and Ørskov (1984) for growing sheep using intragastric infusion of volatile fatty acids into the rumen and microbial protein into the abomasum. More details of the method are given in Chapter 13. A recent alternative approach has been described by Titgemeyer and Merchen (1990a) in which steers were given a semi-purified diet containing little true protein. Abomasal infusions of dextrose and a mixture of crystalline amino acids simulating the non-SAA pattern of casein were followed by infusions of graded levels of L-methionine. Requirements for SAA were estimated from NR and PAA responses and are discussed in more detail later in this chapter.

Early studies on the EAA requirements of the steer involved post-ruminal infusion of the limiting amino acid and measuring NR, PAA or PU responses. These studies included those for methionine (Steinacker *et al.*, 1970; Mathers and Miller, 1979; Schwab *et al.*, 1982), lysine (Boila and Devlin, 1972; Burris *et al.*, 1976) or for lysine, methionine and threonine (Richardson and Hatfield, 1978). Unfortunately these studies failed to measure the quantities of the limiting amino acids in the digesta and it was only possible to estimate the supplementary amino acid requirement. Details of these requirements have been given by Buttery and Foulds (1985) and it is not intended to discuss them further here. Quantitative estimates of EAA requirements can be divided into those based on factorial methods and those based on other, direct methods such as PAA and NR responses.

Factorial estimates

Hutton and Annison (1972) estimated that methionine was the limiting EAA for a 200 kg steer growing at 1.0 kg d^{-1}. Their calculations were based on a comparison of the duodenal requirement of EAA for the steer, estimated by a factorial method using the known EAA requirements of the young pig, with the estimated synthesis of EAA by the rumen bacteria of the steer. Although acknowledging certain weaknesses in their method, notably the appreciable differences in metabolic faecal nitrogen outputs between the pig and the steer, Hutton and Annison (1972) produced estimates for most of the EAA, which are shown in Table 14.4. Unfortunately the cystine requirement was not given and the results were based on the EAA composition of pig carcasses. If the composition of calf carcasses (Williams, 1978) had been used the requirements for certain EAA would have been lower; for instance, the methionine requirement would have been 240 instead of 263 mg kg^{-1} W$^{0.75}$ d^{-1}. Details of the other revised values are given by Buttery and Foulds (1985). Other factorial estimates have been made for similar steers by Burroughs et al. (1974), Griffiths (1977) and Gabel and Poppe (1986). The calculations used by Burroughs et al. (1974) and Gabel and Poppe (1986) provide information for steers of different liveweights and liveweight gains, but for comparative purposes only the EAA requirements for 200 kg steers growing at 1.0 or 1.1 kg d^{-1} are given in Table 14.4. The values marked with an asterisk in Table 14.4 were recalculated from the data of Burroughs et al. (1974) and Griffiths (1977) since these authors only presented results for the SAA, lysine and threonine. Griffiths (1977) used the EAA composition of beef carcass meat and fatty tissues in his calculations. Substitution of the whole body EAA values reported by Williams (1978) resulted in a 15% reduction in the estimated lysine requirements but did not markedly affect the SAA and threonine requirements. Owens and Pettigrew (1989) have also estimated the EAA requirements of a 350 kg steer growing at 1.0 kg d^{-1} by the factorial method and these results are given in Table 14.4. These authors suggested subdividing requirements into two parts, maintenance and performance; by adding the two it is then possible to calculate the requirements for a particular level of performance. The maintenance requirements for a 350 kg steer are also given in Table 14.4 with the sum of these and of those for growth at 1.0 kg d^{-1} also being given. Unfortunately cystine and tyrosine requirements were not reported. The absence of cystine is surprising since the authors considered the EAA requirements for maintenance were highly correlated with the EAA composition of keratin, a protein particularly rich in this amino acid. Since there are few estimates of maintenance requirements, Owens and Pettigrew (1989) calculated them from the sum of amino acids lost as metabolic faecal, endogenous urinary and scurf components. Similar calculations were made by Gabel and Poppe (1986) for the maintenance requirements of a 200 kg steer; these were generally, apart from methionine, isoleucine, threonine and valine, of the same order as those reported by Owens and Pettigrew (1989). Agreement between the different factorial estimates, although for animals of about the same growth rate and size, is not very good, particularly for the SAA, where some values (Griffiths, 1977; Gabel and Poppe, 1986) are quite low. Although the SAA requirements reported by Burroughs et al. (1974) seem to be in excellent agreement with the other factorial and direct estimates in Table 14.4 their estimates for other EAA seem to be twice as high. Recently O'Conner *et al.* (1993) modified the Cornell Net Carbohydrate and Protein System in an attempt to predict the supply of, and requirements for, absorbed essential amino acids in cattle diets. This would have the advantage in being able to predict requirements under different dietary, animal and environmental conditions. However more research was considered necessary in several areas to provide more accurate data for the model. Ainslee *et al.* (1993) have attempted to do this for one area, the amino acid content of the tissue protein of steers.

Direct estimates

The first refereed study on the EAA requirements of any ruminant animal in which the total flow of amino acids into the duodenum was measured was that of Williams and Smith (1974). In this steers of 110–160 kg liveweight, growing at 0.4 kg d^{-1} were given three diets containing flaked maize and straw supplemented with either groundnut meal (diet A, 20 g N kg^{-1} dry matter (DM); diet C, 10 g N kg^{-1} DM) or maize gluten meal (diet B, 20 g N kg^{-1} DM). Steers given diets A and C were abomasally infused with increasing amounts of L-methionine and those given diet B with increasing amounts of L-lysine. Flows of EAA into the duodenum were measured and SAA and lysine requirements estimated from PAA responses. The PU responses, although measured, did not enable requirements to be estimated and production responses were not measured. The SAA and lysine requirements for steers given diets A and B are given in Table 14.4. The SAA requirements of the steers given diet C were lower, 150 g methionine kg^{-1}W$^{0.75}$ d^{-1} and 100 mg cystine kg^{-1}W$^{0.75}$ d^{-1}, probably because of the lower growth rate supported by this diet. No attempt was made to correct these values for the apparent digestibility of EAA in the small intestine or for the net efficiency of utilization of EAA entering the duodenum since such information was not available for the steer at that time. Such corrections are desirable, and Rohr and Leibzien (1991) have suggested mean values of 85 and 65% for the true digestibility and efficiency of utilization of EAA respectively. The protein supplements in this study were selected in an attempt to influence the EAA composition of the duodenal digesta. However, even for maize gluten, a protein of low degradability, selected for its low lysine content, the effect of dietary protein was masked by rumen microbial protein synthesis except for arginine and leucine, which differ considerably between the two protein sources. The lysine requirement on this diet was already satisfied since plasma lysine increased linearly with increasing lysine infusion in contrast to the two-phase response of plasma methionine to methionine infusions. Although the EAA composition of ruminant digesta is considered to be independent of diet since a large proportion is from

Table 14.4. The essential amino acid requirements (mg kg^{-1} W$^{0.75}$ d^{-1}) of the beef steer.

Amino acid/ exptl. details	Hutton and Annison (1972)	Burroughs et al. (1974)	Griffiths (1977)	Gabel and Poppe (1986)	Owens and Pettigrew (1989)		
					(a)	(b)	(a+b)
Factorial methodology							
Methionine	263	225	⎰ 134	86	38	171	209
Cystine	-	150	⎱	41	-	-	-
Lysine	365	750	317	294	37	221	258
Threonine	233	470*	184	157	42	160	202
Valine	259	457*	193*	182	48	173	221
Isoleucine	259	328*	139*	139	41	173	214
Leucine	312	809*	342*	298	44	229	273
Tyrosine	-	293*	124*	112	-	-	-
Phenylalanine	259	422*	178*	157	23	267	300
Histidine	96	200	124*	96	14	74	88
Tryptophan	63	94*	40*	35	5	45	50
Liveweight (kg)	200	200	150-400	200	350	350	
Growth rate (kg d^{-1})	1.0	1.1	1.0	1.0	0	1.0	

	Williams and Smith (1974)	Fenderson and Bergen (1975)	Mathers et al. (1979)	Titgemeyer and Merchen (1990b)	Titgemeyer and Merchen (1991)
Direct methodology					
Methionine	271	249	200	}210	–
Cystine	136	63	70		–
Lysine	<520	<376	–	–	<540
Threonine	–	<252	–	–	⋮
Tryptophan	–	<55	–	–	–
Liveweight (kg)	110–160	274	90–170	294–325	355
Growth rate (kg d^{-1})	0.4	0.7	0.5	1.3	–
Method †	PAA	PAA	PAA	PAA / NR	PAA / NR

* Asterisked values are recalculated from the original data of Burroughs *et al.* (1974) and Griffiths (1977).
† PAA = plasma amino acids; NR = nitrogen retention.
(a) requirement calculated for maintenance
(b) requirement calculated for gain of 1 kg d^{-1}
(a+b) requirement for maintenance and growth.

bacterial protein, the wide range, $9-38 \, g \, kg^{-1}$ amino acids, of rumen bacterial methionine concentrations reported in the literature by Chamberlain *et al.* (1986) is of major concern given the amino acid's nutritional importance.

Fenderson and Bergen (1975) carried out a similar experiment for bigger steers (274 kg) growing at a faster rate ($0.73 \, kg \, d^{-1}$) and given a diet of maize and oats. Abomasal infusions were of L-methionine, L-lysine, L-threonine or L-tryptophan and the total abomasal flows of these amino acids were determined. From PAA responses, of which only methionine infusion resulted in a two-phase response, the requirements for these EAA were estimated and are given in Table 14.4. The lysine, threonine and tryptophan requirements were assumed to have been met by microbial protein synthesis. The results have been corrected to absorbable requirements assuming a digestibility coefficient of 0.70 for amino acids in sheep digesta.

Mathers *et al.* (1979) determined the methionine requirements of $90-170 \, kg$ steers growing at $0.5 \, kg \, d^{-1}$ from the response of plasma methionine to duodenal infusions of methionine. The steers were given a barley/urea diet which supplied mainly microbial protein to the duodenum and were infused with a mixture of amino acids simulating the composition of fish meal but without SAA. The SAA requirements, given in Table 14.4, were calculated from the methionine infused at the plasma methionine break-point plus the methionine and cystine apparently absorbed from the small intestines on the basal diet. In a similar later experiment Mathers and Miller (1983) only observed a break-point in the plasma methionine response curve for one of four steers. No explanation was given for this difference.

Titgemeyer *et al.* (1988) estimated the absorbable lysine requirement of steers given a maize/maize silage diet from plasma lysine responses. For 313 kg and 383 kg steers gaining more than $1 \, kg \, d^{-1}$ the lysine requirement was $<30g \, d^{-1}$ and $44-48 \, g \, d^{-1}$ respectively. The latter values seem excessively high. Attempts to determine the SAA requirements in the same study were unsuccessful, probably because the methionine supplied on this diet was in excess of the requirements. However, the authors concluded that D-methionine could be utilized by the ruminant steer and that lysine and methionine, protected from degradation in the rumen, were effective in supplying these EAA post-ruminally. Because of the difficulty in producing a methionine deficiency with normal diets Titgemeyer and Merchen (1990a) developed the procedure using a semi-purified diet described earlier in this section. Quantitative estimates of the total absorbable SAA requirements of $294-325 \, kg$ steers growing at $1.3 \, kg \, d^{-1}$ based on PAA and NR responses were reported by Titgemeyer and Merchen (1990b) and are given in Table 14.4. These results are considerably lower than those reported by Williams and Smith (1974), Fenderson and Bergen (1975) and Mathers *et al.* (1979), particularly since the steers in these studies were smaller and growing at much slower rates. Titgemeyer and Merchen (1991) have recently reported the absorbable lysine requirement of 355 kg steers using the same procedure used for estimating SAA requirements except that NR responses alone were used. The growth rate of the steers was not given but would presumbly have been about $1.3 \, kg \, d^{-1}$. The result, shown in Table 14.4, like that of Williams and Smith (1974), is a maximum value since the lack of response in NR showed that lysine was not the first-limiting EAA under these conditions.

Conclusions

Estimates of the EAA requirements for the beef steer, whether by factorial or direct methods, strongly support the view that, under most practical conditions, only the quantities of methionine and cystine synthesized in the rumen, together with those surviving from the diet, are insufficient for the needs of the animal. There remains the practical problem of supplementing the ruminant diet with either individual amino acids or proteins protected from degradation in the rumen. Since experimentation of this type is very expensive and given the present economic climate, it is unlikely that further studies will be commissioned. An alternative approach suggested by Smith (1980) might be to consider the total EAA supply since diets, calculated according to new protein rationing schemes for ruminants, generally provide a marked excess of total EAA for the growing steer. This view is supported by the study of Williams *et al.* (1988) using veal calves as model animals since their amino acid intakes could be precisely controlled. From this the optimum EAA: total AA ratio at the duodenum was found to be 0.37, which is close to that found in the whole body of the calf and much lower than the value of 0.48 found in ruminant duodenal digesta.

References

Ainslee, S.J., Fox, D.J., Perry, T.C., Ketchen, D.J. and Barry, M.C. (1993) Predicting amino acid adequacy of diets fed to Holstein steers. *Journal of Animal Science* 71, 1312–1319.

Asplund, J.M. (1986) Somatic nutrient requirements of ruminants. In: Olson, R.E., Beutler, E. and Broquist, H.P. (eds) *Annual Review of Nutrition 6.* Annual Reviews, Palo Alto, pp. 95–112.

Best, P. (1991) New market, new meat. *Feed International* 12, 6–10.

Blaxter, K.L. and Wood, W.A. (1952) The nutrition of the young Ayrshire calf. 7. The biological value of gelatin and of casein when given as a sole source of protein. *British Journal of Nutrition* 6, 56–71.

Boila, R.J. and Devlin, T.J. (1972) Effects of lysine infusion per abomasum of steers fed continuously. *Canadian Journal of Animal Science* 52, 681–687.

Borg, B.S. and Wahlstrom, R.C. (1989) Species and isomeric variation in the utilization of amino acids. In: Friedman, M. (ed) *Absorption and Utilization of Amino Acids II*. CRC Press, Boca Raton, pp. 155–172.

Burris, W.R., Boling, J.A., Bradley, N.W. and Young, A.W. (1976) Abomasal lysine infusion in steers fed a urea supplemented diet. *Journal of Animal Science* 42, 699–705.

Burroughs, W., Trenkle, A. and Velter, R.L. (1974) A system of protein evaluation for cattle and sheep involving metabolizable protein (amino acids) and urea fermentation potential of feedstuffs. *Veterinary Medicine/Small Animal Clinician* 69, 713–722.

Buttery, P.J. and Foulds, A.N. (1985). Amino acid requirements of ruminants. In: Haresign, W. and Cole, D.J.A. (eds) *Recent Advances in Animal Nutrition*. Butterworths, London, Boston, pp. 257–271.

Buttery, P.J., Essex, C., Foulds, A.N. and Soar, J.B. (1984) Methionine to cystine conversion in cattle. *Proceedings of the Nutrition Society* 43, 56A.

Chamberlain, D.G., Thomas, P.C. and Quig, J. (1986) Utilization of silage nitrogen in sheep and cows: amino acid composition of duodenal digesta and rumen microbes. *Grass and Forage Science* 41, 31–38.

Degussa (1988) *Amino Acids in Animal Nutrition*. Degussa Ltd, Bonn, pp. 3–52.

Ericson, L.-E. (1961) A criticism of the carcass analysis procedure for the determination of amino acid requirements. *Acta Physiologica Scandinavica* 52, 90–98.

Fenderson, C.L. and Bergen, W.G. (1975) An assessment of essential amino acid requirements of growing steers. *Journal of Animal Science* 41, 1759–1766.

Foldager, J., Huber, J.T. and Bergen, W.G. (1977) Methionine and sulfur amino acid requirements in the preruminant calf. *Journal of Dairy Science* 60, 1095–1104.

Fuller, M.F. (1991) Present knowledge of amino acid requirements for maintenance and production: non-ruminants. In: Eggum, B.O., Boisen, S., Børsting, C., Danfær, A. and Hvelplund, T. (eds) *Proceedings of the 6th International Symposium on Protein Metabolism and Nutrition*. National Institute of Animal Science, Foulom, Denmark, pp. 116–126.

Gabel, M. and Poppe, S. (1986) Intestinal amino acid supply in relation to the amino acid requirements in growing bulls. *Archiv für Tierernährung* 36, 227–234.

Gill, C. (1989) Evolution of milk replacers. *Feed International* 10, 27–29.

Griffiths, T.W. (1977) Amino acid composition of beef carcass meat and amino acid requirements of growing cattle. *Proceedings of the Nutrition Society* 36, 82A.

Hutton, K. and Annison, E.F. (1972) Control of nitrogen metabolism in the ruminant. *Proceedings of the Nutrition Society* 31, 151–158.

Jenkins, K.J. and Emmons, D.B. (1982) Evidence for beneficial effect of chymosin-casein clots in abomasum on calf performance. *Nutrition Reports International* 26, 635–643.

Kolar, C.W. and Wagner, T.J. (1991) Alternative protein use in calf milk replacers. In: Metz, J.H.M. and Groenestein, C.M. (eds) *Proceedings of the International Symposium on Veal Calf Production*. PUDOC, Wageningen, pp. 211–216.

Komarek, R.J. and Jandzinski, R.A. (1978) Evidence indicating the need for methionine by dairy cows and the metabolic relationship of D-methionine and L-methionine. *Journal of Dairy Science* 61, 876.

Lewis, A.J. (1992) Determination of the amino acid requirements of animals. In Nissen, S. (ed.) *Modern Methods in Protein Nutrition and Metabolism.* Academic Press, London, pp. 67–85.

Lewis, D. and Mitchell, R.M. (1976) Amino acid requirements of ruminants. In: Cole, D.J.A., Boorman, K.N., Buttery, P.J., Lewis, D., Neale, R.J. and Swan, H. (eds) *Proceedings of the 1st International Symposium on Protein Metabolism and Nutrition.* Butterworths, London, pp. 417–424.

Mathers, J.C. and Miller, E.L. (1979) The determination of amino acid requirements. In: Buttery, P.J. (ed.) *Protein Metabolism in the Ruminant.* ARC, London, pp. 3.1–3.11.

Mathers, J.C. and Miller, E.L. (1983) Effects of varying methionine supply on methionine and leucine metabolism in growing calves. In: Pion, R., Arnal, M. and Bonin, D. (eds) *Proceedings of the 4th International Symposium on Protein Metabolism and Nutrition, II.* INRA, Paris, pp. 29–32.

Mathers, J.C., Miller, E.L. and Lerman, P.M. (1979) Methionine supply and requirement in growing calves. *Annales des Recherches Veterinaires* 10, 310–313.

Moughan, P.J., Stevens, E.V.J., Reisima, I.D. and Rendel, J. (1989) The effect of avoparacin on the ileal and faecal digestibility of nitrogen and amino acids in the milk-fed calf. *Animal Production* 49, 63–71.

O'Conner, J.D., Sniffen, C.J., Fox, D.J. and Chalupa, W. (1993) A net carbohydrate and protein system for evaluating cattle diets: IV. Predicting amino acid adequacy, *Journal of Animal Science* 71, 1298–1311.

Otterby, D.E. and Linn, J.G. (1981) Advances in nutrition and management of calves and heifers. *Journal of Dairy Science* 64, 1365–1377.

Owens, F.N. and Pettigrew, J.E. (1989) Subdividing amino acid requirements into portions for maintenance and growth. In: Friedman, M. (ed.) *Absorption and Utilization of Amino Acids I.* CRC Press, Boca Raton, pp. 15–30.

Patureau-Mirand, P., Prugnaud, J. and Pion, R. (1973a) Influence of sulfur amino acid supplementation of a milk replacer on blood-free amino acid levels. Estimation of the methionine requirement of the pre-ruminant calf. *Annales de Biologic animale, Biochimie, Biophysique* 13, 225–246.

Patureau-Mirand, P., Prugnaud, J. and Pion, R. (1973b) Influence of lysine supplementation of a milk replacer on the content of free lysine in the blood and muscle of the pre-ruminant calf. *Annales de Biologie animale, Biochimie, Biophysique* 13, 683–689.

Patureau-Mirand, P., Toullec, R., Paruelle, J.L., Prugnaud, J. and Pion, R. (1974) Influence of the nature of proteins in milk replacers on blood levels of free amino acids in the preruminant calf. I. Milk, whey, fish and alkane yeast proteins. *Annales de Zootechnie* 23, 343–348.

Patureau-Mirand, P., Grizard, J., Prugnaud, J. and Pion, R. (1976) Utilization of a diet rich in starchy products by the rapidly growing preruminant calf. I. Influence on the free amino acid content of blood and muscle. *Annales de Biologie animale, Biochimie, Biophysique* 16, 579–592.

Richardson, C.R. and Hatfield, E.E. (1978) The limiting amino acids in growing cattle. *Journal of Animal Science* 46, 740–745.

348   A.P. Williams

Rohr, K. and Lebzien, P. (1991) Present knowledge of amino acid requirements for maintenance and production. In: Eggum, B.O., Boisen, S., Børsting, C., Danfær, A. and Hvelplund, T. (eds) *Proceedings of the 6th International Symposium on Protein Metabolism and Nutrition.* National Institute of Animal Science, Foulum, pp. 127–137.

Scanff, P., Yvon, M., Pelissier, J.-P., Guilloteau, P. and Toullec, R. (1991) Effect of some technological treatments of milk on amino acid composition of in vivo effluents during gastric digestion. *Journal of Agricultural and Food Chemistry* 39, 1482–1487.

Schwab, C.G., Muise, S.J., Hylton, W.E. and Moore, J.J. III (1982) Response to abomasal infusion of methionine of weaned dairy calves fed a complete pelleted starter ration based on by-product feeds. *Journal of Dairy Science* 65, 1950–1961.

Smith, R.H. (1980) Comparative amino acid requirements. *Proceedings of the Nutrition Society* 39, 71–78.

Steinacker, G., Devlin, T.J. and Ingalls, J.R. (1970) Effect of methionine supplementation posterior to the rumen on nitrogen utilization and sulfur balance of steers on a high roughage ration. *Canadian Journal of Animal Science* 50, 319–324.

Storm, E. and Ørskov, E.R. (1984) The nutritive value of rumen micro-organisms in ruminants 4. The limiting amino acids of microbial protein in growing sheep determined by a new approach. *British Journal of Nutrition* 52, 613–620.

Titgemeyer, E.C. and Merchen, N.R. (1990a) The effect of abomasal methionine supplementation on nitrogen retention of growing steers postruminally infused with casein or nonsulfur-containing amino acids. *Journal of Animal Science* 68, 750–757.

Titgemeyer, E.C. and Merchen, N.R. (1990b) Sulfur-containing amino acid requirement of rapidly growing steers. *Journal of Animal Science* 68, 2075–2083.

Titgemeyer, E.C. and Merchen, N.R. (1991) Lysine requirement of growing steers. *Archiv für Tierernährung* 41, 71–76.

Titgemeyer, E.C., Merchen, N.R., Berger, L.L. and Deetz, L.E. (1988) Estimation of lysine and methionine requirements of growing steers fed corn silage-based or corn-based diets. *Journal of Dairy Science* 71, 421–434.

Tzeng, D. and Davis, C.L. (1980) Amino acid nutrition of the young calf. Estimation of methionine and lysine requirements. *Journal of Dairy Science* 63, 441–450.

Van Kempen, G.J.M. and Huisman, J. (1991) Introductory remarks; some aspects of skim-milk replacement by other protein sources in veal-calf diets. In: Metz, J.H.M. and Groenestein, C.M. (eds) *New Trends in Veal Calf Production.* Pudoc, Wageningen, pp. 201–205.

Van Weerden, E.J. and Huisman, J. (1977) The amino acid requirements of milk-fed calves. *Landbouwkundig Tijdschrift* 89, 206–213.

Van Weerden, E.J. and Huisman, J. (1980) Amino acid requirement of the milk-fed veal calf. In: Oslage, H.J. and Rohr, K. (eds) *Proceedings of the 3rd EAAP Symposium on Protein Metabolism and Nutrition II.* FAL, Braunschweig, pp. 825–831.

Van Weerden, E.J. and Huisman, J. (1985) Amino acid requirement of the young veal calf. *Zeitschrift für Tierphysiologie, Tierernährung und Futtermittelkunde* 53, 232–244.

Walker, D.M. and Kirk, R.D. (1975) The utilization by preruminant lambs of isolated soya bean protein in low protein milk replacers. *Australian Journal of Agricultural Research* 26, 1037–1052.

Williams, A.P. (1978) The amino acid, collagen and mineral composition of preruminant calves. *Journal of Agricultural Science, Cambridge* 90, 617–624.

Williams, A.P. (1984) The amino acid requirements of the preruminant calf. In: Zebrowska, T., Buraczewska, L., Buraczewski, S., Kowalczyk, J. and Pastuszewska, B. (eds) *Proceedings of the VI International Symposium on Amino Acids.* Polish Scientific Publishers, Warsaw, pp. 330–334.

Williams, A.P. and Hewitt, D. (1979) The amino acid requirements of the preruminant calf. *British Journal of Nutrition* 41, 311–319.

Williams, A.P. and Smith, R.H. (1974) Concentrations of amino acids and urea in the plasma of the ruminating calf and estimation of the amino acid requirements. *British Journal of Nutrition* 32, 421–433.

Williams A.P. and Smith, R.H. (1975) Concentrations of amino acids and urea in the plasma of the preruminant calf and estimation of the amino acid requirements. *British Journal of Nutrition* 33, 149–158.

Williams, A.P., Smith, R.H. and Cockburn, J.E. (1988) Essential amino acid requirements of young bovines growing at different rates. In: Elsner, L., Friemel, H., Heitz, G. *et al.* (eds) *Proceedings of the 5th International Symposium on Protein Metabolism and Nutrition.* Wilhelm Pieck University, Rostock, pp. 35–37.

Williams, H.H., Curtin, L.V., Abraham, J., Loosli, J.K. and Maynard, L.A. (1954) Estimation of growth requirements for amino acids by assay of the carcass. *Journal of Biological Chemistry* 208, 277–286.

.

Amino Acid Nutrition of the Dairy Cow

15

J.D. OLDHAM

Genetics and Behavioural Sciences Department,
The Scottish Agricultural College, Bush Estate, Penicuik,
Midlothian EH26 0QE, UK.

Introduction

It seems reasonable to state at the outset that the concepts of amino acid nutrition, as applied to dairy cows, are the same as those which hold for the other animal species which have already been dealt with in previous chapters. As far as tissue metabolism is concerned cows use the same range of amino acids in their metabolism as do other mammalian species, and the concept of some amino acids being essential (indispensable) and others non-essential (dispensable) also has the same general meaning with dairy cattle as it does with other animals.

In ruminant animals the impact of feeding on the metabolism of rumen microorganisms is always an issue to be considered when relating the provision of nutrients in food to performance of the animal. Except for the fact that the relatively high planes of feeding which high-yielding dairy cows achieve differs from lower performing ruminants, and has consequent effects on the dynamics of rumen metabolism, it also seems reasonable to think that any influences of feed composition on rumen microbial metabolism will be similar in the rumen of dairy cows to that of other ruminants.

The basis for presenting this discussion on amino acid nutrition of dairy cows is therefore that cows are no different in concept from other animals as regards the manner in which amino acids are used in the body, and no different, at least qualitatively, from other ruminants in the way that rumen fermentation influences, or is influenced by, the provision of amino acids (or amino acid precursors) in feeds. The differences which are of interest, and which deserve separate treatment, arise from the large effect which the protein secretion from the mammary gland has on the quantities of amino acids which are needed by cows, and on the issues of amino acid partition between the mammary gland and other tissues when amino acid supply is inadequate.

Quantitative aspects of amino acid nutrition in dairy cows do, obviously, differ from other species for reasons of animal size, the nature of the protein products produced (with a very heavy emphasis on the production of milk protein) and perhaps also on other competing demands for amino acids for the sustenance of effective lacatation; in the latter instance one might think particularly of the possibility that the high glucose demand of the animal imposes specific constraints on amino acid use.

Any consideration of amino acid nutrition is largely concerned with two main issues:

1. The matching of amino acid supply to animal demand.
2. The consequences for animal performance and amino acid metabolism when supply is less than perfect in either, or both, quantity or quality (i.e. amino acid balance).

Matching Supply to Demand

Amino acids which are made available to the animal by being absorbed from the small intestine can be referred to as 'metabolizable amino acid supply'. It is more common in the sphere of ruminant nutrition for this supply of a mixture of amino acids to be referred to in terms of its equivalent protein mass – hence the term 'metabolizable protein', which is the mixture of amino acids presented, by absorption from the gastrointestinal tract, to the animal for its metabolism.

It has become conventional in recent years to adopt schemes for matching metabolizable protein (MP) supply in ruminant animals to their demands for MP by accounting, separately, for the needs of rumen microbes for nitrogenous precursors and the demands of the animal, at tissue level, for total metabolizable protein. The amounts of amino acids which are supplied to the animal in the form of microbial protein, once the needs of the rumen microbes for nitrogen have been met, may, or may not, be sufficient to satisfy the animal's demand for metabolizable protein. In general, rumen microbial protein alone might be expected to be sufficient to meet the needs of animals whose demand for metabolizable protein per unit metabolizable energy (ME) does not exceed $6.5\,g$ $MP\,MJ^{-1}$ ME. This is because the yield of metabolizable protein per unit metabolizable energy is likely to be around this value (i.e. $6.5\,g\ MP\,MJ^{-1}$ ME; AFRC, 1992). When the animal's demand for protein (in relation to energy) is greater than this, then it becomes necessary for food protein to be supplied directly to the animal tissues in addition to that which is degraded and incorporated into net microbial protein synthesis.

Figure 15.1 shows that, in dairy cows, the MP:ME ratio as a requirement exceeds 6.5 at around 20 kg milk per day. So, it is above this yield level that a supply of digestible 'undegraded dietary protein' (DUP) becomes necessary to allow the animal's demands to be met. A number of consequences for the amino acid nutrition of dairy cows follow from Fig. 15.1. The first is that below the point at which microbial protein alone is likely to be sufficient to meet the animal's needs, issues of amino acid nutrition are largely determined by the first-limiting amino acid in microbial protein (i.e. the relative proportions of amino acids in rumen microbial protein). Above the point at which microbial protein supply alone is insufficient to meet the animal's demand, opportunities arise to minimize the need for supplemental protein by identifying supplemental amino acid mixtures which will exactly match the relative imbalances in the amino acid composition of rumen microbial protein. By so doing, perhaps one could sustain substantially higher rates of milk production than are indicated in Fig. 15.1, on the basis of microbial protein supplies alone. By enhancing microbial protein yields per MJ ME one can also see, from Fig. 15.1, that quite substantial rates of lactational performance can be sustained, at least in theory, on the basis of microbial protein supplies alone. And indeed the early work of Virtanen (1966) confirms that yields

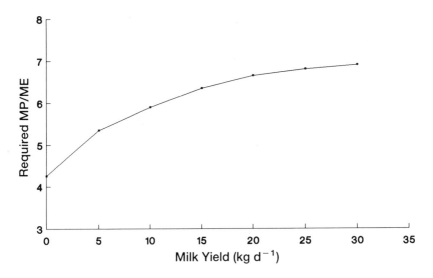

Fig. 15.1. Estimated animal requirements for metabolizable protein (MP; g) per MJ metabolizable energy (ME) for dairy cows of different milk yield. Calculation of required MP/ME ratio is based on the simple assumption that ME requirement for maintenance = 60 MJ ME d^{-1}; ME required per kg milk = 5 MJ kg^{-1}; MP requirement for maintenance = 250 g d^{-1}; MP requirement per kg milk = 40 g MP kg^{-1} milk.

of milk of up to 5000 l in a lactation can be sustained on purified diets in which the sole source of nitrogen is urea. This is consistent with Fig. 15.1.

The rationale for relating dietary (crude) protein intake to animal needs for metabolizable protein is shown in Fig. 15.2. This scheme is based on the particular system described by AFRC (1992) but is similar to the approaches which have been adopted by various other groups around the world (see AFRC, 1992, for references). A major weakness of this scheme for dealing with amino acid nutrition is that it does not account, explicitly, for the use of individual amino acids and, in particular, for the amino acid which is first limiting for production. In this scheme amino acid nutrition is described within the term which accounts for the partial efficiency with which the mixture of amino acids supplied as metabolizable protein is used for particular processes. Thus:

$$\text{Efficiency of MP use} = k_n = k_{aai} \times \text{RV} \tag{15.1}$$

where $k_{aai} =$ the efficiency with which an ideally balanced amino acid mixture is used; RV is the relative value of the amino acid mixture supplied and has a maximum value of 1, when the amino acid mixture is ideal for the purpose.

An alternative approach for incorporating the idea of amino acid balance ('quality') into an MP scheme for ruminants was included in ARC (1984). The concept suggested was that the relative proportions of *total* essential and *total* non-essential amino acids in the MP supply could be related to their proportions in the protein product (milk protein, tissue protein) to allow an assessment of amino acid balance from which influences of imbalance on the efficiency of use of absorbed amino acids could be calculated. Such an approach is so clearly inadequate in the context of well-established principles of amino acid nutrition that it should be no surprise that the approach has never been adopted into practice. The scheme as presented by AFRC (1992) is helpful in matching amino acid supply to demand, only if RV is known. Unfortunately it has proved so technically difficult to estimate RV in the ruminant animal, because of the interference of rumen fermentation in the provision of amino acids to the post-ruminal gut, that good estimates of RV are not yet available. It is, however, possible to estimate from first principles, established elsewhere, the balance of amino acids which might be considered to be ideal for sustaining lactational performance in dairy cows.

Theoretical assessment of ideal amino acid needs in dairy cows

The amounts of each essential amino acid which are needed to sustain a particular level of (lactational) performance can, in theory, be calculated factorially by adding together the estimated requirement for maintenance

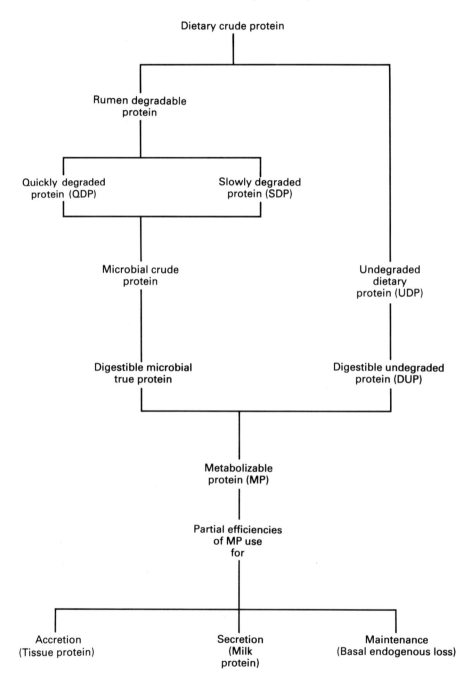

Fig. 15.2. Flow diagram to represent the broad relationships linking dietary (crude) protein intake to metabolizable protein yield and its partition between various animal functions.

and for milk protein secretion with due allowance also for tissue protein changes. The latter may add to requirement if there is net tissue gain, in the short term, or reduce the need for the supply of amino acids of immediate dietary origin if there is net tissue protein mobilization.

Requirements for maintenance include the urinary loss of unmodified amino acids, the obligatory use of amino acids as precursors for other essential metabolites (e.g. creatine, taurine, glutathione, catecholamines, carnitine), the obligatory oxidation of amino acids, those amino acids which are lost from gastrointestinal epithelia (desquamated mucosal cells, mucus, digestive enzymes) and those that are lost from integuments and epidermal structure (Moughan, 1989; Fuller, 1991). In pigs the profile of amino acids which is ideal to satisfy maintenance needs differs a little from that which is ideal for tissue protein accretion (Fuller et al., 1989). As there are no direct estimates of the ideal amino acid profile for maintenance in ruminant animals, it would seem reasonable to accept the profile which applies to pigs as being equally applicable to ruminants, including dairy cows. The absolute amounts of each amino acid which are required for maintenance in ruminants may, however, be greater than that for pigs. Basal endogenous nitrogen (N) losses for growing pigs seem to be substantially lower (at around $0.15\,g\,N\,kg^{-0.75}$; ARC, 1981) than in ruminant animals, where estimates of about $0.35\,g\,N\,kg^{-0.75}$ (ARC, 1984; AFRC, 1992) are accepted as reasonable. The reason for this apparent difference in the scale of basal endogenous nitrogen losses between pigs and ruminants is not clear. It may be inappropriate to scale this demand to $W^{0.75}$ rather than to protein mass and relative degree of maturity, as suggested by Emmans and Fisher (1986). Alternatively the difference might reflect an obligatory use of amino acids for gluconeogenesis to meet maintenance need for glucose in the ruminant animal, which has no implications for nitrogen metabolism in pigs. To satisfy a maintenance need for glucose of $2\,g\,kg^{-0.75}\,d^{-1}$ (Girdler et al., 1984), with a maximum conversion rate of 100 g amino acid to yield 55 g glucose, would require $\sim0.6\,g$ amino $N\,kg^{-0.75}\,d^{-1}$ to be generated through amino acid catabolism. The difference between pigs and ruminants of $0.2\,g\,N\,kg^{-0.75}\,d^{-1}$ in basal endogenous nitrogen excretion could therefore readily be accounted for in terms of an obligatory contribution of amino acids to gluconeogenesis at maintenance – though firm supporting evidence is lacking. For the present purpose it is assumed that the amino acid profile for maintenance is as suggested by Fuller et al. (1989) with the amounts scaled for the apparent difference in basal endogenous nitrogen loss rates (per $kg^{0.75}$) between ruminants and pigs (i.e. adjusted by a factor of 0.35/0.15).

There are no good measurements of the ideal amino acid mixture to meet the demand for secretion of milk proteins. The ideal profile for lean tissue deposition (i.e. tissue protein accretion) has often been taken to be close to that of average tissue protein (ARC, 1981; Fuller et al., 1989). So a sensible starting point for identifying an ideal amino acid balance for milk protein secretion would be the average composition of milk proteins. Fuller *et al.* (1989) have argued that, for growing pigs, the concentration of essential amino acids in ideal protein is a little higher than the concentration of these amino acids in net protein accretion and that it is more reminiscent of the amino acid profile to be found in sow's milk protein. The higher concentration of essential amino acids in the ideal mixture is perhaps needed to accommodate inevitable inefficiencies of incorporation of these amino acids into net protein gain. The amino acid groups released through the catabolism associated with this inevitable inefficiency may be substantially conserved by contributing to the freely exchanging amino N pool – hence the essential:non-essential amino acid ratio in the ideal mixture which is required to be supplied, would be somewhat greater than that found in the deposited tissue.

The extent to which similar arguments apply to ideal amino acid mixtures for milk protein is uncertain. There is a long-standing view (Mepham, 1982) that some essential amino acids (phenylalanine, tyrosine, methionine and tryptophan) are extracted from the blood supply to the mammary gland in amounts which match very closely the quantities of these amino acids secreted in milk protein, whilst other essential amino acids (arginine, branched-chain amino acids, threonine, lysine and histidine) are extracted from blood in amounts which substantially exceed the quantities secreted in milk. By comparison, the remaining non-essential amino acids are generally not taken up from blood in sufficient quantities to account for their secretion in milk (see also De Peters and Cant, 1992). These observations are consistent with the view expressed above, in relation to tissue protein accretion, that the ideal amino acid mixture (as supplied) might differ somewhat from the mixture of amino acids in the secreted milk protein to take into the account the fact that some essential amino acids are necessarily catabolized in the mammary gland and that their amino groups contribute to the *de novo* synthesis of non-essential amino acids within the gland. However, it is far from clear if the circumstances under which the various reported mammary amino acid extraction studies satisfied the conditions that amino acid supply (i) was limiting for milk protein syntheis; and (ii) was close to ideal in its composition (balance).

De Peters and Cant (1992) pointed out a number of technical issues involved in the various reported measurements of amino acid uptake by the mammary gland which would suggest that the existing data on amino acid uptake are not necessarily indicative of the 'efficiency of metabolism' which would apply under amino acid-limiting conditions. There is some evidence that the branched-chain amino acids are subject to quite high rates of oxidation in the mammary gland of cattle (Wohlt *et al.*, 1977). This might apply particularly to isoleucine, for which mammary uptake from both plasma and red blood cells is high compared with its output in milk (Hanigan *et al.*, 1992). The relatively high uptake of leucine observed by Wohlt *et al.* (1977) may be a technical artefact as there appears to be counter-current transport of this amino acid by red blood cells which, when taken into account, brings leucine uptake measurements (by arterio-venous difference measurement) reasonably closely into line with measured rates of output in milk (Hanigan *et al.*, 1992). It may therefore be appropriate to consider that the ideal amino acid mixture for cow's milk protein production has at least an elevated content of isoleucine relative to the composition of milk protein. If arginine is truly essential for dairy cows, then its apparent high rate of catabolism within the mammary gland (Mepham, 1982) might also be indicative of a higher concentration of arginine in the ideal amino acid mixture required for supply than is present in the secreted protein. One of the major reasons for a higher rate of arginine catabolism is that it is a precursor of proline, and post-ruminal supplementation with proline has been shown to reduce arginine uptake across the mammary gland (Bruckental *et al.*, 1991). Peculiarly, though, the main effect of post-ruminal supplementation of proline appeared to be an enhancement of milk fat secretion rather than of milk protein.

The consequence of these considerations is that there are no clear quantitative guidelines to show that the balance of amino acids required in the ideal amino acid supply for milk protein synthesis differs from that of the secreted protein. As a starting point it therefore seems reasonable to stick with the view that the ideal mixture of amino acids for milk protein synthesis is represented by the profile to be found in average milk proteins – although this could particularly underestimate the concentrations of isoleucine and arginine and perhaps also of leucine, valine, histidine and threonine required in an ideal amino acid mixture.

When there is tissue protein gain, it would seem reasonable to accept that, as with pigs, the composition of an ideal amino acid mixture for tissue protein accretion is close to that of average tissue protein (Fuller *et al.*, 1989) – though the point, already made, that there may be some differences because of inevitable inefficiencies in the use of some amino acids, must not be forgotten. There are some, though generally small, differences between the estimated amino acid composition of pig carcasses (ARC, 1981; Fuller *et al.*, 1989) and that found in cattle (Williams, 1978). The limiting mass efficiencies with which individual amino acids in an ideal mixture are used for tissue protein accretion have been precisely estimated, for growing pigs, by Fuller *et al.* (1989). It would seem appropriate to assume that these values also apply for ruminant animals and that one can estimate the ideal amino acid requirements for tissue protein gain by applying those efficiencies, for pigs, with the amino acid composition of cattle carcass protein, estimated by Williams (1978).

When there is net tissue protein mobilization during lactation, it may be assumed that the released amino acids all contribute to the metabolizable amino acid pool with an efficiency of 1.0. But the amino acid composition of the mixture which is released on mobilization must be taken into account. There appear to be no data which describe precisely the amino acid content of mobilized protein as being any different from that for the average amino acid composition of the tissue. (Although Owens (1987) stated that the amino acid composition of protein reserves does not match perfectly the amino acid composition of mobilized protein; the foundation for this remark was not revealed.) The situation in which milk protein secretion proceeds while body tissue protein undergoes net catabolism might therefore best be assumed to be dealt with simply by adjusting the input profile of amino acids for the contribution from net mobilized tissue assuming that the amino acid composition of mobilized tissue is the same as the average composition of body tissue protein.

Theoretical estimates of amino acid requirements for any particular combination of maintenance plus milk protein secretion, plus tissue protein change, can be made factorially by combining the elements of the information given above. Table 15.1 lists theoretical estimates of the amounts of amino acids in the ideal amino acid supply which are required to meet the needs of maintenance, milk protein secretion and tissue protein change (accretion). Two profiles are given for the ideal amino acid mixture for milk protein; one (Milk protein A) is based on the assumption that the efficiency of use of an ideally balanced amino acid mixture is the same, at 0.85 (Oldham, 1987) for all essential amino acids; the second (Milk

Table 15.1. Theoretical amounts of essential amino acids, in an ideal amino acid mixture, which are required for maintenance, milk protein secretion and tissue protein accretion in dairy cows. Values are mg amino acid kg$^{-0.75}$ d^{-1} for maintenance, and g kg^{-1} protein for milk and tissue protein production. Two estimates are given for milk protein - see text for details.

Amino acid	Maintenance	Milk protein A	Milk protein B	Tissue protein accretion
Threonine	124	49	52	53
Valine	47	72	77	47
Methionine + cystine	114	39	37	38
Methionine	21	28	26	20
Isoleucine	37	61	74	37
Leucine	54	105	112	80
Phenylalanine	42	55	52	40
Lysine	84	84	79	69

protein B) assumes that the efficiency of conversion for phenylalanine, lysine and methionine is higher (at 0.9) than applies to threonine, valine and leucine (at 0.8) and that isoleucine has a low efficiency of 0.7. It should be noted that the assumed efficiencies are grossly speculative.

By convention the relative proportions of amino acids in these ideal mixtures are presented in relation to lysine, with a nominal content for lysine of 100 units. Using this convention of a reference amino acid we can calculate the profile of amino acids which is needed to sustain any particular rate of performance as:

$$\text{Amount of amino acid i required per unit of lysine} =$$
$$([AA_{im}]/[Lys_m])^*Lys_m + ([AA_{il}]/[Lys_l])^*Lys_l \qquad (15.2)$$
$$+ (([AA_{it}]/[Lys_t])^*Lys_t) / (Lys_m + Lys_l + Lys_t)$$

where $[AA_{im}]$, $[AA_{il}]$ and $[AA_{it}]$ are the concentrations of amino acid i in the ideal amino acid mixture for maintenance, lactation (milk protein secretion) and tissue protein accretion respectively; $[Lys_m]$, $[Lys_l]$ and $[Lys_t]$ are the equivalent concentrations for lysine (which is conventionally used as the reference amino acid in the ideal mixture because it is used almost exclusively for protein synthesis in its non-oxidative metabolism) and Lys_m, Lys_l and Lys_t are the amounts of lysine needed for the different functions.

Some examples of theoretical calculations of the amounts of essential amino acids which are needed to sustain low and high rates of milk protein secretion in cows of average size, are given in Table 15.2. From such information, likely limiting amino acids in particular dietary circumstances can be estimated if the profile of amino acids in the overall amino acid supply (from the GI tract) is known.

Interactions between Metabolizable Amino Acids and Energy-Yielding Nutrients

As already stated, amino acid supply may be inadequate for reasons of insufficiency or imbalance. A lack of supply (compared with demand) of non-amino nutrients whose role can be substituted by amino acids, can have a bearing on the way in which even an ideally balanced amino acid mixture is used. Thus, for example, the supply of total energy-yielding nutrients, if inadequate, might influence the efficiency of use of an ideally balanced amino acid mixture. This has been clearly demonstrated in growing pigs where the efficiency of use of an ideally balanced amino acid mixture falls below its maximum value when the metabolizable energy: digestible crude protein ratio in the diet falls below about $72 \, MJ \, kg^{-1}$ (Kyriazakis and Emmans, 1992). From simple considerations of the

Table 15.2. Amounts (g d^{-1}) of amino acids required to meet the needs of a 600 kg cow for maintenance (M) or when she is secreting 500, 1000 or 1500 g milk protein per day. The impact of body protein mobilization at the rate of 150 g d^{-1} on the needs of a cow secreting 1500 g milk protein per day is also shown (1500 B).

Amino acid	Performance level				
	M	500	1000	1500	1500 B
Threonine	14	66	118	170	165
Valine	5	81	157	233	228
Methionine + cystine	5	46	87	128	124
Methionine	2	32	62	92	90
Isoleucine	4	68	132	196	192
Leucine	6	116	226	336	327
Phenylalanine	5	63	121	179	174
Lysine	9	53	97	141	132

amounts of ME and limiting quantities of MP which are needed to sustain a milk yield of, say, 40 kg (1300 g protein) d^{-1}, in a cow, one can calculate that the efficiency of use of an ideal amino acid mixture would fall when the ME:MP ratio drops below around 140 MJ kg^{-1} (if there was no body fat mobilization). At this high level of milk yield, such a cow might be losing body fat at a high rate – perhaps even 1.5 kg d^{-1} (Konig *et al.*, 1979). At this rate of body fat loss the ratio of ME:MP below which the efficiency of use of an ideal amino acid mixture might be expected to fall would be closer to 96 MJ kg^{-1}, which is sufficiently close to the value found experimentally by Kyriazakis and Emmans (1992) to make this a matter of interest. The issue of principle is that the efficiency of amino acid use can be expected to fall when the ratio ME:MP falls below a certain level. Again, in theory, one could estimate what the consequence would be if, for our example of the 40 kg milk yield dairy cow, dietary ME:MP were to fall from 96 MJ kg^{-1} (still assuming a body fat loss of 1.5 kg d^{-1}) to 50 MJ kg^{-1}. To make the calculation it is simply assumed that below 96 MJ ME kg^{-1} MP, protein is catabolized to maintain an available ME:MP ratio of 96. Thus at 50 MJ kg^{-1} MP there should only be sufficient energy to use, productively, 0.52 kg MP, i.e. maximum efficiency of ideal amino acid use of 0.42 (if ideal amino acid mixture is used with a limiting efficiency of 0.82).

These theoretical effects of changing ME:MP ratio on the efficiency of ideal amino acid use are shown in Fig. 15.3, superimposed on the data, from pigs, as reported by Kyriazakis and Emmans (1992). The

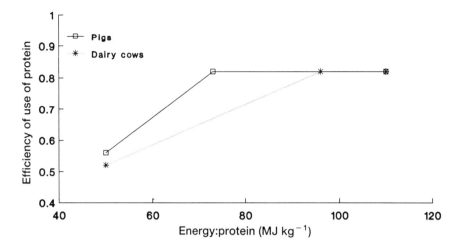

Fig. 15.3. The efficiency of use of an ideally balanced amino acid mixture for growth in pigs (data of Kyriazakis and Emmans, 1992) and for dairy cows (as calculated here; see text for details) as a function of the energy:protein ratio as supplied to the animal's tissue (ME:DCP for pigs; ME:MP for cows).

similarity in the slope of response to diminishing ME:MP ratio for pigs (as found by Kyriazakis and Emmans, 1992) and for cows, as calculated here, is striking. Reference to Fig. 15.1, though, will show that low values for ME:MP in the supply of nutrients to dairy cows are unlikely because the yield of metabolizable protein from rumen microbial protein, at 6.5 g MJ^{-1} ME, corresponds to an ME:MP supply from this source of 154 MJ kg^{-1}. The higher amounts of MP required per MJ ME at higher milk yields (Fig. 15.1) would still only equate to a required ME:MP ratio of around 140 MJ kg^{-1} as previously mentioned.

At the margin, however, increments in the supply of MP which are achieved by altering the amount of digestible undegradable protein which is supplied to the cow are frequently likely to be achieved with little or no change in ME intake, apart from the ME value of the protein itself (this is different from the situation in which MP supply is altered by supplying more or less rumen-degradable protein where there frequently are closely associated changes in ME intake; Oldham, 1984). If the response to such increases in MP supply is achieved in the form of an increase in milk yield (at approximately constant fat and protein composition), then the ME:MP ratio required to achieve that response would be of the order of 5 MJ ME 33 g^{-1} MP, or an ME:MP ratio of about 125 MJ kg^{-1}. If the increment of MP was used both to supply the ME required for extra milk synthesis as well as the amino acids required to secrete extra milk protein, then the efficiency of use of even ideal amino acid mixture *at the margin* would be expected to be around 0.2 – and this does indeed seem to be the case in quite a number of instances (AFRC, 1992; Webster, 1992).

Thus, the principle of energy:protein ratio having an influence on the efficiency of amino acid use in the nutrition of cows would, at least in theory, apply (as might be expected from other species). In particular, ME:MP ratio may be important in determining the responses to increases in DUP supply when ME intake does not change and also where there is no influence on body fat mobilization.

In addition to the foregoing discussion about the possible impact of ME:MP ratio on amino acid use in dairy cows, there is also the possibility that the high rate of glucose utilization in lactating cows (for milk lactose synthesis) which has to be met largely by gluconeogenesis, could have implications for amino acid metabolism and hence nutrition. When lactational performance is truly limited by amino acid supply the net yield of glucose from amino acid catabolism can only be small because the available amino acid supply is being used for protein synthesis with a relatively high efficiency. Thus, in the data summarized by Webster (1992) the conversion of metabolizable amino acid, by catabolism into glucose, above maintenance requirements for protein, would, at a maximum, have been capable of yielding only sufficient glucose to supply 17% of the lactose secretion of those cows. As the partition of metabolism of amino acid carbon between oxidation and net glucose synthesis is very strongly in favour of oxidation (Lindsay, 1976) the net contribution of catabolized amino acid to the animal's glucose economy when protein is limiting is probably only 2 or 3% of the glucose demands for lactose synthesis, as was argued by Bruckental *et al.* (1980).

This is not to say that there are no implications for glucose metabolism in the catabolism of amino acids. Indeed, in the tissues of the gut which appear to use glucose preferentially as an energy substrate, oxidation of amino acids in gut tissues may have a sparing effect on glucose metabolism (Seale and Parker, 1991). Also, at the margin one might argue, as with energy:protein relationships (above), that the translation of increments of metabolizable amino acids into milk protein might vary according to the glucose status of the animal. If we imagine the situation in which the lactational response to an increased supply of metabolizable amino acid is in the form of an enhanced milk yield with no change in milk protein concentration, then each increment in milk protein secretion has associated with it an increase in milk lactose (and possibly also fat) for which an enhancement in glucose usage is required. Where the metabolizable amino acid supply increases with no change in the supply of other energy-yielding nutrients, so that any extra glucose demand to maintain a constant milk protein:lactose ratio comes from amino acid catabolism, then the maximum rate of conversion of metabolizable amino acid (metabolizable protein) to milk protein would be around 0.25 because the majority of the remainder would have to be catabolized to allow the generation of glucose to support the enhanced lactose synthesis. So, as with the discussion on protein:energy ratios above, one can imagine a scenario in which the marginal response (in milk protein yield) to increments in total amino acid supply could be expected to be low, at around 0.25. This provides an alternative hypothesis to account for the low marginal rates of return to metabolizable protein which have been observed (AFRC, 1992; Webster, 1992).

Such thoughts raise some interesting ideas about the likely responses of dairy cows to supplementation with particular limiting amino acids. The technology to provide individual amino acids to the post-ruminal gut by protecting them from degradation in the rumen is now available (Papas et al., 1984; Rulquin and Vérité, 1993). If we take the view that methionine or lysine is most likely to be limiting performance in dairy cows under protein-limiting circumstances (Rulquin and Vérité, 1993) and that neither of these amino acids is glucogenic to any marked degree, then we can speculate on the kinds of response which are to be expected when one of these amino acids is limiting.

If the profile of nutrients from the basal diet is such that high rates of gluconeogenesis can be maintained without the need to call on amino acid catabolism (i.e. strongly glucogenic diets) then the performance response to the provision of a first-limiting amino acid could be expected either to be in the form of enhanced milk yield and/or a change in the protein concentration of milk. Alternatively, if the basal diet provides nutrients such that glucose supply is also limiting for performance, then any response in milk protein output to the correction of supply of a first-limiting amino acid could only be realized through an increase in milk protein concentration. Perhaps the goal of increasing milk protein concentration through nutrition would therefore best be achieved by using supplemental protected amino acids (which are known to be limiting) with diets which are of low glucogenic potential.

If the correction of an amino acid deficiency is achieved through supplying increases in total amino acid (i.e. metabolizable protein) then different response scenarios might be expected to hold. In animals which have substantial fat reserves to mobilize and which also have an adequate non-amino acid glucogenic nutrient supply, responses in milk protein output could be achieved either through enhanced milk volume and/or an increase in milk protein concentration and the marginal response rate could be expected to be reasonably high. Alternatively, in situations where animals have little fat to mobilize and/or are being offered a diet of low glucogenic potential, low rates of response to supplemental metabolizable protein would be expected and the route for response might be more likely to be an increase in concentration than an increase in yield (though both would be allowed). Such hypotheses are open to test.

Lactational responses to metabolizable protein and individual amino acids

The infusion of nutrients directly into the abomasum or duodenum is the most direct way of ensuring an increase in the metabolizable protein or amino acid supply to a cow independently of changes in ME supply (other than that which comes with the protein or amino acids).

Responses, in milk protein output, to the infusion of proteins into the abomasum (usually casein) have frequently been found to conform to some sort of diminishing return pattern (Fig. 15.4). These responses in the yield of milk protein have generally appeared as a combination of an increase in milk volume yield and an associated increase in milk protein concentration. In every case shown in Fig. 15.4 a small increase in abomasal casein supply yielded a high fractional rate of return in terms of enhanced milk protein output, although incremental responses were not maintained at a high rate across the full range of supplement levels of casein, because in the different circumstances other factors would have come into play to limit performance. Milk protein responses to abomasal casein supplements were

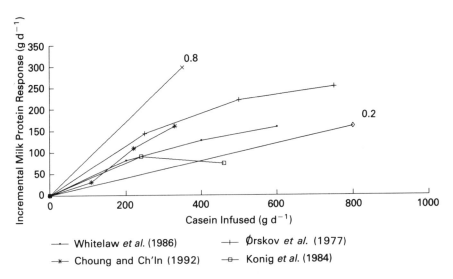

Fig. 15.4. Patterns of response in milk protein output to the infusion of casein directly into the abomasum of dairy cows. Lines of slope 0.2 and 0.8 are shown for reference. (Data from: Whitelaw *et al.*, 1986; Ørskov *et al.*, 1977; Konig *et al.*, 1984; Choung and Chamberlain, 1992.)

maintained over the greatest range of supplementation in the work reported by Ørskov *et al.* (1977) in which, interestingly, increments of casein supplement were infused with amounts of glucose which were designed to maintain constant rates of ME supplementation across the range of infusions used. In each of the experiments represented in Fig. 15.4, the responses to casein infusion over at least part of the supplemented range were associated with an increased output of energy in milk which far outweighed the additional ME supplied as casein (or casein plus glucose, for Ørskov *et al.*, 1977). This phenomenon is not restricted to early lactation as Choung and Chamberlain (1992) achieved it with cows which were at least 20 weeks into their lactations at the start of the experiment. Such responses imply a repartitioning of nutrient use between body tissues, milk synthesis and catabolism. Any shift in the balance between synthetic and catabolic activity would, of course, imply a change in the marginal efficiency of use of metabolizable energy for milk (k_l). Calorimetric studies of the effects of protein supply on k_l would not suggest that such influences are significant (see Choung and Chamberlain, 1993) so that a repartitioning effect of a change in protein supply on the use of energy-yielding substrates is probably implied. The regulation of such a response might be associated with increases in circulating levels of growth hormone (GH), which can follow the enhancement of protein status in ruminants (Oldham *et al.*, 1978, 1982) and sometimes also with reduced insulin concentrations in the circulating blood (Choung and Chamberlain, 1992). Such effects would be compatible with the stimulation of acetate utilization for milk fat synthesis at the expense of adipose tissue synthesis, which is implied in the data of Choung and Chamberlain (1992) from their observations on changes in milk fatty acid profile, following casein infusion.

Changes in total amino acid nutrition (metabolizable protein) which have been achieved in feeding and production studies are more difficult to interpret quantitatively because they depend on predictions of increments in metabolizable amino acid or MP supply rather than on direct measurements. But if we accept that predictions of metabolizable amino acid or protein supply can be made and consider lactational responses to changes in total amino acid nutrition in production trials, then it is quite common for marginal rates of response, as already mentioned, to be rather low (Webster, 1992). Marginal rates of response which comply with expectations from systems such as AFRC (1992) can be found (e.g. Gordon and Small, 1990), but frequently these are identified in studies where both ME supply and metabolizable amino acid supply have changed together. Usually milk protein yield is primarily influenced by an increase in milk volume with little or no change in milk protein concentration under such circumstances, which would be consistent with hypotheses previously aired. When production studies have been designed such that the basal diet might be expected to vary in 'glucogenic potential', either by changing the form of dietary carbohydrate (Lees *et al.*, 1990) or simply forage:concentrate ratio (Mayne and Gordon, 1985) responses to protein supplementation have still largely been achieved through increases in milk volume yield with only small adjustments to milk protein percent (if at all). In the case of Lees *et al.* (1990), changing the nature of carbohydrate in the feed did interact with the response to protein, in that there was a greater response in milk energy output to protein supplementation (because of an increase in milk fat content) when the feed had low rather than high 'glucogenic potential'. Such responses are consistent with the effects of changing amino acid supply on partition of energy-yielding nutrients mentioned previously. Less extreme variations between diets (as regards differences in source of dietary carbohydrate) as were applied by Lees *et al.* (1990), yielded differential responses to protein supplementation of a similar but less exaggerated kind (Mayne and Gordon, 1985). Gordon and Small (1990) found that the marginal response (in milk protein output) to changing protein intake was greater with diets of high concentrate:silage ratio than those with a low ratio – although in each instance the calculated marginal response to metabolizable amino acid supply was very low indeed (at 0.06 or less).

Rulquin and Vérité (1993) have summarized many studies in which supplements of individual amino acids have been given to cows either by direct infusion post-ruminally, or intravenously, or by inclusion in the feed in the protected form. When no single amino acid is limiting performance, then no responses to supplementation are expected. When diets are based substantially on maize (corn) products then, not too surprisingly, lysine can often be identified as a likely first-limiting amino acid. Responses reported recently by Schwab *et al.* (1992) (Fig. 15.5) show a classical pattern of response to the provision of a first-limiting amino acid (lysine in this case). For the first increment of lysine (with methionine), the response was a substantial enhancement of milk protein concentration, but thereafter the response was largely by an alteration in milk volume yield. This must suggest that the range over which the milk protein:lactose ratio can be varied is in some way constrained in cattle, although there is wide variation over the ruminant species in general (Gibson, 1987). If so, this is an added complication in predicting responses to individual amino acids in that resource limitation on milk lactose synthesis might modify 'conventional' responses to supplemental amino acids in terms of an alteration in the net secretion of the protein product.

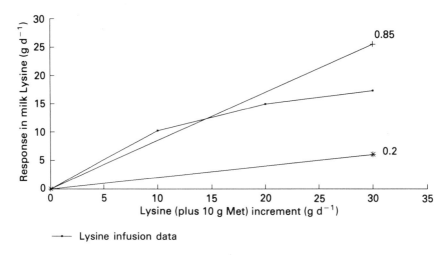

—•— Lysine infusion data

Fig. 15.5. Response of dairy cows (as milk lysine yield, $g\,d^{-1}$) to abomasal infusion of graded amounts of lysine in the presence of a standard supplement of methionine. Lines of slope 0.2 and 0.85 are shown for reference. (Data from Schwab *et al.*, 1992.)

Amino Acids and Protein Partition

A number of issues already dealt with are in fact issues of amino acid, or protein, partition. The partition of amino acids between use for protein synthesis and their oxidation is obviously variable, depending on the relative imbalances between different amino acids when one is limiting production. It also depends on the total amounts of amino acids which are supplied when protein does not limit performance; on energy:protein ratio in the supply of 'nutrients' (energy is not a nutrient!) and perhaps also on the supply of non-amino acid glucose precursors. Such issues are common features of amino acid nutrition and metabolism in general.

Another important question is what determines the partition of limiting amino acids between alternative protein synthetic paths. In the lactating cow (and indeed in any lactating animal) amino acids may be used for net synthesis of secreted milk protein, or possibly also net synthesis of tissue proteins (tissue protein accretion). Tissue protein may be subject to net catabolism whilst milk protein secretion progresses; this is usually referred to as net tissue protein mobilization to support lactation. Under extreme circumstances around 20% of fully replete maternal protein mass might be available for mobilization (Botts *et al.*, 1978) although under more conventional production circumstances such loss might not be so great (Gibb *et al.*, 1992). From studies with rats it would seem that labile maternal protein can be lost at quite a high rate when there is a substantial dietary deficiency of protein. Fractional rates of loss may be as high as 0.03 d^{-1} (Pine *et al.*, 1992). Once the potentially labile protein mass has been lost from the maternal body, in rats at least, there is a severe consequent reduction in lactational performance (Pine *et al.*, 1993). Where maternal protein mass is not fully replete, supplements of amino acids may be partitioned between milk protein synthesis and the replenishment of maternal protein 'reserves'. This phenomenon shows clearly in the data of Whitelaw *et al.* (1986) where, under severely protein-limiting conditions for lactational performance, supplements of casein given by abomasal infusion were partitioned approximately equally between enhanced protein secretion in milk and diminished tissue nitrogen deficit at low levels of casein supplementation. As the response to supplemental casein in milk protein secretion diminished (Fig. 15.5) the proportionate partition towards enhanced nitrogen balance (which may, of course, have been overestimated) increased. The regulation of this partition is not well understood, but for a complete understanding of amino acid nutrition, especially in the sense of response prediction, it is an issue which needs to be addressed.

Supply-side Issues

Strictly speaking, amino acid nutrition refers to the provision of amino acids for their subsequent metabolism, although inevitably the discussion of the subject leads quickly to a consideration of the metabolism of amino acids within the body. The measurement of amino acid supply to the post-ruminal gut is technically difficult and despite the apparent progress of some years ago towards overcoming these difficulties (MacRae, 1975; Faichney, 1975; Sutton and Oldham, 1977) there are still technical difficulties in at least some laboratories (Oldham *et al.*, 1988; Ortigues *et al.*, 1990) which must create some uncertainty about the absolute reliability of measures of post-ruminal nutrient flow reported in the literature. As a simple example from recent literature, Table 15.3 shows the amino acid compositions of duodenal amino acid supply in cows offered diets of very similar apparent constitution in two experiments from two different laboratories (King *et al.*, 1991; Waltz *et al.*, 1989). For the relatively small differences in diet composition the difference in duodenal amino acid profile is remarkable. From these same published papers the post-ruminal supply of non-ammonia nitrogen (per kg DM intake) was 23.1 and 28.3 g kg^{-1} respectively – a 22% difference which in one circumstance (Waltz *et al.*, 1989) was associated with a post-ruminal non-ammonia nitrogen flow 8% less than nitrogen intake, and in the other (King *et al.*, 1991) suggested a 12% greater flow of non-ammonia nitrogen post-ruminally than intake. Such variation obviously confuses rather than

Table 15.3. Relative proportions (g 100 g^{-1} lysine) of essential amino acids in post-ruminal digesta of cows offered diets of very similar composition (approx. 10% alfalfa; 30-40% corn silage; 40-30% ground corn; 3% molasses and ~10% blood meal) in two experiments (King *et al.*, 1991; Waltz *et al.*, 1989). The variability in ratio of values between the two reports shows that the contrast was not merely a reflection of differences in lysine between the two.

Amino acid	King *et al.* (1991)	Waltz *et al.* (1989)	Ratio King:Waltz
Histidine	69	45	1.53
Threonine	87	69	1.26
Arginine	134	45	2.98
Valine	144	77	1.87
Isoleucine	70	57	1.23
Leucine	196	142	1.38
Phenylalanine	108	72	1.5
Lysine	100	100	1

clarifies the interpretation of apparent quantitative studies of post-ruminal nutrient passage. Perhaps the best that one can hope for is to be able to make comparisons in a qualitative way between treatments within experiments; but caution is needed in making quantitative statements which embrace data from different experiments and different laboratories. It is for these reasons that the majority of the discussion in this chapter has centred on theory or the interpretation of results from post-ruminal nutrient supplementation or from production trials.

There is much current interest in the role which peptides play in amino acid nutrition. Short-chain peptides represent a sizeable proportion of the total amino acids which are absorbed from the lumen of the gut (Webb *et al.*, 1992). Those observations which suggest that there might also be substantial peptide absorption from the reticulo-rumen and abomasum (Webb *et al.*, 1992) should be treated with great caution as the technical aspects of the measurement are so contentious. If there were to be substantial amino acid absorption prior to the abomasum, of course, then even greater doubt would be placed on the quantitative relevance of apparent measures of post-ruminal amino acid supply. This author suspects that apparent amino acid absorption prior to the abomasum is artefactual.

It has been suggested that peptide uptake by the mammary gland might also play a significant role in providing amino acids for net protein synthesis. Such suggestions must modify the relatively simple view that arterial concentrations of amino acids are major determinants of substrates supplied for mammary protein synthesis (De Peters and Cant, 1992).

Amino acid nutrition can be complicated by the endogenous cycling of amino acids from tissue, through secretion into the gut and subsequent reabsorption. As relationships between supply and demand are most usefully estimated in terms of the relationship between amino acids of immediate dietary origin (i.e. without having gone through any sort of endogenous cycling) and current synthetic demand, the 'correction' of supply to account for the reabsorption of endogenously secreted amino acids from the gastrointestinal tract is required. An interesting observation in this context is that about 5% of microbial amino acids might have their origin in endogenous secretions (Marsden *et al.*, 1988). The supply from this source may frequently therefore be overestimated. Clark *et al.* (1992) have also raised the issue of uncertainty in estimations of the amino acid content of microbial crude protein. From a review of literature sources they identified a mean (from 35 experiments) of 665 g amino acid N kg^{-1} total microbial N with a coefficient of variation amongst reports of 13.5%. They also identified very substantial variations in estimated ratios of diaminopimelic acid (a common bacterial 'marker'):microbial N and purine N:microbial N, which might be a matter for concern amongst those people who are trying to measure microbial amino acid supply to the tissues using non-invasive methods such as the measurement of purine N end-product excretion in urine (e.g. Chen *et al.*, 1990; Susmel *et al.*, 1994).

In short there are quite a number of issues remaining on the supply side of amino acid nutrition and metabolism which are still unresolved and which are largely sensitive to the inadequacies of difficult techniques. These issues have been mentioned here in only a superficial way because space does not allow a full discussion. They are, nonetheless, important, especially in the context of developing quantitative frameworks (models) of amino acid nutrition, and for the design of experiments to test ideas within these models.

Conclusion

No attempt has been made here to survey the detailed literature on amino acid metabolism in dairy cows. Recent reviews (e.g. De Peters and Cant, 1992; Clark *et al.*, 1992; Rulquin and Vérité, 1993) will serve that purpose. I have tried to point out some of the concepts of amino acid nutrition which we might sensibly expect to apply in dairy cows and in other species. From consideration of those issues, and in particular the relationship between energy and protein supply and possibly glucose requirement and protein supply, one can draw together some ideas about the manner in which dairy cows might use amino acids when amino acids are limiting performance.

The availability of quite reliable means to protect amino acids from rumen degradation offers the opportunity for designing experiments which will test certain hypotheses presented here. The great and continuing uncertainty in designing such experiments is the reliability with which one can predict amino acid (and other nutrient) supply to the absorptive areas of the gut from a knowledge of food intake and the composition of foods ingested. It remains a central feature of our lack of understanding of amino acid nutrition of dairy cows that this technical hurdle has not been satisfactorily overcome – but perhaps this merely represents a lack of imagination on the part of those interested in the subject.

References

AFRC (Agricultural and Food Research Council) (1992) Nutritive requirements of ruminant animals: Protein. Technical Committee on Responses to Nutrients, Report No. 9. *Nutrition Abstracts and Reviews, Series B* 62, 787–835.

ARC (Agricultural Research Council) (1981) *The Nutrient Requirements of Pigs. Technical Review.* Commonwealth Agricultural Bureaux, Farnham Royal, Slough.

ARC (Agricultural Research Council) (1984) *The Nutrient Requirements of Ruminant Livestock. Suppl. No. 1. Report of the Protein Group of the ARC Working Party.* Commonwealth Agricultural Bureaux, Farnham Royal, Slough.

Botts, R.I., Hemken, R.W. and Bull, L.S. (1978) Protein reserves in the lactating dairy cow. *Journal of Dairy Science* 62, 433–440.

Bruckental, I., Oldham, J.D. and Sutton, J.D. (1980) Glucose and urea kinetics in cows in early lactation. *British Journal of Nutrition* 44, 33–45.

Bruckental, I., Ascarelli, I., Yosif, B. and Alumot, E. (1991) Effect of duodenal proline infusion on milk production and composition in dairy cows. *Animal Production* 53, 299–303.

Chen, H.B., Hovell, F.DeB., Ørskov, E.R. and Brown, D.S. (1990) Excretion of purine derivatives by ruminants: effect of exogenous nucleic acid supply on purine derivative excretion by sheep. *British Journal of Nutrition* 63, 131–142.

Choung, J.J. and Chamberlain, D.G. (1992) The effects of abomasal infusions of sodium caseinate, soya protein isolate and a water extracted fishmeal on the utilisation of nitrogen by dairy cows. *Journal of the Science of Food and Agriculture* 60, 131–134.

Choung, J.J. and Chamberlain, D.G. (1993) The effects of abomasal infusions of casein or soyabean protein isolate on the milk production of dairy cows in mid-lactation. *British Journal of Nutrition* 69, 103–115.

Clark, J.H., Klusmeyer, T.H. and Cameron, M.R. (1992) Microbial protein synthesis and flows of nitrogen fractions to the duodenum of dairy cows. *Journal of Dairy Science*, 75, 2304–2323.

De Peters, E.J. and Cant, J.P. (1992) Nutritional factors influencing the nitrogen composition of bovine milk: A review. *Journal of Dairy Science* 75, 2043–2070.

Emmans, G.C. and Fisher, C. (1986) Problems in nutritional theory. In: Fisher, C. and Boorman, K.N. (eds) *Nutrient Requirements of Poultry and Nutritional Research*. Butterworths, London, pp. 9–39.

Faichney, G.J. (1975) The use of markers to partition digestion within the gastrointestinal tract of ruminants. In: MacDonald, I.W. and Warner, A.C.I. (eds) *Digestion and Metabolism in the Ruminant*. University of New England Publishing Unit, Armidale, pp. 277–291.

Fuller, M.F. (1991) Present knowledge of amino acid requirements for maintenance and production: non-ruminants. In: Eggum, B.O., Boisen, S., Børsting, C., Danfaer, A. and Hvelplund, T. (eds) *Proceedings of the EAAP 6th International Symposium on Protein Metabolism and Nutrition, Herning, Denmark*. Institute of Animal Science, Foulum, Denmark pp. 116–126.

Fuller, M.F., McWilliam, R., Wang, T.C. and Giles, L.R. (1989) The optimum dietary amino acid pattern for growing pigs. 2. Requirements for maintenance and for tissue protein accretion. *British Journal of Nutrition* 62, 255.

Gibb, M.J., Ivings, W.E., Dhanoa, M.S. and Sutton, J.D. (1992) Changes in body components of autumn calving Holstein-Friesian cows over the first 29 weeks of lactation. *Animal Production* 54, 339–360.

Gibson, J.P. (1987) The option and prospects for genetically altering milk composition in dairy cattle. *Animal Breeding Abstracts* 55, 231–243.

Girdler, C.P., Thomas, P.C. and Chamberlain, D.G. (1984) Exogenous supply of glucose precursors and nitrogen utilisation in sheep. *Proceedings of the Nutrition Society* 45, 43A.

Gordon, F.J. and Small, J.C. (1990) The direct and residual effects of giving fishmeal to dairy cows receiving different levels of concentrate supplementation in addition to grass silage. *Animal Production* 51, 449–460.

Hanigan, M.D., Calvert, C.C., De Peters, E.J., Reis, V.L. and Baldwin, R.L. (1992) Kinetics of amino acid extraction by lactating mammary glands in control and sometribove treated cows. *Journal of Dairy Science* 75, 161–173.

King, K.J., Bergen, W.G., Sniffen, C.J., Grant, A.L., Grieve, D.B., King, V.L. and Ames, N.K. (1991) An assessment of absorbable lysine requirements in lactating cows. *Journal of Dairy Science* 74, 2530–2539.

Konig, B.A., Parker, D.S. and Oldham, J.D. (1979) Acetate and palmitate kinetics in lactating dairy cows. *Annales de Recherches Veterinaires* (2/3), 368–370.

Konig, B.A., Oldham, J.D. and Parker, D.S. (1984) The effect of abomasal infusion of casein on acetate, palmitate and glucose kinetics in cows during early lactation. *British Journal of Nutrition* 52, 319–328.

Kyriazakis, I. and Emmans, G.C. (1992) The effects of varying protein and energy intakes on the growth and body composition of pigs. II. The effects of varying both energy and protein intake. *British Journal of Nutrition* 68, 615–625.

Lees, J.A., Oldham, J.D., Haresign, W. and Garnsworthy, P.C. (1990) The effect of patterns of rumen fermentation on the response by dairy cows to dietary protein concentration. *British Journal of Nutrition* 63, 177–186.

Lindsay, D.B. (1976) Amino acids as sources of energy. In: Cole, D.J.A. (ed.) *Protein Metabolism and Nutrition*. Butterworths, London, pp. 183–197.

MacRae, J.C. (1975) The use of re-entrant cannulae to partition digestive function within the gastrointestinal tract of ruminants. In: MacDonald, I.W. and

Warner, A.C.I. (eds) *Digestion and Metabolism in the Ruminant*. University of New England Publishing Unit, Armidale, pp. 261–276.

Marsden, M., Bruce, C.I., Bartram, C.G. and Buttery, P.J. (1988) Initial studies on leucine metabolism in the rumen of sheep. *British Journal of Nutrition* 60, 161–171.

Mayne, C.S. and Gordon, F.J. (1985) The effect of concentrate to forage ratio on the milk-yield response to supplementary protein. *Animal Production* 41, 269–280.

Mepham, T.B. (1982) Amino acid utilisation by lactating mammary gland. *Journal of Dairy Science* 65, 287–298.

Moughan, P.J. (1989) Simulation of the daily partitioning of lysine in the 50 kg liveweight pig – A factorial approach to estimating amino acid requirement for growth and maintenance. *Research and Development in Agriculture* 6, 7–14.

Oldham, J.D. (1984) Protein-energy interrelationships in dairy cows. *Journal of Dairy Science* 67, 1090–1114.

Oldham, J.D. (1987) Efficiencies of amino acid utilisation. In: Jarrige, R. and Alderman G. (eds) *Feed Evaluation and Protein Requirement Systems for Ruminants*, EUR 10657EN. EEC, Brussels, pp. 171–186.

Oldham, J.D., Hart, I.C. and Bines, J.A. (1978) Effect of abomasal infusions of casein, arginine, methionine or phenylalanine on growth hormone, insulin, prolactin, thyroxine and some metabolites in blood from lactating goats. *Proceedings of the Nutrition Society* 37, 9A.

Oldham, J.D., Hart, I.C. and Bines, J.A. (1982) Formaldehyde-treated proteins for dairy cows – effects on blood hormone concentrations. *British Journal of Nutrition* 48, 543–547.

Oldham, J.D., Napper, D.J., Jacobs, J.L. and Phipps, R.H. (1988) Digestion and use of dietary nitrogen in dairy cows. In: *Proceedings of the 5th EAAP Symposium on Protein Nutrition and Metabolism*. EAAP Publication No. 35. Wiss. Z. WPU, Rostock N-Reihe, 37: 2 65–66.

Ørskov, E.R., Grubb, D.A. and Kay, R.N.B. (1977) Effect of postruminal glucose or protein supplementation on milk yield and composition in Friesian cows in early lactation. *British Journal of Nutrition* 38, 397–405.

Ortigues, I., Oldham, J.D., Smith, T., Courtenay, M.B. de and Siviter, J.W. (1989) Nutrient supply and growth of heifers offered straw-rich diets. I. A comparison between 3 marker systems for measurement of intestinal digesta flows. *Journal of Agricultural Science, Cambridge* 114, 69–77.

Owens, F.N. (1987) Maintenance protein requirement. In: Jarrige, R. and Alderman, G. (eds) *Feed Evaluation and Protein Requirement Systems for Ruminants*, EUR 10657. EEC, Brussels, pp. 187–212.

Papas, A.M., Sniffen, C.J. and Muscato, T.V. (1984) Effectiveness of rumen-protected methionine for delivering methionine post-ruminally in dairy cows. *Journal of Dairy Science* 67, 545–552.

Pine, A.P., Jessop, N.S., Allan, J.F. and Oldham, J.D. (1992) Effects of dietary protein content during lactation on tissue protein synthesis in rats. *Proceedings of the Nutrition Society* 51, 157A.

Pine, A.P., Jessop, N.S. and Oldham, J.D.(1994) Maternal protein reserves and their influence on lactational performance in rats. *British Journal of Nutrition* (in press).

Rulquin, H. and Vérité, R. (1993) Amino acid nutrition of dairy cows: productive effects and animal requirements. In: Garnsworthy, P.C. and Cole, D.J.A. (eds) *Recent Advances in Animal Nutrition - 1993.* Nottingham University Press, Nottingham, pp. 55–77.

Schwab, C.G., Bozak, C.K., Whitehouse, N.L. and Olsen, V.M. (1992) Amino acid limitation and flow to the duodenum at 4 stages of lactation. 2. Extent of lysine limitation. *Journal of Dairy Science* 75, 3503–3518.

Seale, C.J. and Parker, D.S. (1991) Increased plasma free amino acid concentrations and net absorption of amino acids into portal and mesenteric veins with intraruminal propionate infusion into forage-fed steers. In: Eggum, B.O., Boisen, S., Børsting, C., Danfær, A. and Hvelplund, T. (eds) *Proceedings of the EAAP 6th International Symposium on Protein Metabolism and Nutrition, Herning, Denmark.* Institute of Animal Science, Foulum, Denmark pp. 184–186.

Susmel, P., Spanghero, M., Stefanon, B., Mills, C.R. and Plazzotta, E. (1994) Digestibility and allantoin excretion in cows fed diets differing in nitrogen content. *Livestock Production Science* 39, 97–99.

Sutton, J.D. and Oldham, J.D. (1977) Feed evaluation by measurement of sites of digestion in cannulated ruminants. *Proceedings of the Nutrition Society* 36, 203–209.

Virtanen, A.I. (1966) Milk production of cows on protein free feeds. *Science* 153, 1603.

Waltz, D.M., Stern, M.D. and Illg, D.J. (1989) The effect of ruminal protein degradation of bloodmeal and feathermeal on the intestinal amino acid supply to dairy cows. *Journal of Dairy Science* 72, 1509–1518.

Webb, K.E., Matthews, J.C. and Di Rienzo, D.B. (1992) Peptide absorption: A review of current concepts and future perspectives. *Journal of Animal Science* 70, 3248–3257.

Webster, A.J.F. (1992) The metabolisable protein system for ruminants. In: Garnsworthy, P.C., Haresign, W. and Cole, D.J.A. (eds) *Recent Advances in Animal Nutrition - 1992.* Butterworths Heinemann, London, pp. 93–110.

Whitelaw, R.G., Milne, J.S., Ørskov, E.R. and Smith, J.S. (1986) The nitrogen and energy metabolism of lactating cows given abomasal infusions of casein. *British Journal of Nutrition* 55, 537–556.

Williams, A.P. (1978) The amino acid, collagen and mineral composition of preruminant calves. *Journal of Agricultural Science, Cambridge* 90, 617–624.

Wohlt, J.E., Clark, J.H., Derrig, R.G. and Davis, C.L. (1977) Valine, leucine and isoleucine metabolism by lactating bovine mammary tissue. *Journal of Dairy Science* 60, 1875.

Amino Acid Requirements of Finfish 16

R.P. WILSON
*Department of Biochemistry and Molecular Biology,
Mississippi State University, PO Drawer BB,
Mississippi State, MS 39762, USA.*

Introduction

Certain aspects of amino acid nutrition and metabolism in fish appear to differ from those observed in other vertebrates. Fish are normally fed diets containing 2–4 times as much protein as other vertebrates. This is due to the fact that the optimal dietary protein level required for maximum growth of various cultured fish ranges from about 30 to 55% crude protein (Bowen, 1987; Tacon and Cowey, 1985; Wilson, 1989). This observation has led certain investigators, including myself, to suggest that the efficiency of protein utilization is lower in fish than other animals.

Tacon and Cowey (1985) first noted that the dietary protein requirements of fish are not that dissimilar from those of other vertebrates when expressed relative to feed intake (g protein kg^{-1} bodyweight d^{-1}) and liveweight gain (g protein kg^{-1} liveweight gain). Bowen (1987) has compared several parameters relating protein intake to growth of fish and other vertebrates and found very little difference in protein utilization among the species compared (Table 16.1). The data used in making this comparison included median values for 18 studies for fish and eight studies of other vertebrates, i.e. calf, chicken, lamb, pig and white rat. The only parameters that differed significantly were the level of protein in the diet required for maximum growth and feed efficiency. When the protein requirement data were recalculated to correct for differences in relative protein intake and growth rates, as suggested by Tacon and Cowey (1985), the resulting data were very similar for fish and other vertebrates, thus indicating that the efficiency of protein utilization is very similar among the species compared.

The first definitive studies on amino acid nutrition of fish were

Table 16.1. Comparison of various parameters relating protein intake to growth of fish and other vertebrates. (Data from Bowen, 1987.)

Parameter	Fishes	Other vertebrates
Specific growth rate	2.765	2.445
Protein in diet (%)	40.3	20.0
Protein intake at maximum growth (mg protein ingested g^{-1} bodywt d^{-1})	16.5	12.0
Protein retention efficiency (100 × g protein retained g^{-1} protein ingested)	31.0	29.0
Protein growth efficiency (g growth g^{-1} protein ingested)	1.945	1.965
Food conversion efficiency (g growth g^{-1} diet ingested)	0.78	0.26

conducted in late 1950 and early 1960 with chinook salmon, *Oncorhynchus tshawytscha*. The initial amino acid test diets were formulated to simulate the amino acid content of chicken egg protein, chinook salmon egg protein and chinook yolk-sac fry protein (Halver, 1957). The diet with an amino acid profile based on chicken egg protein gave the best growth rate and feed efficiency. This diet was then used to determine the qualitative amino acid requirements of chinook salmon (Halver *et al.*, 1957).

Qualitative Amino Acid Requirements

Halver *et al.* (1957) determined the essentiality of the 18 common protein amino acids by comparing the relative growth rates of chinook salmon fed the basal and specific amino acid-deficient diet over a 10-week period. For each of the ten indispensable amino acids, groups of the deficient fish were split at 6 weeks, with one subgroup being continued on the deficient diet and the other subgroup fed the basal diet. In each of the subgroups shifted to the basal diet, the fish showed an immediate and substantial growth response to the complete diet. The results indicated that the following ten amino acids were indispensable for chinook salmon: arginine, histidine, isoleucine, leucine, lysine, methionine, phenylalanine, threonine, tryptophan and valine. Other species that have been studied by this method and that have shown similar results include the channel catfish, *Ictalurus punctatus* (Dupree and Halver, 1970); common carp, *Cyprinus carpio* (Nose *et al.*, 1974); European eel, *Anguilla anguilla*, and Japanese eel, *Anguilla japonica* (Arai *et al.*, 1972); rainbow trout, *Oncorhynchus mykiss*

(Shanks *et al.*, 1962); red sea bream, *Chrysophrys major* (Yone, 1976); sockeye salmon, *Oncorhynchus nerka* (Halver and Shanks, 1960); and *Tilapia zillii* (Mazid *et al.*, 1978) Additional studies based on the lack of ¹⁴C incorporation into each amino acid following intraperitoneal injection of uniformly labelled [¹⁴C]glucose have indicated that these same ten amino acids are indispensable for plaice, *Pleuronectes platessa*, and sole, *Solea solea* (Cowey *et al.*, 1970), and European sea bass, *Dicentrarchus labrax* (Metailler *et al.*, 1973).

Determination of Amino Acid Requirements

Amino acid test diets

Most investigators have used the method developed by Halver and co-workers (Mertz, 1972) to determine the quantitative amino acid requirements of fish. This procedure involves feeding graded levels of one amino acid at a time in a test diet containing either all crystalline amino acids or a mixture of casein, gelatin and amino acids formulated so that the amino acid profile is identical to whole chicken egg protein except for the amino acid being tested. This procedure has been used successfully with several species including chinook, coho, chum and sockeye salmon, Japanese eel, Nile tilapia and rainbow trout. However, the amino acid test diets must be neutralized with sodium hydroxide for utilization by carp (Nose *et al.*, 1974) and channel catfish (Wilson *et al.*, 1977).

Other investigators have used semipurified and practical diets supplemented with crystalline amino acids to estimate the amino acid requirements of certain fish. The semipurified diets have usually included an imbalanced protein as the major source of the dietary amino acids, e.g. zein (Kaushik, 1979) or corn gluten (Halver *et al.*, 1958; Ketola, 1983), which are deficient in certain amino acids. Practical type diets utilize normal feed ingredients to furnish the bulk of the amino acids. These may be formulated with a fixed amount of intact protein and the remaining protein equivalent is made up of crystalline amino acids (Luquet and Sabaut, 1974; Jackson and Capper, 1982; Walton *et al.*, 1984a). The various problems inherent in using these types of diets to assess the amino acid requirements of fish have been discussed elsewhere (Wilson, 1985).

Growth studies

Most of the amino acid requirement values have been estimated based on the conventional growth response curve or Almquist plot. Replicate groups of fish are fed diets containing graded levels of the test amino acid until measurable differences appear in the weight gain of the test fish. A linear increase in weight gain is normally observed with increasing amino acid intake up to a break-point corresponding to the requirement of the specific amino acid, at which weight gain levels off or plateaus.

Various methods have been used to estimate or calculate the break-point corresponding to the requirement value based on the weight gain data. The requirement values for chinook salmon (reviewed by Mertz, 1972), common carp and Japanese eel (Nose, 1979) were estimated using an Almquist plot without the aid of any statistical analysis, where others have used regression analysis to generate the Almquist plot (Harding *et al.*, 1977; Akiyama *et al.*, 1985a). Wilson *et al.* (1980) used the continuous broken-line model developed by Robbins *et al.* (1979) to estimate the requirement values. Santiago and Lovell (1988) used both the broken-line model and quadratic regression analysis to estimate the requirement values for Nile tilapia (*Oreochromis niloticus*) based on weight gain data. Quadratic regression analysis resulted in the lowest error term for estimating the requirement values, whereas the broken-line model yielded the lowest error term for only three requirement values. Most of the requirement values that have been reported within the last 10 years have been estimated based on the broken-line model.

The various problems involved in the accurate determination of the amino acid requirements of fish based on growth studies have been reviewed by Cowey and Luquet (1983) and Cowey and Tacon (1983). These problems include: (i) a lack of precision in the interpretation of the growth response curves, i.e. determining the break-point of the curve has often been subjective; (ii) the growth rates commonly observed with amino acid test diets have generally been lower than those observed with intact protein diets; and (iii) the possibility exists that some of the crystalline amino acids in the test diets may be leached during the feeding studies.

Serum or tissue amino acid studies

Some investigators have found a high correlation of either serum or blood and muscle free amino acid levels to dietary amino acid intake in fish. The hypothesis is that serum or tissue concentrations of the amino acid should remain low until the requirement for the amino acid is met and then increase to high levels when excessive amounts of the amino acid are fed. This technique has proven useful in confirming the amino acid requirements in only a few cases. For example, of the ten indispensable amino acid requirement studies in the channel catfish, only the serum lysine (Wilson *et al.*, 1977), threonine (Wilson *et al.*, 1978), histidine (Wilson *et al.*, 1980) and methionine (Harding *et al.*, 1977) data were useful in confirming the requirement values estimated based on weight gain data. Serum methionine data of sea bass (Thebault *et al.*, 1985) and serum lysine of hybrid striped bass (*Morone chrysops* × *M. saxatilis*) (Griffin *et al.*,

1992) have been used to confirm the requirement values for these species. Blood and muscle arginine concentrations were found to gradually increase in rainbow trout fed increasing levels of arginine and were not useful for assessing the arginine requirement of this species (Kaushik, 1979). Walton *et al.* (1984b) were unable to use blood tryptophan levels to confirm the tryptophan requirement of rainbow trout. Of the ten amino acids required by Nile tilapia, Santiago and Lovell (1988) were only able to use muscle free lysine, threonine and isoleucine concentrations to confirm the requirement values of these amino acids based on growth studies.

Amino acid oxidation studies

This technique is based on the general hypothesis that when an amino acid is limiting or deficient in the diet, the major portion will be utilized for protein synthesis, and little will be oxidized to carbon dioxide, whereas when an amino acid is supplied in excess, and is thus not a limiting factor for protein synthesis, more of the amino acid will be oxidized. The intake level which produces a marked increase in amino acid oxidation should then be a direct indicator of the requirement value for that specific amino acid.

This technique has been evaluated in rainbow trout with only limited success. Walton *et al.* (1984a) were successful in using this technique to confirm the lysine requirement of rainbow trout based on weight gain data. Following the growth study, three fish from each dietary treatment were injected intraperitoneally with a tracer dose of $[U-^{14}C]$lysine and the respired carbon dioxide was collected over a 20-hour period. The level of $[^{14}C]$carbon dioxide produced was used as a direct measurement of the rate of oxidation of lysine in the fish. The level of oxidation observed was very low in those fish fed low dietary levels of lysine, somewhat higher for the intermediate dietary levels, and much higher for the higher levels of dietary lysine. The break-point of the dose–response curve indicated a dietary requirement of 20 g lysine kg^{-1} diet, which was in close agreement with a value of 19 g lysine kg^{-1} diet obtained from growth data. In a similar study involving tryptophan, these workers (Walton *et al.*, 1984b) found that the requirement value based on oxidation data was lower, 2.0 versus 2.5 g kg^{-1} diet, than the value based on weight gain data. These workers concluded that the oxidation technique is not suitable for use in the absence of growth data because of its lack of precision in determining requirement values from graphical plots.

Kim *et al.* (1992c) were unsuccessful in using phenylalanine oxidation rates to evaluate the phenylalanine requirement of rainbow trout. In this study, the fish were fed diets containing varying levels of phenylalanine plus L-$[1-^{14}C]$ phenylalanine for 10 to 20 days. The expired $^{14}CO_2$ increased gradually with increasing levels of phenylalanine in the diet without any apparent break-point. These workers concluded that this technique is probably not appropriate for determining amino acid requirements in fish.

Quantitative Amino Acid Requirements

The complete quantitative amino acid requirements have been established for only seven of the estimated 300 different fish species being cultured worldwide, namely the catla (*Catla catla*, an Indian major carp), channel catfish, chinook salmon, chum salmon (*Oncorhynchus keta*), common carp, Japanese eel and Nile tilapia. A limited number of requirement values have also been reported for coho salmon (*Oncorhynchus kisutch*), gilthead bream (*Chrysophrys aurata*), hybrid striped bass, milkfish (*Chanos chanos*), mossambique tilapia (*Oreochromis mossambicus*), rainbow trout, red drum (*Sciaenops ocellatus*), European sea bass and sockeye salmon.

Arginine

The arginine requirement values for fish are summarized in Table 16.2. Salmon have the highest requirement of about 6% of dietary protein, whereas the other species require about 4 to 5% of protein. The requirement value of about 4% of protein for rainbow trout appears to be the most reasonable; however, values ranging from 3.3 to 5.9% have been reported. Some confusion exists with respect to the effect of salinity on the arginine requirement of rainbow trout. Kaushik (1979) found that the requirement for arginine decreased as salinity increased. However, Zeitoun *et al.* (1973) found that the protein requirement of rainbow trout was higher at 20 ppt salinity than at 10 ppt salinity.

A dietary lysine–arginine antagonism has been well documented in certain animals. Based upon growth studies, this antagonism has not been demonstrated in channel catfish (Robinson *et al.*, 1981) or rainbow trout (Kim *et al.*, 1992b). Kaushik and Fauconneau (1984) have presented some biochemical evidence which indicates some metabolic antagonism may exist between lysine and arginine in rainbow trout. These workers found that increasing dietary lysine intake affected plasma arginine and urea levels and ammonia excretion. These changes were found to be due to a decrease in the relative rate of arginine degradation as the level of dietary lysine increased.

Histidine

The histidine requirements of fish are presented in Table 16.3. Excellent agreement has been found among the species studied with a range of 1.5

Table 16.2. Arginine requirements (% of protein) for various fish species.

Species	Requirement	Type of diet	Reference
Catla	4.8	Purified	Ravi and Devaraj (1991)
Channel catfish	4.3	Purified	Robinson *et al.* (1981)
Chinook salmon	6.0	Purified	Klein and Halver (1970)
Chum salmon	6.0	Purified	Akiyama (1987)
Coho salmon	5.8	Purified	Klein and Halver (1970)
Common carp	4.3	Purified	Nose (1979)
Gilthead bream	5.0	Semipurified	Luquet and Sabaut (1974)
Japanese eel	4.5	Purified	Arai (Nose, 1979)
Milkfish	5.3	Purified	Borlongan (1991)
Mossambique tilapia	4.0	Practical	Jackson and Capper (1982)
Nile tilapia	4.2	Purified	Santiago and Lovell (1988)
Rainbow trout	3.3	Semipurified	Kaushik (1979)
	3.6	Semipurified	Walton *et al.* (1986)
	4.0	Purified	Kim *et al.* (1992b)
	5.9	Semipurified	Ketola (1983)

Table 16.3. Histidine requirements (% of protein) for various fish species.

Species	Requirement	Type of diet	Reference
Catla	2.5	Purified	Ravi and Devaraj (1991)
Channel catfish	1.5	Purified	Wilson *et al.* (1980)
Chinook salmon	1.8	Purified	Klein and Halver (1970)
Chum salmon	1.6	Purified	Akiyama *et al.* (1985a)
Coho salmon	1.8	Purified	Klein and Halver (1970)
Common carp	2.1	Purified	Nose (1979)
Japanese eel	2.1	Purified	Arai (Nose, 1979)
Nile tilapia	1.7	Purified	Santiago and Lovell (1988)

to 2.5% of protein for the requirement values. Wilson *et al.* (1980) were able to confirm the requirement value by the serum free histidine pattern in channel catfish. There was a significant increase in serum free histidine concentration up to the dietary requirement as determined based on growth data and then the serum histidine remained constant at higher dietary intake.

Muscle carnosine concentration has been shown to be altered by dietary histidine in the chick (Robbins *et al.*, 1977) and chinook salmon (Lukton, 1958). In both cases, the muscle carnosine was depleted when a histidine-deficient diet was fed. In the chick, as the dietary histidine was increased above the requirement, the muscle carnosine level returned to that of the control. Counter to the above, carnosine could not be detected in muscle tissue of the channel catfish regardless of the dietary level of histidine (Wilson *et al.*, 1980).

Isoleucine

The isoleucine requirements of fish are presented in Table 16.4. The requirement appears to be about 2.2 to 3% of protein for those species studied except for the Japanese eel, which has a much higher requirement value.

Wilson *et al.* (1980) determined the effects of dietary isoleucine on serum free isoleucine, leucine and valine in channel catfish. Even though the serum isoleucine increased somewhat with increasing isoleucine intake, these data did not confirm the requirement as determined based on growth data. The serum free leucine and valine concentrations appeared to parallel the serum free isoleucine concentrations. A relatively high mortality rate was observed in the fish fed the isoleucine-deficient diet.

Leucine

The leucine requirement values are presented in Table 16.5. The requirements values agree quite well, ranging from 3.3 to 3.9% of protein, except for the high value of 5.3% of protein reported for the Japanese eel.

Wilson *et al.* (1980) reported that the serum free leucine level in channel catfish remained constant regardless of dietary leucine intake. There was, however, a marked effect of dietary leucine on the serum free isoleucine and valine levels. There was about a six-fold increase in serum free isoleucine and valine concentrations at the 0.7% dietary leucine level, as compared to the 0.6% leucine level. These elevated levels of isoleucine and valine did not return to the baseline levels until a dietary level of 1.2% or above was fed. This observation was interpreted to indicate that leucine may facilitate the tissue uptake of branched-chain amino acids and/or their intracellular metabolism.

Table 16.4. Isoleucine requirements (% of protein) of various fish species.

Species	Requirement	Type of diet	Reference
Catla	2.4	Purified	Ravi and Devaraj (1991)
Channel catfish	2.6	Purified	Wilson *et al.* (1980)
Chinook salmon	2.2	Purified	Chance *et al.* (1964)
Chum salmon	2.4	Purified	Akiyama (1987)
Common carp	2.5	Purified	Nose (1979)
Japanese eel	4.0	Purified	Arai (Nose, 1979)
Nile tilapia	3.1	Purified	Santiago and Lovell (1988)

Table 16.5. Leucine requirements (% of protein) of various fish species.

Species	Requirement	Type of diet	Reference
Catla	3.7	Purified	Ravi and Devaraj (1991)
Channel catfish	3.5	Purified	Wilson *et al.* (1980)
Chinook salmon	3.9	Purified	Chance *et al.* (1964)
Chum salmon	3.8	Purified	Akiyama (1987)
Common carp	3.3	Purified	Nose (1979)
Japanese eel	5.3	Purified	Arai (Nose, 1979)
Nile tilapia	3.4	Purified	Santiago and Lovell (1988)

Valine

The valine requirement values of fish are presented in Table 16.6. Reasonable agreement exists between the values reported for the species studied, indicating that the requirement ranges from about 3 to 4% of protein. Studies showed that serum valine levels in the channel catfish responded to valine intake in a manner similar to that described for isoleucine (Wilson *et al.*, 1980).

Isoleucine–leucine–valine interactions

Some differences appear to exist in the apparent isoleucine–leucine–valine interactions among different fishes. Chance *et al.* (1964) found that the isoleucine requirement in chinook salmon was increased slightly with increasing levels of dietary leucine. This effect was not observed in either

Table 16.6. Valine requirements (% of protein) of various fish species.

Species	Requirement	Type of diet	Reference
Catla	3.6	Purified	Ravi and Devaraj (1991)
Channel catfish	3.0	Purified	Wilson *et al.* (1980)
Chinook salmon	3.2	Purified	Chance *et al.* (1964)
Chum salmon	3.0	Purified	Akiyama (1987)
Common carp	3.6	Purified	Nose (1979)
Japanese eel	4.0	Purified	Arai (Nose, 1979)
Nile tilapia	2.8	Purified	Santiago and Lovell (1988)

the common carp (Nose, 1979) or channel catfish (Robinson *et al.*, 1984). Nose (1979) did, however, observe reduced growth rates in carp fed high dietary isoleucine levels during his leucine requirement study. This reduced growth rate was not observed when the leucine requirement study was repeated at lower isoleucine levels. Robinson *et al.* (1984) concluded that a nutritional interrelationship does exist among the branched-chain amino acids in the channel catfish but the interaction does not appear to be as severe as has been observed in certain other animals.

Lysine

The lysine requirement values for fish are summarized in Table 16.7. In general, lysine appears to be the first-limiting amino acid in feedstuffs commonly used in formulating feeds for warmwater fish (Robinson *et al.*, 1980b) and perhaps other fish as well. Therefore, more requirement values have been reported for this amino acid. The requirement appears to range from 4 to 5% of protein for most fishes. The values of 5.7 for common carp and 6.2 for catla, an Indian major carp, may indicate that carps have a higher lysine requirement than other fishes. The value of 6.1 for rainbow trout appears to be out of line since two other investigators have reported much lower values.

Serum free lysine levels were useful in confirming the lysine requirement in channel catfish originally determined at 24% crude protein (Wilson *et al.*, 1977); however, serum free lysine levels provided little indication of the lysine requirement when re-evaluated at a 30% crude protein level (Robinson *et al.*, 1980b). Walton *et al.* (1984a) observed good agreement between the lysine requirement values determined by either growth studies or amino acid oxidation studies in rainbow trout.

Table 16.7. Lysine requirements (% of protein) of various fish species.

Species	Requirement	Type of diet	Reference
Catla	6.2	Purified	Ravi and Devaraj (1991)
Channel catfish	5.0	Purified	Robinson *et al.* (1980b)
	5.1	Purified	Wilson *et al.* (1977)
Chinook salmon	5.0	Semipurified	Halver *et al.* (1958)
Chum salmon	4.8	Purified	Akiyama *et al.* (1985a)
Common carp	5.7	Purified	Nose (1979)
Gilthead bream	5.0	Semipurified	Luquet and Sabaut (1974)
Hybrid striped bass	4.0	Purified	Griffin *et al.* (1992)
	4.0	Semipurified	Keembiyehetty and Gatlin (1992)
Japanese eel	5.3	Purified	Arai (Nose, 1979)
Milkfish	4.0	Semipurified	Borlongan and Benitez (1990)
Mossambique tilapia	4.1	Practical	Jackson and Capper (1982)
Nile tilapia	5.1	Purified	Santiago and Lovell (1988)
Rainbow trout	3.7	Purified	Kim *et al.* (1992b)
	4.2	Semipurified	Walton *et al.* (1984a)
	6.1	Semipurified	Ketola (1983)
Red drum	4.4	Semipurified	Craig and Gatlin (1992)
Sea bass	4.8	Practical	Tibaldi and Lanari (1991)

Methionine

Methionine and cystine are classified as sulfur-containing amino acids. Adequate amounts of both methionine and cystine are needed for proper protein synthesis and other physiological functions of the fish. Cystine is considered dispensable because it can be synthesized by the fish from the indispensable amino acid methionine. When methionine is fed without cystine, a portion of the methionine is used for protein synthesis, and a portion is converted to the cystine monomer for incorporation into protein. If cystine is included in the diet, it reduces the amount of dietary methionine needed. Thus, to determine the total sulfur amino acid requirement (methionine plus cystine), the dietary requirement for methionine is determined either in the absence of cystine or with test diets containing very low levels of cystine.

The methionine or total sulfur amino acid requirement values are presented in Table 16.8. It appears that most fish have a requirement value of about 2 to 3% of protein, whereas catla, chinook salmon and gilthead bream appear to require higher levels of methionine.

Rainbow trout appear to be unique in that methionine deficiency results in bilateral cataracts (Poston *et al.*, 1977). These workers observed cataracts in rainbow trout fed diets containing isolated soyabean protein. The cataracts were prevented by supplementing the diet with methionine. Cataracts have also been observed in methionine-deficient rainbow trout by Walton *et al.* (1982), Rumsey *et al.* (1983) and Cowey *et al.* (1992). This deficiency sign has not been reported in any other species of fish.

As indicated above, dietary cystine can reduce the amount of dietary methionine required for maximum growth. The cystine replacement value for methionine on a sulfur basis has been determined to be about 60% for channel catfish (Harding *et al.*, 1977) and 42% for rainbow trout (Kim *et al.*, 1992a).

Table 16.8. Methionine or total sulfur amino acid requirements (% of protein) of various fish species.

Species	Requirement	Type of diet	Reference
Catla	3.6	Purified	Ravi and Devaraj (1991)
Channel catfish	2.3	Purified	Harding *et al.* (1977)
Chinook salmon	4.0	Purified	Halver *et al.* (1959)
Chum salmon	3.0	Purified	Akiyama (1987)
Common carp	3.1	Purified	Nose (1979)
Gilthead bream	4.0	Semipurified	Luquet and Sabaut (1974)
Hybrid striped bass	2.9	Semipurified	Keembiyehetty and Gatlin (1993)
Japanese eel	3.2	Purified	Arai (Nose, 1979)
Mossambique tilapia	3.2	Practical	Jackson and Capper (1982)
Nile tilapia	3.2	Purified	Santiago and Lovell (1988)
Rainbow trout	1.9	Semipurified	Cowey *et al.* (1992)
	2.2	Purified	Walton *et al.* (1982)
	2.3	Purified	Kim *et al.* (1992a)
	3.0	Purified	Rumsey *et al.* (1983)
Red drum	3.0	Semipurified	Moon and Gatlin (1991)
Sea bass	2.0	Practical	Thebault *et al.* (1985)

Robinson *et al.* (1978) evaluated the utilization of several sulfur compounds for their potential replacement value for methionine in channel catfish. Growth and feed efficiency data indicated that DL-methionine was utilized as effectively as L-methionine. Methionine hydroxy analogue was only about 26% as effective as L-methionine in promoting growth. No significant growth response was observed when taurine or inorganic sulfate was added to the basal diet. Page *et al.* (1978) were also unable to detect the utilization of taurine and inorganic sulfate as sulfur sources in rainbow trout. D-methionine has been shown to replace L-methionine on an equal basis in rainbow trout (Kim *et al.*, 1992a).

Phenylalanine

A relationship similar to that presented for methionine and cystine exists for phenylalanine and tyrosine, two important aromatic amino acids. Tyrosine is considered dispensable because it can be synthesized by the fish from the indispensable amino acid phenylalanine. If tyrosine is included in the diet, it reduces the amount of phenylalanine needed in the diet. Thus, fish have a total aromatic amino acid requirement.

The phenylalanine or total aromatic amino acid requirement values for fish are presented in Table 16.9. Most requirement values fall within the range of 5 to 6% of protein except for the lower value for rainbow trout and the higher value for common carp.

Since the fish has a metabolic need for both phenylalanine and tyrosine, and only a certain portion of the phenylalanine can be converted

Table 16.9. Phenylalanine or total aromatic amino acid requirements (% of protein) for various fish species.

Species	Requirement	Type of diet	Reference
Catla	6.2	Purified	Ravi and Devaraj (1991)
Channel catfish	5.0	Purified	Robinson *et al.* (1980a)
Chinook salmon	5.1	Purified	Chance *et al.* (1964)
Chum salmon	6.3	Purified	Akiyama (1987)
Common carp	6.5	Purified	Nose (1979)
Japanese eel	5.8	Purified	Arai (Nose, 1979)
Milkfish	5.2	Purified	Borlongan (1992)
Nile tilapia	5.5	Purified	Santiago and Lovell (1988)
Rainbow trout	4.3	Purified	Kim (1993)

to tyrosine and still meet the animal's need for phenylalanine, it is important to determine how much of the total aromatic amino acid requirement can be provided by dietary tyrosine. Growth studies indicate that tyrosine can replace or spare about 60% of the phenylalanine requirement in common carp (Nose, 1979), 50% in channel catfish (Robinson et al., 1980a) and 48% in rainbow trout (Kim, 1993).

Threonine

The threonine requirement values for fish are summarized in Table 16.10. The reported values range from 2 to 5% of protein. It is difficult to offer any explanation for the lack of agreement in these requirement values. Additional research is needed to determine if this wide range of values represents true differences in the threonine requirements or just a difference in the techniques used to determine the requirement values.

DeLong et al. (1962) found the threonine requirement of chinook salmon to be the same when determined at rearing temperatures of 8 and 15°C. These findings were not expected since these workers had previously reported the protein requirement of chinook salmon to increase from 40% at 8°C to 55% at 15°C (DeLong et al., 1958).

Tryptophan

The tryptophan requirement values for fish are presented in Table 16.11. The requirement appears to be about 0.5 to 1% of protein for the various species studied. The high value of 1.4% of protein for rainbow trout may have been overestimated because no dietary levels between 0.25 and 0.50% of diet were fed (Poston and Rumsey, 1983).

Tryptophan deficiency results in several deficiency signs in salmonids that have not been observed in other fish species. Halver and Shanks (1960) observed scoliosis and lordosis in sockeye salmon but not in chinook salmon fed tryptophan-deficient diets. Scoliosis and lordosis have also been observed in tryptophan-deficient rainbow trout (Shanks et al., 1962; Kloppel and Post, 1975; Poston and Rumsey, 1983; Walton et al., 1984b) and chum salmon (Akiyama et al., 1985b). These deformities were found to be reversible in rainbow trout when the fish were fed adequate dietary tryptophan (Shanks et al., 1962; Kloppel and Post, 1975) and appear to be related to a depletion of 5-hydroxytryptophan in the body or brain (Akiyama et al., 1986). Other tryptophan deficiency signs in rainbow trout include renal calcinosis (Kloppel and Post, 1975), caudal fin erosion, cataracts, and short gill opercula (Poston and Rumsey, 1983), and increased liver and kidney levels of calcium, magnesium, sodium and potassium (Walton et al., 1984b).

Table 16.10. Threonine requirements (% of protein) for various fish species.

Species	Requirement	Type of diet	Reference
Catla	5.0	Purified	Ravi and Devaraj (1991)
Channel catfish	2.0	Purified	Wilson *et al.* (1978)
Chinook salmon	2.2	Purified	DeLong *et al.* (1962)
Chum salmon	3.0	Purified	Akiyama *et al.* (1985a)
Common carp	3.9	Purified	Nose (1979)
Japanese eel	4.0	Purified	Arai (Nose, 1979)
Milkfish	4.5	Purified	Borlongan (1991)
Nile tilapia	3.8	Purified	Santiago and Lovell (1988)

Table 16.11. Tryptophan requirements (% of protein) of various fish species.

Species	Requirement	Type of diet	Reference
Catla	1.0	Purified	Ravi and Devaraj (1991)
Channel catfish	0.5	Purified	Wilson *et al.* (1978)
Chinook salmon	0.5	Purified	Halver (1965)
Chum salmon	0.7	Purified	Akiyama *et al.* (1985b)
Coho salmon	0.5	Purified	Halver (1965)
Common carp	0.8	Purified	Nose (1979)
Gilthead bream	0.6	Semipurified	Luquet and Sabaut (1974)
Japanese eel	1.1	Purified	Arai (Nose, 1979)
Nile tilapia	1.0	Purified	Santiago and Lovell (1988)
Rainbow trout	0.5	Semipurified	Walton *et al.* (1984b)
	0.6	Purified	Kim *et al.* (1987)
	1.4	Purified	Poston and Rumsey (1983)
Sockeye salmon	0.5	Purified	Halver (1965)

Other Methods of Estimating Amino Acid Needs

Various investigators have observed improved growth and feed efficiency when experimental diets for salmonids were supplemented with indispensable amino acids to simulate levels found in isolated fish protein or the respective eggs and whole body tissue of the species being studied (Rumsey and Ketola, 1975; Arai, 1981; Ketola, 1982; Ogata *et al.*, 1983). The indispensable amino acid requirements of certain fish have also been shown to correlate well with the indispensable amino acid pattern of the whole body tissue of that fish (Cowey and Tacon, 1983; Wilson and Poe, 1985). Therefore, it seems reasonable to suggest that these types of information may be useful in designing test diets for fish when their amino acid requirements have not been established.

The amino acid composition of whole body tissue of certain fishes is presented in Table 16.12. The amino acid compositions of these five species of fish are surprisingly similar. Of the indispensable amino acids, only slight differences appear to exist with the methionine and possibly tryptophan content. Therefore, it appears that the amino acid needs of these species should be very similar.

The amino acid composition of eggs of various fishes has been summarized by Ketola (1982). In general, the amino acid composition appears to vary more than the whole body composition data presented in Table 16.12. Ketola (1982) also points out that although the amino acid content of the fish eggs appears to differ from the reported dietary requirements of the fish, the composition of the eggs has provided useful data in formulating test diets for Atlantic salmon and rainbow trout.

On the basis of observations in other animals, Cowey and Tacon (1983) suggested that the indispensable amino acid requirements of a fish should be related, or even governed, by the pattern of amino acids present in muscle tissue. They showed a high correlation between the requirement pattern found by feeding experiments for the ten indispensable amino acids as determined for carp by Nose (1979) using amino acid test diets and the pattern of the same amino acids in the whole body tissue of growing carp.

Wilson and Poe (1985) have tested this hypothesis in channel catfish. These workers obtained a regression coefficient of 0.96 when the indispensable amino acid requirement pattern for the channel catfish was regressed against the whole body indispensable amino acid pattern found in a 30 g channel catfish. A lower regression coefficient of 0.68 was found when the requirement pattern was regressed against the channel catfish egg amino acid pattern.

Arai (1981) used A/E ratios [(essential amino acid content/total essential amino acid content including cystine and tyrosine) × 1000] of whole body coho salmon fry to formulate test diets for this fish. Fish fed

Table 16.12. Amino acid composition (g/100 g amino acids) of whole body tissue of certain fishes.

Amino acid	Rainbow trout[1]	Atlantic salmon[1]	Coho salmon[2]	Cherry salmon[3]	Channel catfish[4]
Alanine	6.57	6.52	6.08	6.35	6.31
Arginine	6.41	6.61	5.99	6.23	6.67
Aspartic acid	9.94	9.92	9.96	9.93	9.74
Cystine	0.80	0.95	1.23	1.34	0.86
Glutamic acid	14.22	14.31	15.25	15.39	14.39
Glycine	7.76	7.41	7.31	7.62	8.14
Histidine	2.96	3.02	2.99	2.39	2.17
Isoleucine	4.34	4.41	3.70	3.96	4.29
Leucine	7.59	7.72	7.49	7.54	7.40
Lysine	8.49	9.28	8.64	8.81	8.51
Methionine	2.88	1.83	3.53	3.14	2.92
Phenylalanine	4.38	4.36	4.14	4.63	4.14
Proline	4.89	4.64	4.76	4.33	6.02
Serine	4.66	4.61	4.67	4.48	4.89
Threonine	4.76	4.95	5.11	4.63	4.41
Tryptophan	0.93	0.93	1.40	0.83	0.78
Tyrosine	3.38	3.50	3.44	3.58	3.28
Valine	5.09	5.09	4.32	4.85	5.15

[1] Data from Wilson and Cowey (1985).
[2] Data from Arai (1981).
[3] Data from Ogata *et al.* (1983).
[4] Data from Wilson and Poe (1985).

casein diets supplemented with amino acids to simulate the A/E ratios of whole body tissue showed much improved growth and feed efficiency. Ogata *et al.* (1983) used A/E ratios based on amino acid composition data from cherry salmon (*Oncorhynchus masou*) to design test diets for cherry salmon and amago salmon (*Oncorhynchus rhodurus*) fry. A casein diet supplemented with amino acids to simulate the A/E ratios of cherry salmon resulted in better growth in both species than diets containing casein alone, casein plus amino acids to simulate the A/E ratio of eyed cherry salmon eggs, or white fish meal.

The relationships between whole body amino acid patterns and amino acid requirement patterns discussed above are very similar to the ideal protein concept that has been advocated for use in expressing the amino acid requirements of pigs (Agricultural Research Council, 1981). The ideal protein concept is based on the idea that there should be a direct correlation between the whole body amino acid pattern of the animal and the dietary amino acid requirements of that animal. In addition, since lysine is normally the first-limiting amino acid in most feedstuffs, the requirements for the other indispensable amino acids are expressed relative to the lysine requirement. Thus, if one knows the dietary lysine requirement and the whole body amino acid composition of an animal, then one should be able to estimate the dietary requirement for the remaining indispensable amino acids relative to the lysine requirement. A comparison of the amino acid requirement values as determined by conventional means in our laboratory and as estimated based on the ideal protein concept for channel catfish is presented in Table 16.13. These data show excellent agreement.

We are currently evaluating the use of the ideal protein concept as a means of estimating the amino acid requirements of fish. If this procedure proves satisfactory, then all one will need to do in order to formulate diets based on the amino acid needs of a certain fish species is to determine the whole body amino acid composition and the lysine requirement of that species. This new procedure should be much less time consuming and less costly than determining amino acid requirements of the fish by conventional means.

Table 16.13. A comparison of the amino acid requirement values (% of protein) as determined by conventional means and as estimated based on the ideal protein concept for channel catfish.

Amino acid	Amino acid ratio	Determined requirement	Estimated requirement
Lysine	100	5.1	–
Arginine	78	4.3	4.0
Histidine	25	1.5	1.3
Isoleucine	50	2.6	2.6
Leucine	87	4.4	4.4
Met + Cys	44	2.3	2.2
Phe + Tyr	87	5.0	4.4
Threonine	52	2.0	2.7
Tryptophan	9	0.5	0.5
Valine	61	3.0	3.1

Met = methionine; Cys = cystine; Phe = phenylalanine; Tyr = tyrosine.

References

Agricultural Research Council (1981) Protein and amino acid requirements. In: *The Nutrient Requirements of Pigs.* Commonwealth Agricultural Bureaux, Farnham Royal, Slough, pp. 67–124.

Akiyama, T. (1987) Studies on the essential amino acids and scoliosis caused by tryptophan deficiency of chum salmon fry. Unpublished PhD thesis, University of Kyushu, Japan.

Akiyama, T., Arai, S. and Murai, T. (1985a) Threonine, histidine and lysine requirements of chum salmon fry. *Bulletin of the Japanese Society of Scientific Fisheries* 51, 635–639.

Akiyama, T., Arai, S., Murai, T. and Nose, T. (1985b) Tryptophan requirement of chum salmon fry. *Bulletin of the Japanese Society of Scientific Fisheries* 51, 1005–1008.

Akiyama, T., Murai, T. and Mori, K. (1986) Role of tryptophan metabolites in inhibition of spinal deformity of chum salmon fry caused by tryptophan deficiency. *Bulletin of the Japanese Society of Scientific Fisheries* 52, 1255–1259.

Arai, S. (1981) A purified test diet for coho salmon, *Oncorhynchus kisutch,* fry. *Bulletin of the Japanese Society of Scientific Fisheries* 47, 547–550.

Arai, S., Nose, T. and Hashimoto, Y. (1972) Amino acids essential for the growth of eels, *Anguilla anguilla* and *A. japonica. Bulletin of the Japanese Society of Scientific Fisheries* 38, 753–759.

Borlongan, I.G. (1991) Arginine and threonine requirements of milkfish (*Chanos chanos* Forsskal) juveniles. *Aquaculture* 93, 313–322.

Borlongan, I.G. (1992) Dietary requirement of milkfish (*Chanos chanos* Forsskal) juveniles for total aromatic amino acids. *Aquaculture* 102, 309–317.

Borlongan, I.G. and Benitez, L.V. (1990) Quantitative lysine requirement of milkfish (*Chanos chanos*) juveniles. *Aquaculture* 87, 341–347.

Bowen, S.H. (1987) Dietary protein requirements of fishes – a reassessment. *Canadian Journal of Fisheries and Aquatic Sciences* 44, 1995–2001.

Chance, R.E., Mertz, E.T. and Halver, J.E. (1964) Nutrition of salmonoid fishes. XII. Isoleucine, leucine, valine and phenylalanine requirements of chinook salmon and interrelations between isoleucine and leucine for growth. *Journal of Nutrition* 83, 177–185.

Cowey, C.B. and Luquet, P. (1983) Physiological basis of protein requirements of fishes. Critical analysis of allowances. In: Pion, R., Arnal, M. and Bonin, D. (eds) *Protein Metabolism and Nutrition* 1, pp. 364–384, INRA, Paris.

Cowey, C.B. and Tacon, A.G.J. (1983) Fish nutrition – relevance to invertebrates. In: Pruder, G.D., Langdon, C.J. and Conklin, D.E. (eds) *Proceedings of the Second International Conference on Aquaculture Nutrition: Biochemical and Physiological Approaches to Shellfish Nutrition.* Louisiana State University, Division of Continuing Education, Baton Rouge, pp. 13–30.

Cowey, C.B., Adron, J.W. and Blair, A. (1970) Studies on the nutrition of marine flatfish. The essential amino acid requirements of plaice and sole. *Journal of the Marine Biological Association of the United Kingdom* 50, 87–95.

Cowey, C.B., Cho, C.Y., Sivak, J.G., Weerheim, J.A. and Stuart, D.D. (1992) Methionine intake in rainbow trout (*Oncorhynchus mykiss*), relationship to cataract formation and the metabolism of methionine. *Journal of Nutrition* 122, 1154–1163.

Craig, S.R. and Gatlin, D.M. III (1992) Dietary lysine requirement of juvenile red drum *Sciaenops ocellatus*. *Journal of the World Aquaculture Society* 23, 133–137.

DeLong, D.C., Halver, J.E. and Mertz, E.T. (1958) Nutrition of salmonoid fishes. VI. Protein requirements of chinook salmon at two water temperatures. *Journal of Nutrition* 65, 589–599.

DeLong, D.C., Halver, J.E. and Mertz, E.T. (1962) Nutrition of salmonoid fishes. X. Quantitative threonine requirements of chinook salmon at two water temperatures. *Journal of Nutrition* 76, 174–178.

Dupree, H.K. and Halver, J.E. (1970) Amino acids essential for the growth of channel catfish, *Ictalurus punctatus*. *Transactions of the American Fisheries Society* 99, 90–92.

Griffin, M.E., Brown, P.B. and Brant, A.L. (1992) The dietary lysine requirement of juvenile hybrid striped bass. *Journal of Nutrition* 122, 1332–1337.

Halver, J.E. (1957) Nutrition of salmonoid fishes. IV. An amino acid test diet for chinook salmon. *Journal of Nutrition* 62, 245–254.

Halver, J.E. (1965) Tryptophan requirement of chinook, sockeye and silver salmon. *Federation Proceedings* 24, 229 (abstr).

Halver, J.W. and Shanks, W.E. (1960) Nutrition of salmonoid fishes. VIII. Indispensable amino acids for sockeye salmon. *Journal of Nutrition* 72, 340–346.

Halver, J.E., DeLong, D.C. and Mertz, E.T. (1957) Nutrition of salmonoid fishes. V. Classification of essential amino acids for chinook salmon. *Journal of Nutrition* 63, 95–105.

Halver, J.E., DeLong, D.C. and Mertz, E.T. (1958) Threonine and lysine requirements of chinook salmon. *Federation Proceedings* 171, 1873 (abstr).

Halver, J.E., DeLong, D.C. and Mertz, E.T. (1959) Methionine and cystine requirements of chinook salmon. *Federation Proceedings* 18, 2076 (abstr).

Harding, D.E., Allen, O.W. Jr and Wilson, R.P. (1977) Sulfur amino acid requirement of channel catfish: L-methionine and L-cystine. *Journal of Nutrition* 107, 2031–2035.

Jackson, A.J. and Capper, B.S. (1982) Investigations into the requirements of the tilapia *Sarotherodon mossambicus* for dietary methionine, lysine and arginine in semi-synthetic diets. *Aquaculture* 29, 289–297.

Kaushik, S. (1979) Application of a biochemical method for the estimation of amino acid needs in fish: Quantitative arginine requirements of rainbow trout in different salinities. In: Halver J.E. and Tiews, K. (eds) *Finfish Nutrition and Fishfeed Technology, Vol. 1*. Heenemann, Berlin, pp. 197–207.

Kaushik, S.J. and Fauconneau, B. (1984) Effects of lysine administration on plasma arginine and on some nitrogenous catabolites in rainbow trout. *Comparative Biochemistry and Physiology* 79A, 459–462.

Keembiyehetty, C.N. and Gatlin, D.M. III (1992) Dietary lysine requirement of juvenile hybrid striped bass (*Morone chrysops* × *M. saxatilis*). *Aquaculture* 104, 271–277.

Keembiyehetty, C.N. and Gatlin, D.M. III (1993) Total sulfur amino acid requirement of juvenile hybrid striped bass (*Morone chrysops* x *M. saxatilis*). *Aquaculture* 110, 331–339.

Ketola, H.G. (1982) Amino acid nutrition of fishes: requirements and supplementation of diets. *Comparative Biochemistry and Physiology* 73B, 17–24.

Ketola, H.G. (1983) Requirement for dietary lysine and arginine by fry of rainbow trout. *Journal of Animal Science* 56, 101–107.

Kim, K.I. (1993) Requirement for phenylalanine and replacement value of tyrosine for phenylalanine in rainbow trout (*Oncorhynchus mykiss*). *Aquaculture* 113, 243–250.

Kim, K.I., Kayes, T.B. and Amundson, C.H. (1987) Effects of dietary tryptophan levels on growth, feed/gain, carcass composition and liver glutamate dehydrogenase activity in rainbow trout (*Salmo gairdneri*). *Comparative Biochemistry and Physiology* 88B, 737–741.

Kim, K.I., Kayes, T.B. and Amundson, C.H. (1992a) Requirements for sulfur amino acids and utilization of D-methionine by rainbow trout (*Oncorhynchus mykiss*). *Aquaculture* 101, 95–103.

Kim, K.I., Kayes, T.B. and Amundson, C.H. (1992b) Requirements for lysine and arginine by rainbow trout (*Oncorhynchus mykiss*). *Aquaculture* 106, 333–344.

Kim, K.I., Grimshaw, T.W., Kayes, T.B. and Amundson, C.H. (1992c) Effect of fasting or feeding diets containing different levels of protein or amino acids on the activities of the liver amino acid-degrading enzymes and amino acid oxidation in rainbow trout (*Oncorhynchus mykiss*). *Aquaculture* 107, 89–105.

Klein, R.G. and Halver, J.E. (1970) Nutrition of salmonoid fishes: Arginine and histidine requirements of chinook and coho salmon. *Journal of Nutrition* 100, 1105–1109.

Kloppel, T.M. and Post, F. (1975) Histological alterations in tryptophan-deficient rainbow trout. *Journal of Nutrition* 105, 861–866.

Luquet, P. and Sabaut, J.J. (1974) Nutrition azotée et caroissance chez la daurade et la truite. *Actes de Colloques, Colloques sur L'Aquaculture*, Brest No. 1, 243–253.

Lukton, A. (1958) Effect of diet on imidazole compounds and creatine in chinook salmon. *Nature* 182, 1014–1020.

Mazid, M.A., Tanaka, Y., Katayama, T., Simpson, K.L. and Chichester, C.O. (1978) Metabolism of amino acids in aquatic animals. III. Indispensable amino acids for *Tilapia zillii*. *Bulletin of the Japanese Society of Scientific Fisheries* 44, 739–742.

Mertz, E.T. (1972) The protein and amino acid needs. In: Halver, J.E. (ed.) *Fish Nutrition*. Academic Press, New York, pp. 105–143.

Metailler, R., Febvre, A. and Alliot, E. (1973) Preliminary note on the essential amino-acids of the sea-bass, *Dicentrarchus labrax* (Linne). *Study Review GFCM* (France) 52, 91–96.

Moon, H.Y. and Gatlin, D.M. III (1991) Total sulfur amino acid requirements of juvenile red drum, *Sciaenops ocellatus*. *Aquaculture* 95, 97–106.

Nose, T. (1979) Summary report on the requirements of essential amino acids for carp. In: Halver, J.E. and Tiews, K. (eds) *Finfish Nutrition and Fishfeed Technology, Vol. 1*. Heenemann, Berlin, pp. 145–156.

Nose, T., Arai, S., Lee, D.L. and Hashimoto, Y. (1974) A note on amino acids essential for growth of young carp. *Bulletin of the Japanese Society of Scientific Fisheries* 40, 903–908.

Ogata, H., Arai, S. and Nose, T. (1983) Growth responses of cherry salmon *Oncorhynchus masou* and amago salmon *O. rhodurus* fry fed purified casein diets supplemented with amino acids. *Bulletin of the Japanese Society of Scientific Fisheries* 49, 1381–1385.

Page, J.W., Rumsey, G.L., Riis, R.C. and Scott, M.L. (1978) Dietary sulfur requirements of fish: nutritional and pathological criteria. *Federation Proceedings* 37, 1189 (abstr.).

Poston, H.A. and Rumsey, G.L. (1983) Factors affecting dietary requirement and deficiency signs of L-tryptophan in rainbow trout. *Journal of Nutrition* 113, 2568–2577.

Poston, H.A., Riis, R.C., Rumsey, G.L., and Ketola, H.G. (1977) The effect of supplemental dietary amino acids, minerals and vitamins on salmonids fed cataractogenic diets. *The Cornell Veterinarian* 67, 472–509.

Ravi, J. and Devaraj, K.V. (1991) Quantitative essential amino acid requirements for growth of catla, *Catla catla*. *Aquaculture* 96, 281–291.

Robbins, K.R., Baker, D.H. and Norton, H.W. (1977) Histidine status in the chick as measured by growth rate, plasma free histidine and breast muscle carnosine. *Journal of Nutrition* 107, 2055–2061.

Robbins, K.R., Norton, H.W. and Baker, D.H. (1979) Estimation of nutrient requirements from growth data. *Journal of Nutrition* 109, 1710–1714.

Robinson, E.H., Allen, O.W. Jr, Poe, W.E. and Wilson, R.P. (1978) Utilization of dietary sulfur compounds by fingerling channel catfish: L-methionine, DL-methionine, methionine hydroxy analogue, taurine and inorganic sulfate. *Journal of Nutrition* 108, 1932–1936.

Robinson, E.H., Wilson, R.P. and Poe, W.E. (1980a) Total aromatic amino acid requirement, phenylalanine requirement and tyrosine replacement value for fingerling channel catfish. *Journal of Nutrition* 110, 1805–1812.

Robinson, E.H., Wilson, R.P. and Poe, W.E. (1980b) Re-evaluation of the lysine requirement and lysine utilization by fingerling channel catfish. *Journal of Nutrition* 110, 2313–2316.

Robinson, E.H., Wilson, R.P. and Poe, W.E. (1981) Arginine requirement and apparent absence of a lysine-arginine antagonism in fingerling channel catfish. *Journal of Nutrition* 111, 46–52.

Robinson, E.H., Poe, W.E. and Wilson, R.P. (1984) Effects of feeding diets containing an imbalance of branched-chain amino acids on fingerling channel catfish. *Aquaculture* 37, 51–62.

Rumsey, G.L. and Ketola, H.G. (1975) Amino acid supplementation of casein diets of Atlantic salmon (*Salmo salar*) fry and of soybean meal for rainbow trout (*Salmo gairdneri*) fingerlings. *Journal of the Fisheries Research Board of Canada* 32, 422–426.

Rumsey, G.L., Page, J.G. and Scott, M.L. (1983) Methionine and cystine requirements of rainbow trout. *The Progressive Fish-Culturist* 45, 139–143.

Santiago, C.B. and Lovell, R.T. (1988) Amino acid requirements for growth of Nile tilapia. *Journal of Nutrition* 118, 1540–1546.

Shanks, W.E., Gahimer, G.D. and Halver, J.E. (1962) The indispensable amino acids for rainbow trout. *The Progressive Fish-Culturist* 24, 68–73.

Tacon, A.G.J. and Cowey, C.B. (1985) Protein and amino acid requirements. In:

Tytler, P. and Calow, P. (eds) *Fish Energetics: New Perspectives*. The Johns Hopkins University Press, Baltimore, pp. 155–183.

Thebault, H., Alliot, E. and Pastoureoud, A. (1985) Quantitative methionine requirement of juvenile sea-bass (*Dicentrarchus labrax*). *Aquaculture* 50, 75–87.

Tibaldi, E. and Lanari, D. (1991) Optimal dietary lysine levels for growth and protein utilization of fingerling sea bass (*Dicentrarchus labrax* L.) fed semi-purified diets. *Aquaculture* 95, 297–304.

Walton, M.J., Cowey, C.B. and Adron, J.W. (1982) Methionine metabolism in rainbow trout fed diets of differing methionine and cystine content. *Journal of Nutrition* 112, 1525–1535.

Walton, M.J., Cowey, C.B. and Adron, J.W. (1984a) The effect of dietary lysine levels on growth and metabolism of rainbow trout (*Salmo gairdneri*). *British Journal of Nutrition* 52, 115–122.

Walton, M.J., Coloso, R.M., Cowey, C.B., Adron, J.W. and Knox, D. (1984b) The effects of dietary tryptophan levels on growth and metabolism of rainbow trout (*Salmo gairdneri*). *British Journal of Nutrition* 51, 279–287.

Walton, M.J., Cowey, C.B., Coloso, R.M. and Adron, J.W. (1986) Dietary requirements of rainbow trout for tryptophan, lysine and arginine determined by growth and biochemical measurements. *Fish Physiology and Biochemistry* 2, 161–169.

Wilson, R.P. (1985) Amino acid and protein requirement of fish. In: Cowey, C.B., Mackie, A.M. and Bell, J.G. (eds) *Nutrition and Feeding in Fish*. Academic Press, London, pp. 1–16.

Wilson, R.P. (1989) Amino acids and proteins. In: Halver, J.E. (ed.) *Fish Nutrition*, 2nd edn. Academic Press, San Diego, pp. 111–151.

Wilson, R.P. and Cowey, C.B. (1985) Amino acid composition of whole body tissue of rainbow trout and Atlantic salmon. *Aquaculture* 48, 373–376.

Wilson, R.P. and Poe, W.E. (1985) Relationship of whole body and egg essential amino acid patterns to amino acid requirement patterns in channel catfish, *Ictalurus punctatus*. *Comparative Biochemistry and Physiology* 80B, 385–388.

Wilson, R.P., Harding, D.E. and Garling, D.L. Jr (1977) Effect of dietary pH on amino acid utilization and the lysine requirement of fingerling channel catfish. *Journal of Nutrition* 107, 166–170.

Wilson, R.P., Allen, O.W. Jr., Robinson, E.H. and Poe, W.E. (1978) Tryptophan and threonine requirements of fingerling channel catfish. *Journal of Nutrition* 108, 1595–1599.

Wilson, R.P., Poe, W.E. and Robinson, E.H. (1980) Leucine, isoleucine, valine and histidine requirements of fingerling channel catfish. *Journal of Nutrition* 110, 627–633.

Yone, Y. (1976) Nutritional studies of red sea bream. In: Prince, K.S., Shaw, W.N. and Danbert, K.S. (eds) *Proceedings of the First International Conference on Aquaculture Nutrition*. Lewes/Rehoboth, University of Delaware, pp. 34–39.

Zeitoun, I.H., Tack, P.I., Halver, J.E. and Ullrey, D.E. (1973) Influence of salinity on protein requirements of rainbow trout (*Salmo gairdneri*) fingerlings. *Journal of the Fisheries Research Board of Canada* 30, 1867–1873.

Index

effect of
  age, hens 271, 274–275
  amino acid imbalance 73, 227–231
  dietary crude protein level 65, 67,
    212–213, 227–231
  feeding frequency 38, 68, 122
  heat processing of feeds 125–127
  efficiency 170–172, 209, 215–235,
    354–369
  of precursors 37–51
Aminotransferase reactions *see*
  Transamination
Aminotransferases *see* Transaminases
Ammonia 2, 282–283, 285, 287, 301, 308,
  382
  requirement of rumen microbes 285
Analogues
  hydroxy 43–45, 49–51
  keto 37–41, 49–51
  of amino acids 39–41, 80–88
Antagonisms (interactions) 64, 74–88,
    231–235, 248–249, 382, 385–386
  branched-chain
    (leucine–isoleucine–valine) 64,
    74–78, 234–235, 248–249
  lysine–arginine 64, 74, 78–80, 84–86,
    231–234, 248, 382
  non-protein amino acids 64, 80–88, 90,
    231, 233
Arginine 2, 4, 20, 40, 47–48, 64, 74, 78–80,
    84–86, 104–106, 110, 155–156,
    179, 210, 215, 221–223, 225,
    231–234, 266, 322–323,
    332–333, 357–358, 370,
    382–383, 394
  analysis 20
  antagonism by
    canavanine 84–88, 231, 233
    lysine 64, 74, 78–80, 84–86, 231–234,
    248, 382
  isomers 47
  metabolism 5
  in microbial protein 322–323
  for pigs 155–156
  in post-ruminal digesta, cows 370
  requirements
    broiler chickens 104, 221–223, 231
    calves, veal 332–333
    ducks 105, 222
    fish 110, 382–383, 394
    goslings 222
    laying hens 104, 266
    pigs 179
    turkeys 105–106, 221–222
  responses
    broiler chickens 221, 223, 225–226,
    231–234
    turkeys 221, 223, 225

synthesis 2, 47–48
uptake
  mammary gland 5, 357–358
  utilization 221–223, 225, 231–234
Arginase 2, 80, 84, 86, 87
ATP 286–287, 296
Available lysine 136, 148
  see also FDNB lysine

Bacteria, in hind gut 188, 193–196
  recycling role 194
Bacterial marker 310, 371
Bacterial nitrogen 282–283
Bacterial yield 283–286
Basal endogenous nitrogen, losses in
  ruminants 356
Biotechnology 8–9
Biochemical mechanisms underlying adverse
    effects of
  non-protein amino acids 86–88
  amino acid imbalance 69–71
  branched-chain amino acid
    antagonisms 77–78
  canavanine 86–88
  lysine–arginine antagonism 80
  mimosine 87–88, 90
  S-methylcysteine sulfoxide 87–88, 90
Body fat 235–238, 318–319, 361
  in broiler chickens 235–238
  loss, in dairy cows 361
  in sheep 318–319
Branched-chain amino acids 49–50, 74–78,
    210, 214, 234–236, 248–249,
    357–358, 384–386
  antagonisms in
    chicks 74–76, 234–235
    fish 385–386
    laying hens 76, 248–249
    pigs 77
      pre-ruminant lambs 77
      rats 74, 77
      turkey poults 76, 234–235
  biochemical effects 77–78
  effects on
    brain concentrations of
      neurotransmitters 77–78
    enzymes 77
    food intake 77–78
    growth 74–76, 234–235
    plasma amino acid concentrations 75–77
  extraction by mammary gland 357
  see also Isoleucine, Leucine and Valine
Branched-chain keto acids 77
Brassica crops 82, 84, 90

Calf, veal
  amino acid requirements 329–338